Environmental Effects of Afforestation in North-Western Europe

PLANT AND VEGETATION

Volume 1

Series Editor: M.J.A. Werger

Environmental Effects of Afforestation in North-Western Europe

From Field Observations to Decision Support

Edited by

Gerrit W. Heil
Utrecht University,
The Netherlands

Bart Muys
KU Leuven,
Belgium

and

Karin Hansen
Forest & Landscape,
Hørsholm, Denmark

Springer

A C.I.P. Catalogue record for this book is available from the Library of Congress.

ISBN-10 1-4020-4567-0 (HB)
ISBN-13 978-1-4020-4567-7 (HB)
ISBN-10 1-4020-4568-9 (e-book)
ISBN-13 978-1-4020-4568-4 (e-book)

Published by Springer,
P.O. Box 17, 3300 AA Dordrecht, The Netherlands.

www.springer.com

Printed on acid-free paper

TABLE OF CONTENTS

PREFACE

Although the Common Agricultural Policy (CAP) of the European Union is still largely driven by socio-economic factors, environmental concerns are increasingly integrated into it. With the reforms of 1992, 2000 and 2003, measures have indeed been introduced or fine-tuned aiming at enhancing the environmental sustainability of the agricultural sector in a farm perspective as well as in a regional one. This is illustrated by the fact that, by 2003, farmers must maintain their cultivated and set aside land in good agricultural and environmental condition in order to be eligible for income support under the CAP's so-called 1st pillar, dealing with the agricultural market organisation. Under the 2nd pillar, focusing rural development, the agri-environmental schemes are considered to be the most useful tools for making agriculture more environment-friendly. These schemes allow farmers to be compensated for income losses due to the adherence to practices with an added-value for the environment.

Afforestation of agricultural land can be considered under the 1st pillar as an alternative for set-aside of agricultural land in order to cope with economically unsustainable surplus-production. However, afforestation is also supported by the 2nd pillar as one of the measures contributing to rural development, avoidance of land abandonment and preservation of the environment.

Forestry is known to generate less job opportunities and to contribute less to the gross domestic product than agriculture when practiced under equal site conditions. However, in a situation of subsidized agricultural surplus-production, wood or biomass production may be economically interesting alternatives at the macro-economical level. In addition, the extensivation of land use through afforestation may create opportunities for new types of economic activities in the sphere of recreation and tourism. New forests may also have other, less tangible benefits for society. Examples are the buffering of noise-generating or visually unattractive human activities and the creation of conditions in which biodiversity can be preserved or enhanced. Afforestation of today is therefore supposed to serve multiple purposes, whereas afforestation of the past primarily aimed at increased wood production.

It is generally assumed that afforestation can play a substantial role in meeting the greenhouse gas emission reduction targets under the Kyoto-protocol by increased carbon (C) sequestration in biomass and soil. Other expected environmental effects of afforestation as compared to the agricultural situation are a decreased hydrological recharge to water bodies and a decrease of nitrate losses to these bodies. However, there is limited scientific understanding about the influence on the hydrological, C and nitrogen (N) cycles as a function of the physical conditions of the sites designated for afforestation, of the former agricultural management of these sites and of the conducted afforestation management. As a result, predictions of the environmental effects of afforestation remain largely uncertain and are of limited use for afforestation planners and managers.

Against this background, the AFFORESTproject was conducted within the EU 5th Framework Programme for Research & Technological Development (Energy, Environment and Sustainable Development theme). The project ran for four years from May 2000 onwards. Six partners from four countries participated in the AFFOREST project. These were Belgium (Division Forest, Nature and Landscape of the Katholieke Universiteit Leuven), the Netherlands (Alterra Green World Research, Institute of Environmental Sciences Energy Research and Process Innovation (TNO-MEP), and Institute of Environmental Biology, Department of Biology at University of Utrecht), Sweden (Department of Forest Soils at the Swedish University of Agricultural Sciences) and Denmark (Department of Applied Ecology at Forest & Landscape Denmark) who co-ordinated the project.

The major objective of the project was to strengthen the knowledge regarding environmental effects of afforestation of agricultural land in north-western Europe with focus on the water, C and N cycles and their interrelationships. Furthermore, the ambition was to disseminate the improved knowledge by means of guidelines and a computerised system capable of providing support for decisions regarding 'where, how and how long to afforest?' in order to reach one or more environmental targets. This system should be able to assist managers to optimise the location of new forests according to specified environmental criteria.

All partners want to acknowledge the European Commission for financing AFFOREST (project no. EVK1-CT-1999-00020). We also thank our own institutions for co-financing the project. We are very grateful to Dr. Richard Skeffington and Dr. Miko Kirschbaum for giving us good and valuable scientific advice all through the project period. All laboratory and administrative help from our institutions are likewise highly acknowledged.

Also thanks to the external members of the reviewing board of the book chapters Annemarie Bastrup-Birk, Ingeborg Callesen, An De Schrijver, Wim de Vries, Juan Garcia, Per Gundersen, Griet Heuvelmans, Olivier Honnay, Linda Meiresonne, Mats Olsson, Richard Skeffington and some anonymous reviewers.

A special thanks goes to Sofie Bruneel who managed the whole review process and edited the text as a camera ready print proof.

Karin Hansen, Bart Muys and Gerrit Heil

CHAPTER 1

INTRODUCTION: DEMAND FOR AFFORESTATION MANAGEMENT IN NORTH-WESTERN EUROPE

G.W. HEIL[1], K. HANSEN[2], B. MUYS[3] and J. VAN ORSHOVEN[3]

[1]*Faculty of Science, Institute of Environmental Biology, Sorbonnelaan 16, 3584 CA Utrecht, the Netherlands, E-mail: g.w.heil@bio.uu.nl,* [2]*Forest and Landscape Denmark, KVL, Hørsholm Kongevej 11, 2970 Hørsholm, Denmark,* [3]*Katholieke Universiteit Leuven, Division Forest, Nature and Landscape, Celestijnenlaan 200E, 3001 Leuven, Belgium*

Abstract. The current landscapes in north-western Europe are the result of natural processes and human-induced land use and land use changes. Whereas during past centuries, the natural ancient forest cover disappeared and was -sometimes abruptly- converted in intensively used cultural landscapes, an opposed trend has become apparent recently. This trend reversal is at least partly related to the decision of the European Union and several of its member states to actively promote afforestation of –mainly- farmland, in order to respond to surplus agricultural production and to contribute to the international efforts to reduce the nitrate pollution of water bodies and the emission of greenhouse gases. However, many questions pertain regarding the implementation and evaluation of environmental effectiveness and efficiency of this afforestation policy.

The ultimate goal of this book is to assist stakeholders, such as forest and landscape planners and policy makers, in scheming and planning new forests in an environmentally sound way with as many positive and as few negative effects on the environment as possible. The book focuses on the influence of afforestation on carbon sequestration, nitrogen deposition, nitrate leaching, water recharge and potential biodiversity. It deals specifically with the ex-ante evaluation of afforestation measures based on i) field analyses in chronosequences of forest stands planted on former arable land, ii) modelling of these systems using mechanistic models describing the water, C- and N-cycles as well as the understory, iii) simplification of the process-based models into a metamodel (METAFORE) and iv) integration of data and models in a computerized spatial decision support system (the AFFOREST-sDSS).

G.W. Heil et al. (eds.), Environmental Effects of Afforestation in North-Western Europe, 1–18.

1. FOREST AND TREE SPECIES DISTRIBUTION IN NORTH-WESTERN EUROPE

The dominant natural vegetation cover in Europe is forest. The conditions for forest growth are very favorable. In the north, the Gulf Stream and the North Atlantic Drift warm the continent. In temperate Europe, mixed forests with both deciduous and coniferous trees dominate. Changes in distribution and spread of forest species in Europe occur continuously. In north-western Europe, the natural succession of vegetation will develop to mixed stands of cool temperate forests with oak and beech as the predominant tree species (Figure 1.1 and 1.2).

Modelling the present and past range limits of forest trees has suggested that climate change is the major driving force for range changes (Claussen & Esch 1994; Prentince et al. 1998). Anthropogenic activity did catalyze the spread of e.g. Beech (*Fagus sylvatica)* by increasing the rates of forest disturbance (Bradshaw 2004). Fossil pollen data from sites across Europe have been used to reconstruct the location of refugia of Beech and other species, and the spread from these refugia into their current ranges (Pott 1993) (Figure 1.3).

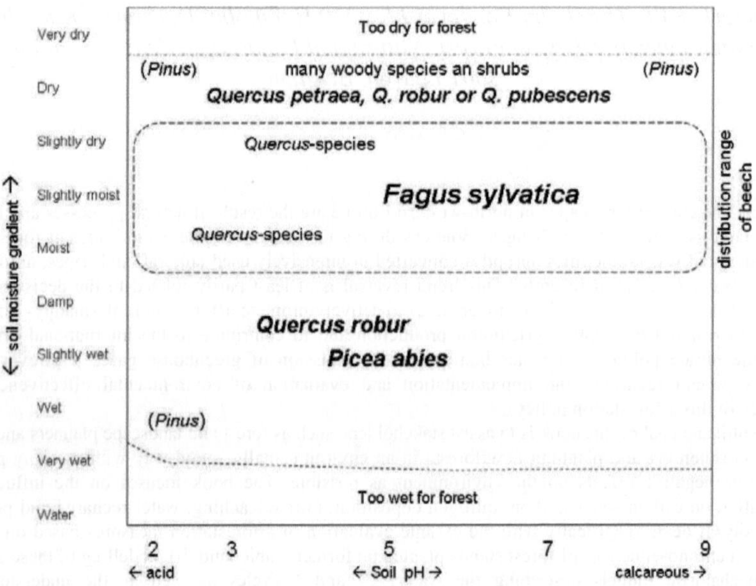

Figure 1.1. *Eco-diagram of oak (Quercus), beech (Fagus), pine (Pinus) and spruce (Picea), under (sub-)oceanic climate conditions and under different soil moisture – soil pH conditions. Letter size and boldness of the names of the tree species represent the strength of (co-)dominance in this moisture-pH domain (simplified after Härdtle et al. 2004).*

Three areas of southern Europe have been identified as refugia for Beech. These areas are the southern French coastal area, the Italian peninsula, and the southern Balkan. The spread of Beech took place in two steps. First in late glacial period (7.000-5.000 BC) Beech spread to central European mountains from these refugia. Second, with the stabilization of a climate favorable to deciduous trees species in

Holocene, beech spread into northern Europe, rapidly into north-western Europe, and more slowly into central and eastern Europe, due to physical barriers. The earlier distribution changes are strongly correlated with shifts in climate whereas the later changes are mostly controlled by competition between species, landscape topography, unfavorable late-glacial stages (Pott 1993). Similar records are observed for oak (*Quercus*) (Brewer et al. 2002).

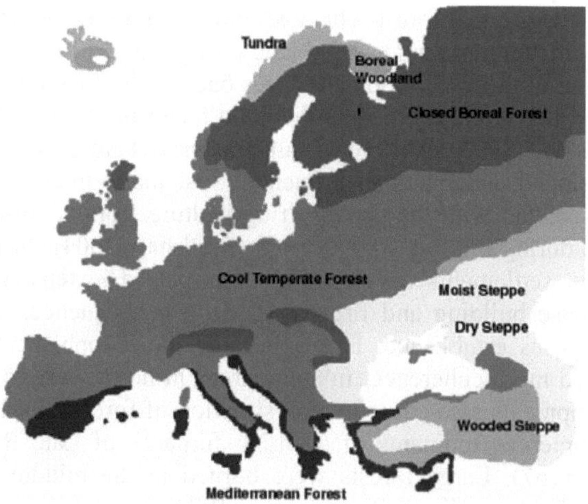

Figure 1.2. *Present potential distribution of forests (Kuusela 1994).*

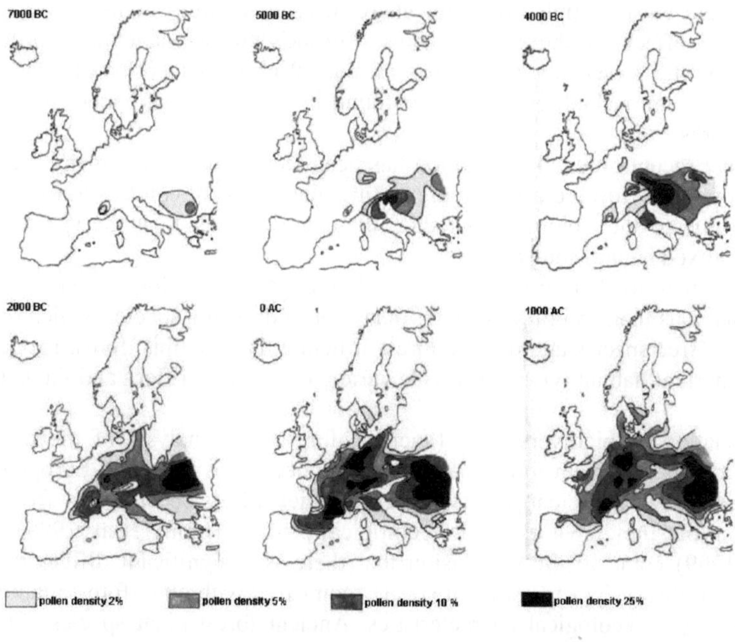

Figure 1.3. *Spread of beech (Fagus sylvatica) from its refugia to north-western Europe (after Pott 1993).*

At the time when large human impact took place, the forests were not in a static stage. They were dynamic and far from being in climax stage (Pott 1993). Glaciation during the most recent ice age and the presence of man affected the distribution of European flora. Pine and Beech stands, for example, were not completely developed and the formation of these species to mixed needle forests, to beech, to mixed beech- and to oak-Beech forests is still going on. The intensification of agriculture and urbanization changed the surface area and the structure of the relatively untouched natural forests. The original forests changed into a cultural landscape during distinct periods of intensification of land use and land changes under heavy human impact.

The beginning of these developments goes back to the first agricultural activities in history. Since then mankind became the main factor of the outcome of vegetation and landscape (Pott 1993). Different multi-factorial landscapes with mosaics of vegetation developed under human influence. Systematic interference with forests was closely associated with the spread of agriculture from south-eastern to north-western Europe during the last 10,000 years (Bradshaw 2004). In the early Middle Ages, permanent settlements developed. In this period, extensive areas of forest were cut for house building and fire wood. As a consequence, larger areas with agricultural purposes established. Farms became more sophisticated and villages grew and showed more coherence. In some areas in north-western Europe, ancient industrial developments also caused the destruction of forests. In the northern parts of Germany, numerous remnants of smelting furnaces of Late Roman Age were found (Richter 1967). Large forests were burned in the middle Ages (Graebner 1925; Tüxen 1968). According to Gimingham (1972) forest destruction took place in England and Wales during the period 1100 to 1700 A.C. Apparently, large fellings of especially oak took place in order to build ships during wartimes in Europe. Also, forests were cut to supply wood for charcoal used for iron-smelters, which had a destructive effect on the forests. Thus, Europe's forests have been profoundly affected by the presence and activities of man. With the exception of Scandinavia, few areas of untouched wilderness are found in Europe today, except for different natural parks.

During recent times, deforestation has been slowed down and many areas have been reforested. However, often monoculture plantations of conifers have been used because due to their pioneer character, these grow faster than the species of the original mixed natural forest.

Recent man-made forests represent the bigger part of that forested land, but offer poorer habitats than natural forests for many forest dwelling species, which require a mixture of tree species and diverse forest structure. Despite this, forest is one of the most important habitat types for a wide variety of species of flora and fauna (Figure 1.4).

In relation to biodiversity of (ancient) forests, an analysis of the ecological characteristics of ancient forest plant species in deciduous forests of Europe showed that the affinity for ancient forests of these species differs considerably from region to region, but they have a definite ecological profile (Hermy et al. 1999). Hermy et al. (1999) come to the conclusion that there is a significant difference in the response of the ancient forest plant species compared with other forest plant species for a variety of ecological characteristics. Ancient forest plant species tend to be more shade-tolerant than the other forest plant species; dry and wet sites are avoided. They are typical of forest sites with an intermediate pH and nitrogen (N) availability. Geophytes and hemi cryptophytes are more frequent amongst ancient forest plant species. The stress-tolerant plant strategy type is significantly more abundant under the ancient forest species when compared with other forest plant species and vice versa for the competitive plant strategy. This distinct ecological profile suggests that ancient forest plant species may be considered as a guild. The poor ability of these species to colonize new forest sites may be attributed to a complex of interacting variables such as limited dispersal abilities (many species have a short-distance dispersal strategy), low diaspore production and recruitment problems (e.g. low competitive ability). Due to their distinct ecological profile and low colonizing abilities, ancient forest plant species may be considered as important biodiversity indicators for forests.

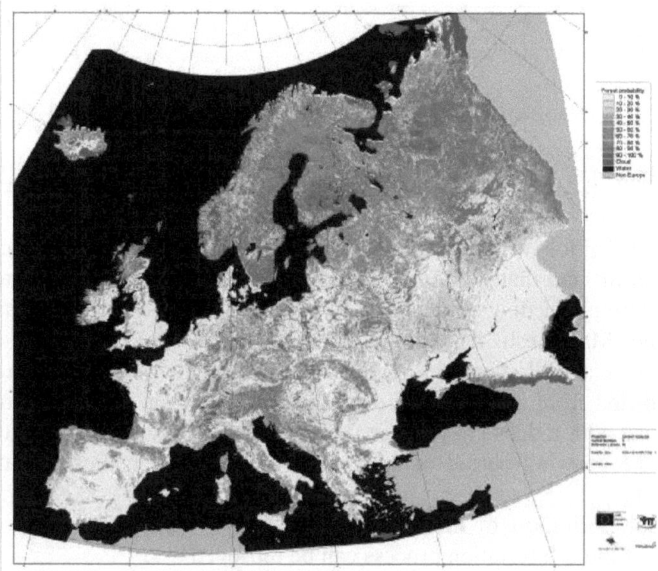

Figure 1.4. *Map of European forests produced for the pan-European area by Kennedy & Folving (1999). An image interpretation was carried out using an AVHRR satellite image compiled from the red and near infra-red channels, for the entire pan-European area. Legend: light = low probability of forest, dark grey = high probability of forest, black = water.*

2. AFFORESTATION IN A NORTH-WESTERN EUROPEAN CONTEXT

Forest is an essential element of the European cultural landscapes. It covers about one third of Europe's land area (27% in central, 32% in southern and 50% in northern Europe according to the FAO's Forest Resources Assessment Report (FAO 2000) or around 195 million ha as estimated by Kuusela (1994). Out of this, the amount of natural forest in western Europe is just 2-3% or less. Presently, the area of forest in Europe is increasing, owing predominantly to the abandonment of marginal farmland and afforestation. Forest expansion in Europe (excluding former USSR) occurred at a rate of 0.4 million ha per year (or 0.3% of the forest area) in the 1990ies (Mather 2000). Opposed to this, the global trend is –0.2% per year (according to FAO). Also, in north-western Europe, afforestation of agricultural land is an important land use change (Table 1.1).

Table 1.1. *Absolute and relative area of agricultural land afforested in Sweden, Denmark, the Netherlands and Belgium (Flanders) between 1992 and 2004.*

Country	Agricultural Area (1992, Million Hectares)	Afforested Area (1992-2004, Thousands of Hectares)	Afforested Area (1992-2004, %)
Sweden	3.4	43.0	1.26
Denmark	3.1	27.3	0.88
The Netherlands	2.5	4.0	0.16
Belgium, Flanders	0.8	1.5	0.19

A general definition of afforestation is the direct human-induced conversion to forested land of land that was not covered by trees for a period of at least 50 years through planting, seeding and/or the human-induced promotion of natural seed sources (the Marrakesh Accords, UNFCCC 2002). In this book, we define afforestation as a well specified human activity by which the use of a given land area is modified from agriculture, i.e. a land use without a tree cover into a forest, i.e. a land use with a dense tree cover. The actions taken at the time of afforestation and along the growing period of the trees are integrated in the term afforestation management.

From our definition of afforestation four major elements for the conceptual description of the afforestation process can be retrieved:

- The initial system, i.e. a land unit under agriculture before the time of afforestation, defined by given climate and soil conditions, a certain land cover (cropland or grassland) and specific agricultural practices (e.g. fertilization level);
- The afforested system, i.e. the same land unit or part of it, but now covered with trees;
- Afforestation management, i.e. the set of choices regarding forest management, including the tree species choice and site preparation applied to the initial system, and the stand tending applied to the developing afforested system;
- The temporal dimension or afforestation time, i.e. the moment in time after afforestation, at which the state of the afforested system is evaluated.

Both the initial system and the afforested system can be characterised by state variables informing about their performance in social, economic and environmental terms. In this book, we focus on four environmental state variables:

- the annual amount of carbon (C) stored in soil and biomass;
- the annual amount of nitrogen (N) leaving the soil profile by leaching;
- the annual amount of water leaving the system through runoff or percolation and ultimately recharging groundwater and surface water bodies, also called precipitation surplus or water yield;
- the field layer biodiversity.

These environmental state variables inform us about the environmental effect/performance of the initial system, and of the afforested system as a function of time. We consider this multi-dimensional environmental performance as a fifth conceptual element of the afforestation process.

What we are particularly interested in is the change of the environmental performance, i.e. the difference in environmental performance between the initial system and the afforested system at a given time for each of the environmental state variables, separately or in combination. Information on these changes (how much?) provides the basis for supporting decisions on where to plant the trees (which initial system?), which species to plant and how to manage it (which afforestation management?) and how long to keep the trees in rotation before reaching a certain effect (what time?).

With regard to its contribution to safeguarding and improving the quality of the environment, it is generally assumed that afforestation can play a substantial role in improving the environment (Powlson et al. 1998; IPCC 2000; Lal 2004). However, there is limited scientific understanding about the influence on the water, carbon and nitrogen cycles, of the physical conditions of the sites designated for afforestation, of the former agricultural management of these sites and of the conducted afforestation management. As a result, predictions of the environmental effects of afforestation remain largely uncertain and are of limited use for afforestation planners and managers. In addition, afforestation of today is supposed to serve multiple purposes, whereas afforestation of the past primarily aimed at increased wood production.

In regions dominated by agricultural activities, N is recognised as a major pollutant of the environment (Figure 1.5). Intensified management of farmland has caused modern arable soils to carry large pools of N bound in organic matter, and soils have a high nitrifying capacity (Jussy et al. 2000). Fertilizer application is the dominant source of groundwater nitrate contamination (van der Voet et al. 1996). In contrast to agricultural soils, old existing forests are characterised by a more tight N cycle. The rate of net nitrification is also rather low. Water from old forest land is, therefore, generally of good quality with a relatively low concentration of dissolved N compared to other land uses (Thornton et al. 2000). Nitrogen leaching is at risk when ecosystems become saturated with N, i.e. when the availability of inorganic N exceeds the demand from plants and micro-organisms (Aber et al. 1989; Gundersen 1991). Thus, an effective measure to reduce leaching of nitrate to the groundwater and hereby create possibilities for a better quality of groundwater could be the conversion of agricultural land to forest. However, because of the high N status of former arable soils, retention of N in afforested ecosystems may be less efficient than in old forest land, resulting in an enhanced risk of nitrate leaching.

Afforestation could also constitute a potential threat for groundwater quality since trees are highly efficient scavengers of pollutants (Allen & Chapman 2001). The scavenging leads to enhanced supply of N, which in combination with high mineralization may potentially result in N saturation and increased nitrate leaching (Allen & Chapman 2001). However, nitrate leaching may still be less than from the former agricultural land use.

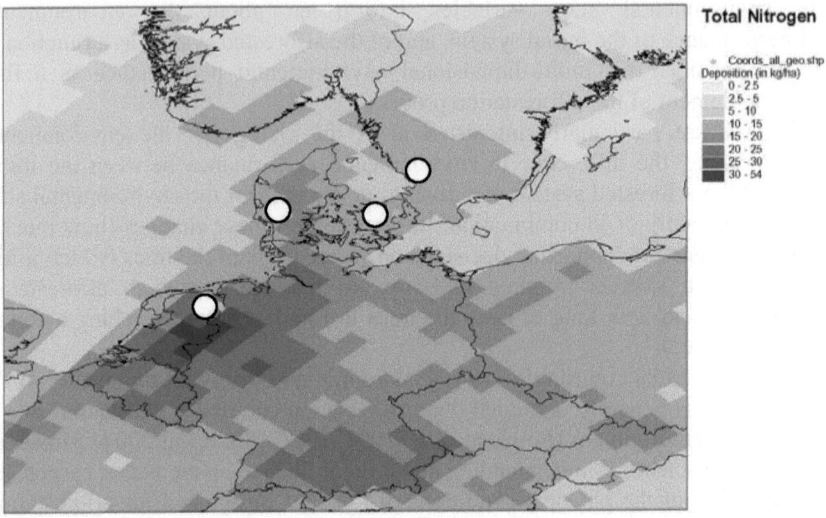

Figure 1.5. *Nitrogen deposition measured at different sites in Sweden, Denmark, and the Netherlands (light circles) (kg ha*$^{-1}$*), and calculated according to the modeling methods explained in Chapter 5.*

The coming into force of the Kyoto Protocol in 2005 has brought attention to the multiple positive roles of forests and wood in C cycles: mitigation of the negative effects of climate change, sequestration of C, provision of a renewable C neutral energy source, substitution for non-renewable fuels or raw materials, and contribution to energy conservation. Sequestration of C has been recognised as an important environmental effect of afforestation and in a European perspective afforestation may provide the greatest potential for C sequestration in agricultural soils (Powlson et al. 1998). Following the ratification of the Kyoto Protocol, afforestation of former arable land has been acknowledge as a land use change that contributes to the mitigation of increasing atmospheric CO_2-concentrations (IPCC 2000). The most evident effect of afforestation on C sequestration is the net sink of atmospheric CO_2 in the growing biomass, however, the soil C pool may also change slowly after afforestation. A basis for reporting and modelling C changes after afforestation is needed to fulfil the Kyoto commitments.

Afforestation may have an adverse environmental effect through the reduction of water recharge to ground- and surface water due to higher transpiration of forest compared to agricultural land. Studies indicate that water recharge declines with an increase in forest cover. The decline in recharge is generally larger for coniferous forest compared to deciduous forests. The reductions in recharge strongly differ from site to site due to differences in climate, site characteristics and field layer vegetation. Thus, to make decisions on afforestation of former agricultural areas in Europe more information on the impact of afforestation on hydrological fluxes is needed. In order to increase the knowledge on the long-term effects of afforestation a series of forests of different age planted on comparable soils (chronosequences) were investigated in the AFFOREST project.

3. THE AFFOREST PROJECT – BACKGROUND AND OBJECTIVE

A surplus of grain has caused decisions and changes in the EU Common Agricultural Policy (CAP) and important areas of agricultural land have been taken out of production. Today, it is estimated that approximately 12.2 million hectares of agricultural land are available for alternative land use compared with the situation in 1985. Much of this land is suitable for afforestation. The governments of several EU countries have decided to actively promote an increase in the forest area planted on former farmland within the next decades. Apart from improving the economic structure of agricultural systems, the main aims of this change in land use are to create a better quality of groundwater, to increase the stock of C in the phytomass and soil of the growing forest as a contribution to the lowering of the large atmospheric emissions of greenhouse gases, and to improve the recreational possibilities for the public. The decisions on afforesting farmland in the EU in the coming decades are reasonably fresh and as such have generated a rather new area for applied research. There is a major need for research on the implications of the decisions in relation to the environment, especially on the effect of the afforestation on the leaching of nitrate and the groundwater recharge and on reducing the net emissions of carbon dioxide through increasing the carbon pools in the new forests.

Several biogeochemical research projects have been performed in the past in forest ecosystems, including nitrate leaching and to some extent carbon sequestration. This research has been very beneficial for basic research and monitoring on a forest stand level and can be used as a reference for measurements in afforested systems. However, forests on former agricultural land will be subject to different processes and may have different environmental impacts since the soils have quite different properties. In addition, the range of possible design and management options create a huge number of degrees of freedom which have hardly ever been studied. Therefore, there is a need for plot measurements in newly afforested systems.

Afforestation is a shift in land use which may have significant effects on the environment. It is advisable, already in an early phase of the afforestation process, to evaluate if the afforestation investments will lead to the preferred environmental effects. When planning and designing the new forests of the future there is a great opportunity to reduce the environmental effects and develop a sustainable management of afforested systems by taking into account the potential of the soil (site classification), the size and structure of the forest, the selection of different tree species, and the future management. Against this background, the AFFOREST project was conducted within the EU 5th Framework Programme for Research & Technological Development (Energy, Environment and Sustainable Development theme) - contract no. EVK1-CT1999-00020. The project ran for four years during the period 2000-2004 and included partners from four countries: Belgium, the Netherlands, Sweden and Denmark.

The major objective of the AFFOREST project was to strengthen the knowledge regarding environmental effects of afforestation of agricultural land in north-western Europe with focus on the water, carbon and nitrogen cycles, their interrelationships and potential biodiversity. Furthermore, the ambition was to disseminate the improved knowledge by means of guidelines and a computerized system capable of providing support for decisions regarding 'where, how and how long to afforest?' in order to reach one or more environmental targets. This system should be able to assist managers to optimize the location of new forests according to specified environmental criteria. Policy elaboration for afforestation areas and the physical planning, design, and management of these forest systems can seldom be targeted to one single objective. The complexity of environmental and ecological systems requires planning and designing for multiple objectives. With this project, at least 4 impact categories are studied, the results of which are integrated in a multiple objective decision support system (AFFOREST-sDSS). Up-scaling of research results from detailed afforestation plot studies to the effect on a higher level at the regional landscape level in AFFOREST is performed using knowledge rules, transfer functions and process-based models linked to Geographical Information Systems.

4. STAKEHOLDER INVOLVEMENT

Afforestation is based on decisions driven by multiple objectives. When multiple-objective demands are involved it is difficult and involves many pitfalls may emerge if one tries to plan a new forest in a proper way realizing the array of natural, political, and socio-economic conditions. Maximizing a single objective will often cause trade-offs for another objective. Addressing more than one objective is therefore a complex optimization challenge. Even when the objective of afforestation only concerns the various environmental effects it may involve conflicts and trade-offs.

Decisions on how and where to afforest and how much these decisions will affect the environmental impacts is a compromise to meet the goals set by managers and stakeholders. A stakeholder analysis was performed in AFFOREST which gave an idea of potential stakeholders to contact in north-western Europe. Furthermore, a questionnaire was developed exploring the interests and the wishes of the potential stakeholders (Figure 1.6).

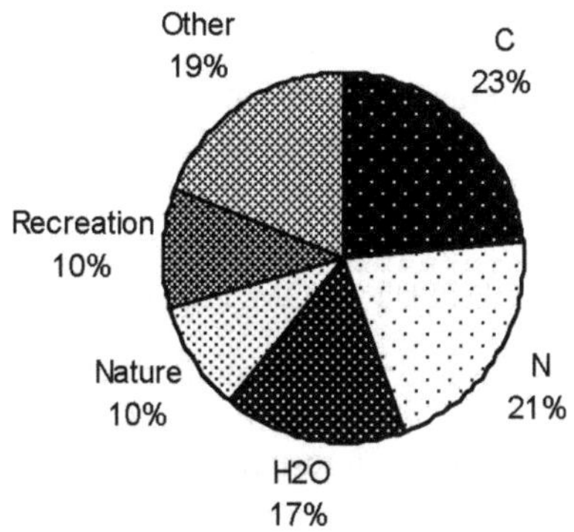

Figure 1.6. *Stakeholder preferences for afforestation objectives. C = carbon sequestration, N = nitrogen leaching, H_2O = water storage capacity, Nature = potential nature value, and Recreation = forest quality for recreation.*

Stakeholder meetings and interviews were organized to analyze the preference for different afforestation measures. In the first year of the AFFOREST project, two separate stakeholder meetings, i.e. a Swedish-Danish and a Flemish-Dutch, were organized during which participants from several public and private forest management organizations and planning institutions on national and regional scales participated. Several questionnaires were returned and formed the basis for further planning of activities.

Afforestation objectives and goals vary from one country to another and even within countries. The specific conditions should always be considered in the entire afforestation process, from policy decisions through location of the new forest, establishment and management, and the final utilization of the forests. In addition to the latter, the interest of stakeholders for the impact scale of afforestation measures is mainly at a local scale as shown by the results from the questionnaires (Figure 1.7).

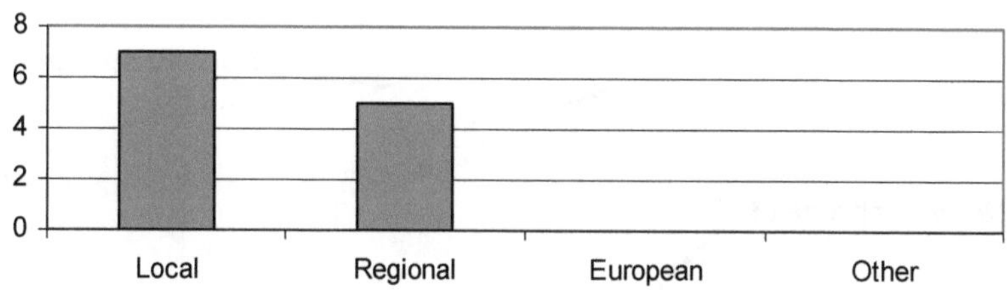

Figure 1.7. *Stakeholder interest for a particular scale in an afforestation DSS.*

Important issues that came out of the stakeholder meetings and the questionnaires:

- Stakeholders should be able to prioritize functions by weights and they should be given possibilities to setting quantitative norms. The latter point implies high data reliability and model reliability;

- The AFFOREST-sDSS should work on two different scale levels (low and high). There should be flexibility between the scales;
- The stakeholders want an open system with the possibility to bring in own data layers and with a high transparency, so that the stakeholder can see what happens within the AFFOREST-sDSS.

5. OUTPUT FROM THE PROJECT

The research in AFFOREST was organised along three lines, based on a common conceptual framework and terminology. Throughout the book, we exclusively use the term afforestation' to designate 'afforestation of agricultural land'.

5.1. Field data

Along the entire project period considerable effort has been put in the collection of field data in chronosequences of afforested land in order to better document, quantify and understand changes in the C, N and H_2O pools and fluxes when agricultural land is afforested according to some well specified afforestation practice (Chapters 2, 3 and 4). Chronosequences of oak and spruce stands, planted on former arable land during the latest 0-90 years, were selected in Denmark, southern Sweden and the Netherlands (Figure 1.8 and Table 1.2). Measurements were conducted in order to obtain data on:

- Effects of afforestation on deposition, nitrate leaching, groundwater recharge, and carbon sequestration in field experiments of new forests established on agricultural land
- Temporal and spatial dynamics to create data for the different models and a decision support system.

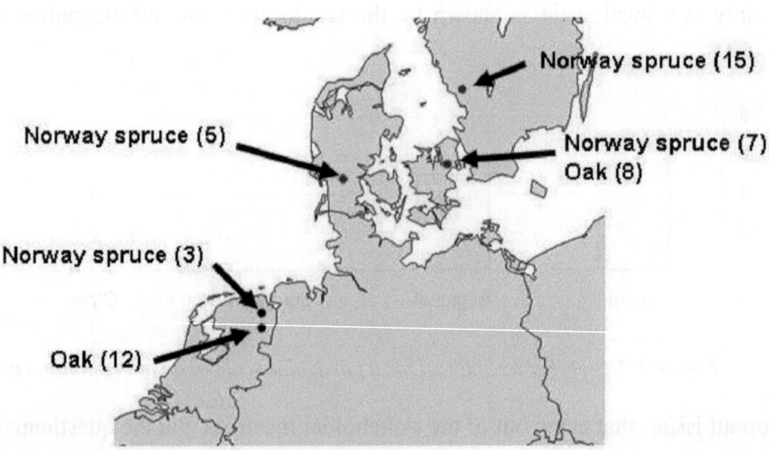

Figure 1.8. *Locations of the AFFOREST chronosequences within the north-west European region.*

Chronosequences form an excellent experimental setup to evaluate effects of afforestation over time. They are an important source of information on hot topics like carbon sequestration, precipitation surplus, and biodiversity effects following land conversions to forest. Chronosequences are based on two main assumptions, 1) the assumption of constant site quality and 2) the assumption of known age. The first assumption means that all forest stands are situated on the same site in terms of growth conditions, including climate, elevation, aspect, soil texture, stoniness and drainage class. Most often this assumption is met by choosing stands in each other's vicinity, but it must be checked in the field. For the second assumption the age of the trees can be easily determined using tree ring analysis. Most often the study will not only be interested in the effects of afforestation as such, but also in the effects of a particular species or management practice. It should be emphasized that it is intuitively believed that the species effect is bigger than the management effect, but in reality it might be the other way around. The tree species choice is visible in the field, but the management practices that were used must be checked.

5.2. Models

The field data and additional data from published experiments have been used in a second part to calibrate, validate, integrate and simplify mechanistic models describing the H_2O-, C- and N-cycles in a coupled way so that the models can be applied to simulate the effects of afforestation in other circumstances than those for which calibration has been done, at the condition of course that valid model input data are available (Chapters 7 and 8). The results of the deposition modelling are described in Chapter 5. The final result of the model simplification is the metamodel METAFORE (Figure 1.9), which is described in Chapter 9.

In addition, an field layer model has been developed to test the possibilities for the development of the forest field layer after afforestation of former agricultural land, and to show the interactive effect of light and nitrogen on the potential field layer plant performance (Chapter 6).

Table 1.2. *Characteristics of the different field sites and chronosequences.*

Country	Denmark	Denmark	Denmark	The Netherlands	The Netherlands	Sweden
Name	Vestskoven 1	Vestskoven 2	Gejlvang	Sellingen	Drenthe	Tönnersjöheden
Closest city	Copenhagen	Copenhagen	Vejle	Groningen	Assen	Halmstad
Tree species	Norway spruce	Oak	Norway spruce	Oak	Norway spruce	Norway spruce
Number of stands	5	6	5	3	3	5
Year of planting	1997 1990 1988 1973 1969	1993 1988 1977 1979 1970 before 1800	1997 1994 1981 1976 1960	1998 1995 1991	1997 1994 1990	1983 1972 1937 1928 1910
Soil type (USDA/FAO)	Hapludalf	Hapludalf	Durorthod/ Quartzipsamment	Gleyic Podzol	Gleyic Podzol	Arenosol
Texture	Sandy loamy	Sandy loamy	Sandy	Sandy	Sandy	Sandy
Avg. precipitation, (mm/yr)	640	640	900	745	745-830	800-1050
Avg. temperature (°C/yr)	7.7	7.7	7.7	9	9	6.1-7.3

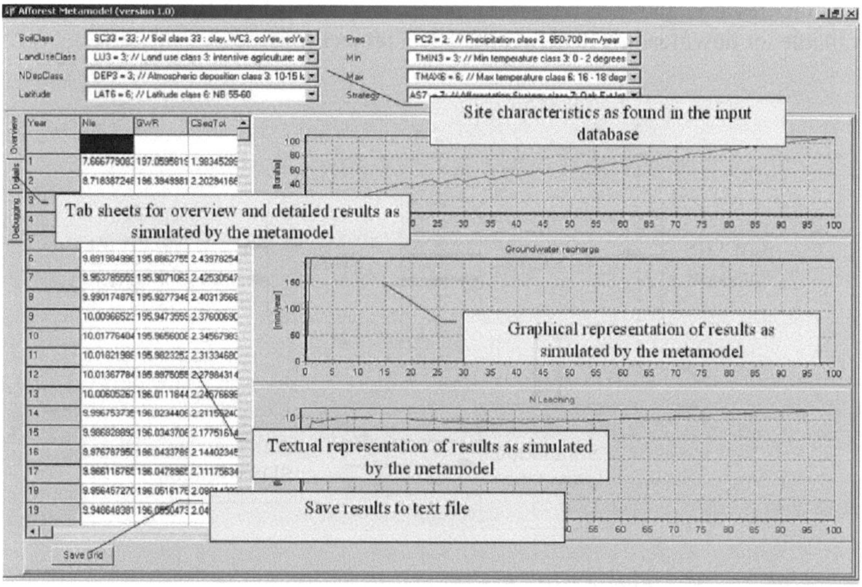

Figure 1.9. *Output interface of the metamodel METAFORE.*

5.3. The decision support system

Model input datasets covering the four studied countries were compiled and METAFORE was applied to these input datasets. The model output was captured in a database which can be queried, i.e. which allows questions to be formulated, in order to obtain useful answers. The model output database, the query tools and a user interface are the key components of the spatial AFFOREST-decision support system (AFFOREST-sDSS) (Chapter 10). The spatial decision support system allows scientific results to be translated into policy-relevant guidelines, and supports decision-making about afforestation strategies in particular places. The AFFOREST-sDSS is schematically presented in Figure 1.10 and the graphical user interface of the AFFOREST-sDSS is shown in Figure 1.11.

5.4. Guidelines and AFFOREST website

Results and evidence from chronosequence experiments, literature and operational knowledge has together with the AFFOREST-sDSS contributed to the elaboration of guidelines for environmentally sound afforestation (Chapter 11). A separate booklet of the guidelines, and other results is available at the AFFOREST website (http://www.sl.kvl.dk/afforest/) and on a CD-Rom (Figure 1.12). All major AFFOREST products, which do not form part of this book, mainly the AFFOREST literature review, the METAFORE model and the AFFOREST-sDSS are also available for downloading at the AFFOREST project website.

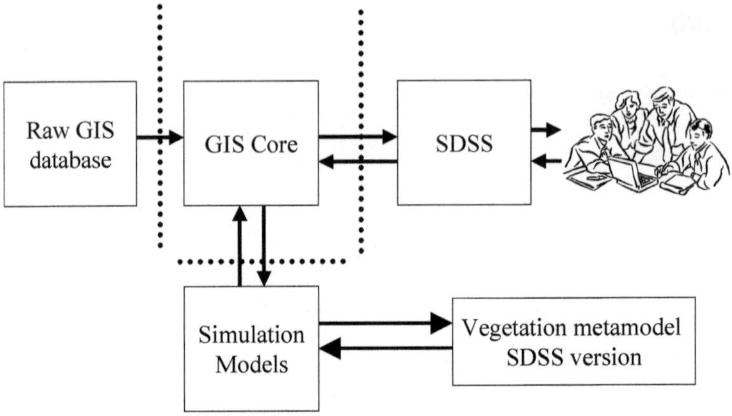

Figure 1.10. *Overview of the AFFOREST system.*

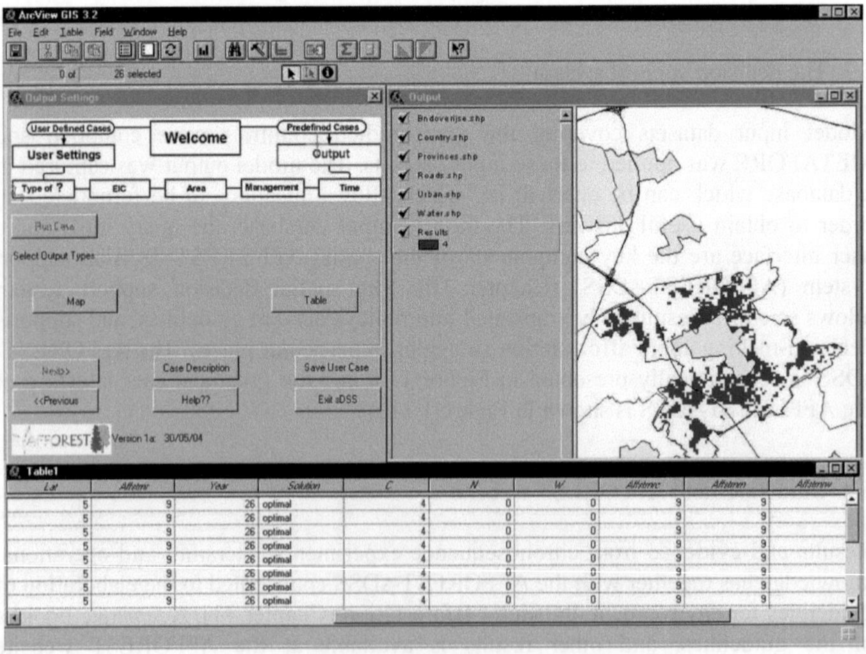

Figure 1.11. *Example of the user interface of the sDSS.*

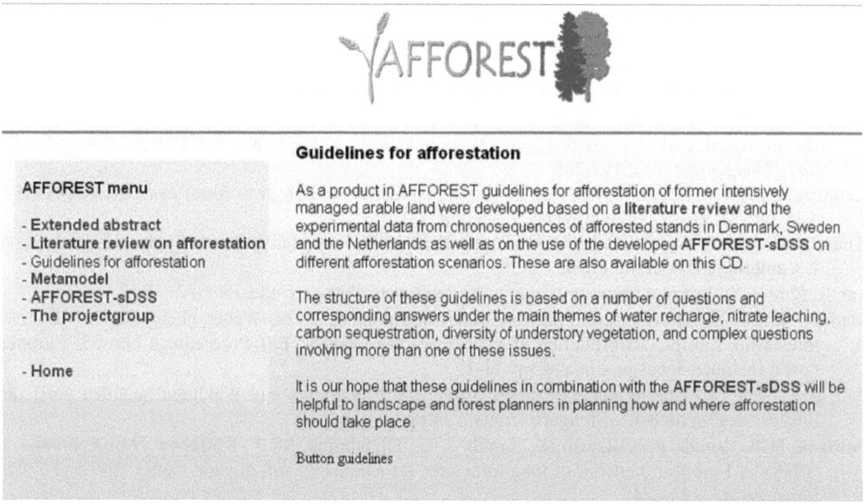

Figure 1.12. *Front page of the guidelines on the AFFOREST CD-ROM . The guidelines are also available at the AFFOREST website* http://www.sl.kvl.dk/afforest/.

6. CONCLUSION

In this book, we present some fundamental views on afforestation in north-western Europe and we have given access to a tool for planning and environmental impact assessment. It is our hope that this book along with the AFFOREST products will be helpful and inspire landscape and forest planners in planning new forests on abandoned arable land in north-western Europe. We also hope to offer inspiration to other researchers in the area of afforestation and decision support. The research of this book has been carried out in the 5th framework of the EU, contract no. EVK1-CT1999-00020.

7. REFERENCES

Aber, J. D., Nadelhoffer, K. J., Steudler, P., & Melillo, J. M. (1989). Nitrogen saturation in northern forest ecosystems. *BioScience, 39,* 378-386.

Allen, A., & Chapman, D. (2001). Impacts of afforestation on groundwater resources and quality. *Hydrogeology Journal, 9,* 390-400.

Bradshaw, R. H. W. (2004). Past anthropogenic influence on European forests and some possible genetic consequences. *Forest Ecology and Management, 197,* 203-212.

Brewer, S., Cheddadi, R., de Beaulieu, J. L., & Reille, M. (2002). The spread of Quercus throughout Europe since the last glacial period. *Forest Ecology and Management, 156,* 27-48.

Claussen, M., & Esch, M. (1994). Biomes computed from simulated climatologies. *Climate Dynamics, 9,* 235-243.

FAO. (2000). Forest Resources of Europe, CIS, North America, Australia, Japan, and New Zealand - main report. United Nations Publications no. 99-II-E-36.

Gimingham, C. H. (1972). Ecology of heathlands. London: Chapman and Hall.

Graebner, P. (1975). Die Heide Norddeutschlands. Leipzig. 272 pp.

Gundersen, P. (1991). Nitrogen deposition and the forest nitrogen cycle: role of denitrification. *Forest Ecology and Management, 44,* 15-28.

Härdtle, W., Ewald. J., & Hölzel, N. (2004). Wälder des Tieflande und der Mittelgebirge. Stuttgart: Ulmer. 252 pp.

Hermy, M., Honnay, O., Firbank, L., Grashof-Bokdam, C., & Lawesson, J. E. (1999). An ecological comparison between ancient and other forest plant species of Europe, and the implications for forest conservation. *Biological Conservation, 91,* 9-22.

IPPC. (2000). Special Report on Land Use, Land-Use Change and Forestry. UNEP-WMO.

Jussy, J. H., Colin-Belgrand, M., & Ranger, J. (2000). Production and root uptake of mineral nitrogen in a chronosequence of Douglas-fir (Pseudotsuga menziesii) in the Beaujolais Mounts. *Forest Ecology and Management, 128,* 197-209.

Kennedy, P., & Folving, S. (1999). Sub-pixel area estimation of European forest using NOAA-AVHRR data. http://rogow99.sggw.waw.pl/07_session_3/03/

Kuusela, K. (1994). Forest Resources in Europa 1950-1990. European Forest Institute. Research Report 1. Cambridge University Press.

Lal, R. (2004). Soil carbon sequestration to mitigate climate change. *Geoderma, 123,* 1-22.

Mather, A. (2000). Afforestation: Progress, trends and policies. In N. Weber, (Ed.), NEWFOR – New forests for Europe: Afforestation at the turn of the century, EFI Proceedings No. 35, European Forest Institute, Joensuu, Finland, pp. 11-19.

Pott, R. (1993). Farbatlas Waldlandschaften: ausgewählte Waltypen und Waldgesellschaften unter dem Einfluss des Menschen – Stuttgart: Ulmer, 224 pp.

Powlson, D. S., Smith, P., Coleman, K., Smith, J. U., Glendining, M. J., Körschens, M., & Franko, U. (1998). A European network of long-term sites for studies on soil organic matter. *Soil and Tillage Research, 47,* 263-274.

Prentince, C, Harrison, S. P., Jolly, D., & Guiot, J. (1998). The climate and biomes of Europe at 6000yr BP: comparison of simulations and pollen based reconstructions. *Quaternary Science Review, 17,* 659-668.

Richter, R. (1967). Kaiserzeitliche Waldverwusting in der Schleswiger Geest. Ein Beitrag zum Heideproblem in Schleswig-Holstein. Mitt. Flor. Soz. Arbeitsgem. N.F. 11/12: 223-229. Stolzenau-Weser.

Thornton, K. W., Holbrook, S. P., Stolte, K. L., & Landy, R. B. (2000). Effects of forest management pratices on Mid-Atlantic streams. *Environmental Monitoring and Assessment, 63,* 31-41.

Tüxen, R. (1968). Die Luneburger Heide. Wrden und Vergehen der nordwestdeutschen Heidelandschaft. In A. Kell, (Ed.), Nuzeitliche Biologie, pp 9-55. Hannover.

UNECE. (2005). Forest certification—Do governments have a role? ECE/TIM/05/N01.

UNFCCC. (2002) Report of the Conference of the Parties on its Seventh Session, Held in Marrakech from 29 October to 10 November 2001. Addendum Part Two: Action Taken by the Conference of the Parties.

Van der Voet, E., Klein, R. K., & Udo de Haes, H. K. (1996). Nitrogen pollution in the European Union – origins and proposed solutions. *Environmental Conservation, 23,* 120-132.

CHAPTER 2

CARBON SEQUESTRATION IN SOIL AND BIOMASS FOLLOWING AFFORESTATION: EXPERIENCES FROM OAK AND NORWAY SPRUCE CHRONOSEQUENCES IN DENMARK, SWEDEN AND THE NETHERLANDS

L. VESTERDAL[1], L. ROSENQVIST[2], C. VAN DER SALM[3], K. HANSEN[1], B.-J. GROENENBERG[3] and M.-B. JOHANSSON[2]

[1]Forest and Landscape Denmark, KVL, Hørsholm Kongevej 11, 2970 Hørsholm, Denmark, E-mail: lv@kvl.dk, [2]Swedish Agricultural University, Box 7001, 750 07 Uppsala, Sweden, [3]Alterra Green World Research, P.O. Box 47, 6700 AA Wageningen, the Netherlands

Abstract. There is limited knowledge of the contribution of afforested arable land to mitigation of greenhouse effects. In the AFFOREST project we evaluated the rate and magnitude of carbon (C) sequestration in biomass and soils following afforestation of cropland. Two oak (*Quercus robur*) and four Norway spruce (*Picea abies*) afforestation chronosequences (age range 1 to 90 years) were studied with respect to C sequestration in Denmark, Sweden and the Netherlands.

Biomass C sequestration ranged between 2.7 and 4.6 Mg C ha^{-1} yr^{-1} for stands younger than 45 years with no clear influence of different site characteristics. Such effects were probably masked by the soil enrichment, which is a legacy of former agriculture. Biomass C sequestration differed more between sites after 40-50 years owing to different management, tree species-specific growth patterns and less influence of former fertilization.

For the total soil compartment studied, i.e. forest floor and mineral soil 0-25 cm, afforestation of cropland as a minimum resulted in unchanged soil C contents and in most cases led to net C sequestration. Rates of soil C sequestration ranged from being negligible in two of the Danish chronosequences to 1.3 Mg C ha^{-1} yr^{-1} for the Dutch chronosequence. The allocation of sequestered soil C was also quite different among chronosequences. While forest floor development consistently led to C sequestration, there was no general pattern in mineral soil C sequestration. In the short term (30 years), tree species had little influence on total soil C sequestration. Afforestation of nutrient-poor sandy soils seemed to result in larger C sequestration in forest floors and the whole soil than afforestation of nutrient-rich, clayey soils.

G.W. Heil et al. (eds.), Environmental Effects of Afforestation in North-Western Europe, 19–51.

For the afforested ecosystem as a whole, the general contribution of soils to C sequestration (i.e. to a net gain in C stock) was about one third of the total C sequestration. The contribution of soil varied among the chronosequences from none to 31%, which is not far from reported contributions of soil in similar studies. In the short term (30-40 years), total C sequestration was higher in Norway spruce than in oak whereas soil type did not clearly influence the rate of C sequestration.

The work in AFFOREST has improved the knowledge of C sequestration in afforested cropland. The new results may help to bridge the gap between existing knowledge and policy demands.

1. INTRODUCTION

Sequestration of carbon (C) has recently been recognized as an important environmental effect of afforestation, and in a European perspective afforestation may provide the greatest potential for C sequestration in agricultural soils (Powlson et al. 1998). A new challenge in the context of climate change mitigation is the management of terrestrial ecosystems to conserve existing carbon stocks and to remove carbon from the atmosphere by increasing the existing stocks (Malhi et al. 1999). Afforestation of former arable land has been acknowledged under the Kyoto Protocol (article 3.3) as an eligible activity that contributes to the mitigation of increasing atmospheric CO_2 concentrations. As a specific change in land use, increases in terrestrial C stocks due to afforestation may serve as one of the measures to meet national reduction commitments.

Forests are generally characterized by a higher density of carbon than arable land, mainly due to the presence of perennial vegetation with a high biomass. Soils under forest are often considered to contain larger stocks of C than cropland soils, but the literature is not consistent (Rodriguez-Murillo 2001; Krogh et al. 2003; Lettens et al. 2004). What happens when cropland is converted to forest? Based on comparisons of C stocks in forests and arable land afforestation is expected to result in significant sequestration of C due to accumulation of woody biomass and further accumulation of organic matter in the soil. Already Billings (1938) reported quantitative data on soil C sequestration in former arable soils based on a study of shortleaf pine (*Pinus echinata* Mill.) afforestation in North Carolina. It is nevertheless still debated whether soil C stocks generally will increase following afforestation. For instance, the previous arable land use may be important for predicted increases in soil carbon, as soil C differ between permanent pasture, annually tilled cropland and no-till management (Römkens et al. 1999; Dick 1983; Denef et al. 2004).

There are, in addition, large uncertainties in terms of the rate of C sequestration and also the allocation of carbon to soil and biomass following afforestation. The contribution of soils to C sequestration is especially uncertain and it is a significant future challenge to quantify the potential soil sinks for CO_2 (Smith 1999; Garcia-Oliva & Masera 2004). In a review Johnson (1992) concluded that the reversion of former agricultural land to forest usually results in substantial increases in soil C, and Bouwman & Leemans (1995) suggested that 50 Mg C ha^{-1} would be sequestered in afforested tropical soils in 30 years. On the other hand some studies indicate that these expectations for soils are far too high (Hamburg 1984; Jug et al. 1999; Richter et al. 1999; Paul et al. 2002). The aggrading biomass of forest trees accounts for much of the C sequestered after afforestation. However, the relative contributions of biomass and soil are uncertain and must be expected to vary between forest, soil and climate types. Few studies included concurrent measurements of soil and biomass C sequestration, but these studies suggested that about 25% of the total C can be sequestered in the soil (Richter et al. 1999; Hooker & Compton 2003).

While C stored in forest biomass is strongly influenced by changing forest management or by disturbances such as clearcutting and replanting, C stored in mineral soils is less susceptible to such changes (Dewar & Cannell 1992). Soils may be a more permanent sink for C and the question remains to what extent they contribute to the total C sequestration of an afforested ecosystem. The quality of C stores with respect to permanency should be considered together with the quantity of C. Better quantitative information on C sequestration is needed in several European countries in order to evaluate the potential contribution of terrestrial systems and to meet the obligations following ratification of the Kyoto Protocol.

Direct measurements of C sequestration would require flux towers or the reassessment of C pools over a considerably long time period to assess changes. In order to save time, sampling of paired plots of afforested and arable land has often been undertaken. Many of the previous studies on afforestation were done on marginal land, on former pastures or on land abandoned before frequent fertilization and liming were introduced in agriculture. Such studies may not be representative for the intensively managed arable soils afforested today. Also, there is a need for studies that provide a basis for reporting and modeling national C sinks in the short-term, e.g., during the Kyoto commitment period 2008-2012.

This chapter evaluates the general effect of afforestation of former cropland on C sequestration in a synthesis of AFFOREST chronosequence experiments in three European countries. The results from the AFFOREST project are compared to previous studies of afforested cropland. The specific objectives were i) to estimate the rates of total C sequestration, ii) to determine the relative contributions to C sequestration of the soil and biomass components of the new forest ecosystems, iii) to study the possible differences in C sequestration between deciduous (oak) and coniferous (Norway spruce) tree species, and iv) to explore C sequestration at contrasting soil types.

2. MATERIALS AND METHODS

2.1. General approach

To study the effect on carbon sequestration following afforestation with oak and spruce, chronosequences of afforestation stands were selected in Denmark, southern Sweden and the Netherlands (Chapter 1). Measurements were conducted in each chronosequence stand to assess the changes in C storage. Biomass was measured for estimation of above- and belowground biomass C storage, and forest floors and mineral soils were sampled to assess soil C storage at different points in time following afforestation of former arable land.

2.2. Study sites

The study included six chronosequences of which two were differently aged oak stands and four were differently aged Norway spruce stands. In Denmark, one oak chronosequence and one Norway spruce chronosequence were assessed within the same forest with clay-rich and nutrient-rich soil in Vestskoven close to Copenhagen. Just outside Vestskoven, a 200-year-old afforested oak stand, Ledøje Plantage, was included for comparison. In contrast to all other stands Ledøje Plantage had a multi-layered stand structure, i.e. with beech and sycamore maple forming a subcanopy. Another spruce chronosequence in a contrasting environment was studied on sandy, poor soil at Gejlvang west of Vejle in southern Jutland. In the Netherlands, one chronosequence of oak and one chronosequence of spruce were studied on similar sandy soil close to Sellingen. The last chronosequence of spruce was situated in south-western Sweden, at Tönnersjöheden east of Halmstad. A map of the locations is found in Chapter 1 (Figure 1.8), and detailed site information is given on the AFFOREST web site (www.sl.kvl.dk/afforest). Briefly, annual temperatures ranged from 6-7°C in Sweden over 7.7°C in Denmark to 9°C in the Netherlands. Annual precipitation ranged from about 640 mm at Vestskoven, Denmark over ca. 800 mm in the Netherlands and to ca. 1000 mm at Gejlvang, Denmark and in Sweden. Soil types ranged from loamy Hapludalfs at Vestskoven, Denmark, to sandy Spodosols in the other chronosequences (Soil Survey Staff 1992). All sites were former cropland with annual tillage. The soils are mainly well drained except in the Netherlands where groundwater levels may be within 50 cm depth during wet periods of the winter. Nitrogen deposition was lowest to the Swedish spruce chronosequence (ca. 20 kg ha^{-1} yr^{-1}) and highest to the Danish spruce chronosequence at Gejlvang and the Dutch oak chronosequence (16-33 kg ha^{-1} yr^{-1}) (Chapter 4).

In all three countries, soil and biomass C stocks were measured once whereas rates of litterfall C were measured for 1-2 years. Table 2.1 gives an overview of the number of stands in each of the chronosequence for litterfall C and biomass and soil C stocks. In Denmark, all chronosequence stands were characterized in terms of biomass and soil C stocks, whereas biomass C was only estimated for a subset of the Swedish and Dutch stands. In the Swedish and Dutch chronosequences, baseline data on soil C data were included from fields still in arable use. Rates of litterfall C were only measured in selected chronosequence stands in all three countries.

Table 2.1. *The number of stands and their range in age with respect to litterfall, biomass and soil C measurements in each chronosequence. Age ranges deviate slightly between the C pools as they were not assessed in the same year.*

Chronosequence	Litterfall C		Biomass C		Soil C	
	Stands	Age range, (years)	Stands	Age range, (years)	Stands	Age range, (years)
Oak, Vestskoven, DK	5	8-31	8	5-28, 200	8	5-28, 200
Norway spruce, Vestskoven, DK	4	11-32	7	1-29	7	1-29
Norway spruce, Gejlvang, DK	3	20-41	5	4-41	5	4-41
Norway spruce, Tönnersjöheden, SE	4	20-93	5	19-92	24 (15)*	0-87
Oak, Sellingen, NL	4	4-19		4-19	12 (10)*	0-29
Norway spruce, Sellingen, NL	-	-	3	9-15	3	9-15

**In these two chronosequences, data on arable land were obtained from 9 fields (SE) and 2 fields (NL) within the afforestation area. The number of afforested stands is given in brackets.*

2.3. Biometrics and calculation of carbon stocks in biomass

In Denmark, stand characteristics at Vestskoven and Gejlvang were assessed in June 2001 according to the methods used in the Danish National Forest Inventory (Johannsen et al. 2004). Briefly, the three circular plots per stand were used as inventory plots for stem number, diameter at breast height and tree height. Diameter at breast height was measured for all trees within plots of different radii according to the diameter of trees, whereas height was measured for a subgroup of these trees randomly selected with probability of selection proportional to diameter. Heights of single trees were subsequently estimated by diameter-height regressions (Näslund 1936). Standing volume of merchantable wood (stem and crown wood for oak and stem wood for spruce) was estimated using equations developed by Madsen (1987) and Madsen & Heusérr (1993).

In Sweden, stand mensuration was done in the five intensively studied stands of the Tönnersjöheden chronosequence in September 2002. Stand data (stem number, diameter at breast height and tree height) were gathered on plots of 25 m × 25 m. Mean diameter, basal area and stand density were calculated. Diameter at breast height was recorded for all trees whereas height was measured for a subgroup of trees selected according to a procedure described by Karlsson (1998). Heights of remaining trees were subsequently estimated by diameter-height regressions. Standing volume of merchantable wood was estimated using equations developed by Brandel (1990).

In the Netherlands, aboveground biomass of trees was estimated on the 15 m × 15 m measurement plots in the 4 oak stands and the 3 spruce stands in October 2002. In order to save time per monitoring area and to be able to estimate the variability in the estimated biomass 3 sections of 7 × 7 m^2 were selected within the measurement plots. Within each section diameter at breast height (DBH), tree height (H) and stand density (N) were measured. DBH and N were measured for each individual tree, whereas H was based on an average estimate per 7 m × 7 m plot. The volume per tree (V) was calculated by using the equations for oak and Norway spruce in Jansen et al. (1996).

Standing volume of merchantable wood was subsequently expanded to total above- and belowground biomass and carbon stock of the total biomass. In Denmark this was done using the expansion factors, basic wood densities and carbon content used for oak and Norway spruce in annual Danish reports on LUCF to UNFCCC (Illerup et al. 2005). Briefly, total biomass was estimated by use of expansion factors 1.2 and 1.8 for oak and Norway spruce, respectively. The lower expansion factor for oak is partly due to inclusion of crown biomass in the Danish volume equations. The volume of wood was converted to mass by use of basic densities of 0.56 and 0.38 Mg m^{-3} for oak and Norway spruce, respectively. In Sweden, specific biomass functions were available for estimating total dry mass (Marklund 1988). In the Netherlands, tree volumes were transformed to biomass by using an expansion factor of 1.8 and wood densities similar to those used in Denmark. Carbon content for stands in all three countries was estimated using a common C concentration of 500 mg g^{-1} (IPCC 2003).

2.4. Litterfall

Total litterfall was measured for two years in Denmark and for one year only in Sweden and the Netherlands. In Denmark, five oak stands and four Norway spruce stands were sampled at Vestskoven and three Norway spruce stands were sampled at Gejlvang. Five litter traps (diameter 31 cm) per stand were placed 1 m above ground in the center and four cardinal points of one circular subplot (10 m radius) per stand. Litter from the five traps was collected every month and dried to constant weight (at 60°C) before weighing the litter produced during each of the two years. A subsample of the litter was subsequently ground for chemical analysis. Ground samples of litter material were analyzed for total C by dry combustion (Dumas method) in a Leco CSN 2000 Analyzer (Matejovic 1993). In Sweden, total litterfall was measured in four stands at Tönnersjöheden. Five litter traps (diameter 64 cm) were randomly placed within a 25 m × 25 m plot in each stand, and litter was collected two times during the sampling period. In other respects the methods were similar to those applied in Denmark. In the Netherlands litterfall was measured in four of the oak stands. Three litter traps (100 × 100 cm) were randomly placed 0.3 m above ground in each of the stands. Litter was collected bi-monthly from the traps and dried to constant weight at 40°C before weighing. A subsample of the litter was subsequently ground and analyzed for total C by dry combustion and measurement of the CO_2 gasses using an IR-cell.

2.5. Soil sampling and analyses

Stand sizes were different in the three countries and slightly different sampling designs and sampling methods were used for soil C while analyses and calculations were performed similarly in the three countries. The sampling design and methods are summarized in Table 2.2. More detailed descriptions of the sampling and analysis of soils in Denmark and Sweden may be found in Vesterdal et al. (2002) and Rosenqvist & Johansson (2005), respectively.

In Denmark and Sweden, C concentrations in the forest floor and soil were determined by dry combustion (Dumas method) in a Leco CSN Analyzer (Matejovic 1993). The Dutch soil samples were analyzed for C by wet oxidation with potassium dichromate (Kurmies 1949).

Carbon sequestration is reported as the relationship between C contents and stand age. Forest floor C contents were calculated by multiplying C concentrations with forest floor mass. For the mineral soil C contents for the fraction $\geq$ 2mm (small stones and gravel) were neglected (McNabb et al. 1986; Homann et al. 1995), and soil organic carbon (SOC) contents in [Mg ha^{-1}] for each of the three soil layers were calculated via

$$SOC_i = \rho_i \bullet (1 - (\delta_{i,\,2mm}/100)) \bullet d_i \bullet C_i \qquad (1)$$

where ρ_i is the bulk density of the < 2 mm fraction in g cm^{-3}, $\delta_{i,\,2\,mm}$ is the relative volume of the fraction $\geq$ 2 mm (%), d_i denotes the thickness of layer i in cm, and C_i denotes the C concentration of layer i.

***Table 2.2.** Summary of soil sampling designs for estimation of C stocks.*

	Denmark	Sweden	The Netherlands
Time of sampling	Vestskoven: Sept. 1998 Gejlvang: June 2001	Sept. 1998	May 2003
No. of sampling plots per stand	3 circular plots of 10 m ∅	1 plot 25 × 25 m	1 plot 15 × 20 m
Sampling points per stand	12 (4 cardinal points × 3 plots)	In grid intersections Soil: 20 Forest floor: 16	In grid intersections Soil: 20 Forest floor: 10
Forest floor sampling area	25 × 25 cm	Auger ∅ 10 cm or 25 × 25 cm	10 × 10 cm
Mineral soil sampling method and depth	Auger (50 mm ∅) to 25 cm (Westman 1995)	Auger (44 mm ∅) to bottom of Ap (ca. 25 cm)	Auger (44 mm ∅) to 30 cm
Subdivisions of mineral soil for analysis	0-5, 5-15, 15-25 cm	0-5 cm, 5 cm - bottom of Ap	0-10, 10-30 cm
Bulk density (< 2 mm fraction)	Measured, 6 cores per stand	Measured on composite sample of 16 subsamples	Estimated from equations by Hoekstra & Poelman (1982)

2.6. Calculations and statistics

Soil C stocks were assessed for the forest floor and upper 25 cm of the mineral soil in all chronosequences. Relationships between stand age and soil, biomass and ecosystem C contents were explored by simple linear regression while the influence of tree species in Denmark was tested by analysis of covariance. No transformations were necessary to fulfil the requirements regarding normally distributed residuals and homogeneity of variances. All statistical tests were carried out using the procedure GLM in SAS (SAS Institute 1993). The significant regression models were used to calculate the changes in soil C stores for the time span of the chronosequence. The 200-yr old stand in Denmark was not included in regressions, but was included in figures for comparison.

3. RESULTS

3.1. Carbon in biomass

The basic data for estimation of carbon sequestration in biomass was the accumulated biomass measured as the volume of merchantable wood. Thus, measured heights, stem diameters and stand densities (stem number per ha) determine the amount of carbon sequestered after afforestation. The development in stand height with age was surprisingly similar across all chronosequences (Figure 2.1a). The low mean height of the ca. 200-year-old Ledøje Plantage is due to multi-layered structure, i.e. with beech and sycamore maple forming a subcanopy. The heights of dominant oak trees were within 25-30 m at Ledøje Plantage.

Stand densities were more variable in the younger stands reflecting different planting densities and different thinning intensities (Figure 2.1b). Especially the Dutch oak stands and the Danish spruce stands at Gejlvang had high densities, the latter because they were planted for Christmas tree production.

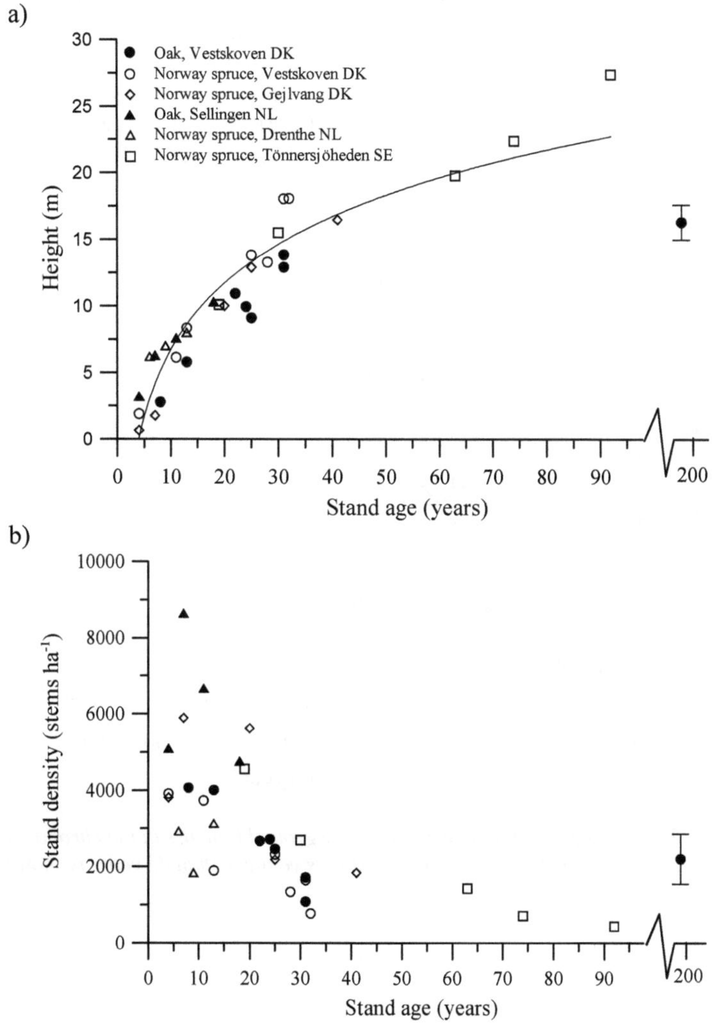

Figure 2.1. *a) Stand height and b) number of trees per hectare as a function of age in chronosequence stands.*

The C content of the total biomass is shown in Figure 2.2. There was a strong relationship with age, and the rates of biomass C sequestration for different chronosequences were quite similar for the first 40-50 years following afforestation (Table 2.3). For all stands younger than 45 years, the rate of C sequestration was about 3.7 Mg C ha^{-1} yr^{-1}. However, the Swedish chronosequence, which included older stands, had a clearly lower rate of biomass C sequestration that was not quite significantly different from 0 (Table 2.3). The older Swedish stands have a relatively lower C storage and do not continue the rate of C sequestration depicted by the Swedish stands younger than age 40. Oak stands in Denmark also tended to have a lower rate of biomass C sequestration, and differed from spruce at the same site. As indicated by stand heights, there was little difference in rate of C sequestration between stands of spruce on contrasting soil types within Denmark. Spruce stands at the nutrient-rich Vestskoven site sequestered carbon only at a slightly higher rate than at the sandy, nutrient-poor Gejlvang site (Table 2.3). The few spruce stands in the Netherlands prevent a general comparison between species.

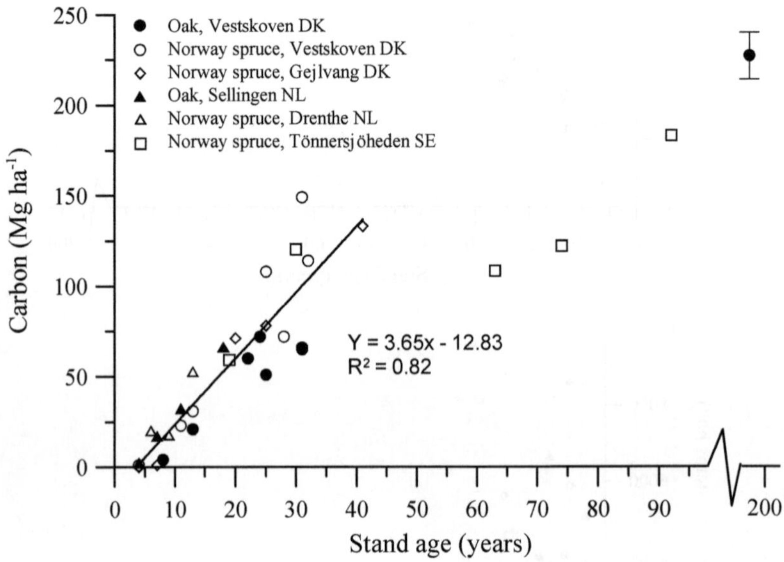

Figure 2.2. *Carbon content in above- and belowground biomass in the chronosequence stands. The general regression for all stands younger than 45 years is shown.*

Table 2.3. *Rates of biomass C sequestration (SE of regression slope) in the AFFOREST chronosequences.*

Chronosequence	Rate (Mg C ha^{-1} yr^{-1})	P*
Oak, Vestskoven, DK	2.72 (0.53)	0.004
Norway spruce, Vestskoven, DK	4.61 (0.82)	0.002
Norway spruce, Gejlvang, DK	3.76 (0.31)	0.001
Norway spruce, SE	1.20 (0.47)	0.082
Oak/Norway spruce, NL	4.55 (0.56)	<0.001
All stands <45 years	3.65 (0.34)	<0.001

**P values < 0.05 indicate whether slopes of regressions ($y = ax + b$) are significantly different from 0.*

3.2. Aboveground input of C to soils with litterfall

The carbon in annual litterfall represents the aboveground input of C to the soil. Litterfall is the source of organic matter for development of forest floors following afforestation. There was no clear relationship between stand age and the rate of litterfall C over the age span of the chronosequences. However, especially for the Danish stands there were increasing rates of litterfall C during the first 20 years until the annual C content of litterfall appear to level off. Litterfall tended to be higher in oak than in spruce stands in Denmark, and Dutch oak stands had comparable litterfall C contents to Danish oak stands. However, litterfall C contents representative of closed stands were reached at a younger age than in Denmark. In the oldest Swedish stands, litterfall C contents appear to decrease.

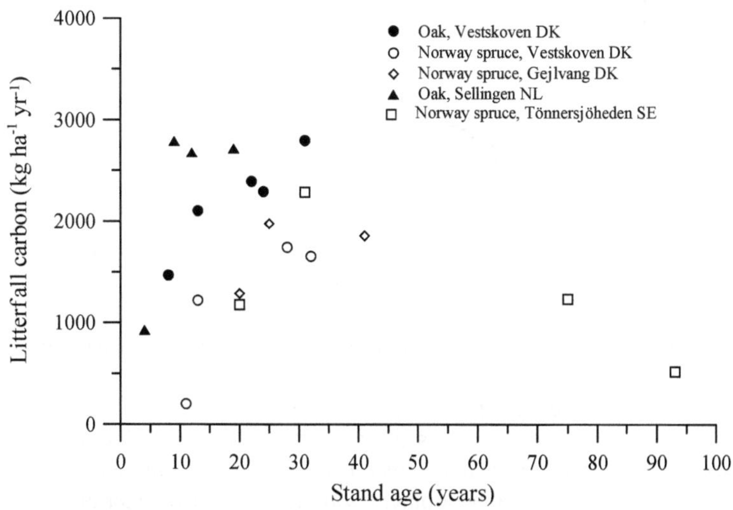

Figure 2.3. *Rates of litterfall C measured over two years in Denmark and for one year in Sweden and the Netherlands.*

3.3. Carbon in soils

The forest floor is the layer of dead organic matter, i.e. leaves, needles, twigs, branches and fruits, that blankets the mineral soil of a forest. Immediately after afforestation there is no forest floor present, but it developed rapidly (Figure 2.4) when the input of C from litterfall increases concurrently with canopy closure (Figure 2.3). The development in forest floor C sequestration was at a first glance quite similar between tree species and sites. However, this similarity is probably an artefact of the different age spans represented by the chronosequences. For instance oak stands in Denmark are in the low range of forest floor C storage at 25-30 years, and the carbon storage in forest floors of the 200-year-old Ledøje Plantage is comparable to that of the 25-30-year-old oak stands. Rates of C sequestration ranged from 0.08 (oak at Vestskoven, DK) to 0.65 Mg C ha^{-1} yr^{-1} (spruce in Sweden, Table 2.4.). Oak stands in Denmark had significantly lower ($P=0.005$) rate of C sequestration than spruce at the same site. Danish spruce stands and Dutch oak and spruce stands were quite comparable in forest floor C sequestration rate (0.34-0.4 Mg C ha^{-1} yr^{-1}, Figure 2.4 and Table 2.4).

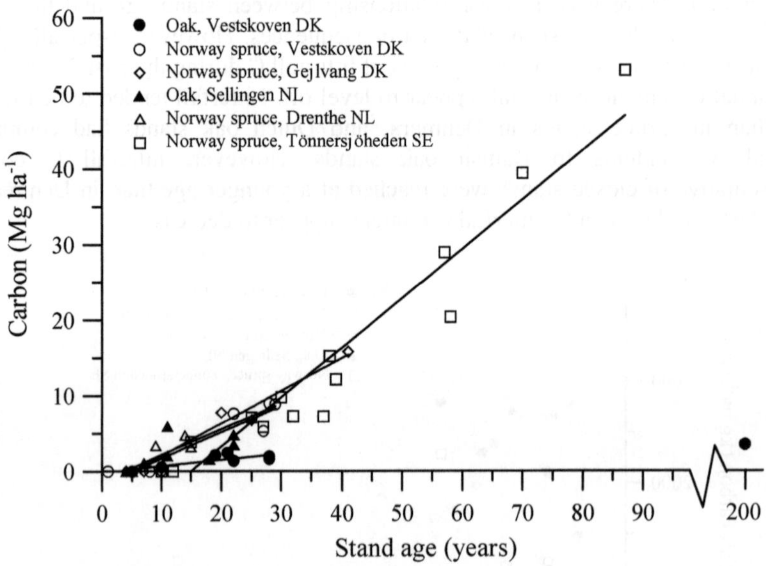

Figure 2.4. *The C content of forest floors in the chronosequence stands.*

In the mineral soil, i.e. the former plow layer 0-25 cm, the pattern was more diverse and the relative change less dramatic than for forest floors. The different chronosequence sites were clearly showed different behaviour with respect to the initial level of soil C and its development with age since afforestation (Figure 2.5a). There was no difference between the Danish spruce and oak chronosequences at Vestskoven, so all data from Vestskoven were combined in the analysis of soil C sequestration rates (Vesterdal et al. 2002). Similarly the few Dutch spruce stands did not differ from the pattern of the oak chronosequence and these data were consequently combined. Dutch sites had the highest C contents in the mineral soil and the Danish site Gejlvang was lowest in mineral soil C content. Only in the Dutch oak chronosequence and to some extent also the Danish Gejlvang chronosequence did the mineral soil C contents increase significantly with increasing stand age. For the studied soil compartments as a whole, i.e. the forest floor and the plow layer, there was a general pattern of constant or increasing soil C stores over 30-90 years after afforestation (Figure 2.5b).

The allocation of sequestered soil C was quite different for the studied chronosequences (compare Figures 2.5a-b). For the Swedish spruce chronosequence and the Danish spruce and oak chronosequences on fertile soils, the change in C was solely due to development of forest floors on top of the mineral soil. The amount of C in the mineral soil was unchanged (SE) or even decreased slightly (Vestskoven, DK). However, in the Danish spruce chronosequence on poor soil in Gejlvang the former plow layer also increased significantly in C content. The allocation of sequestered C within the soil was again different for the Dutch oak chronosequence. Here the plow layer sequestered almost four times more C than the forest floor.

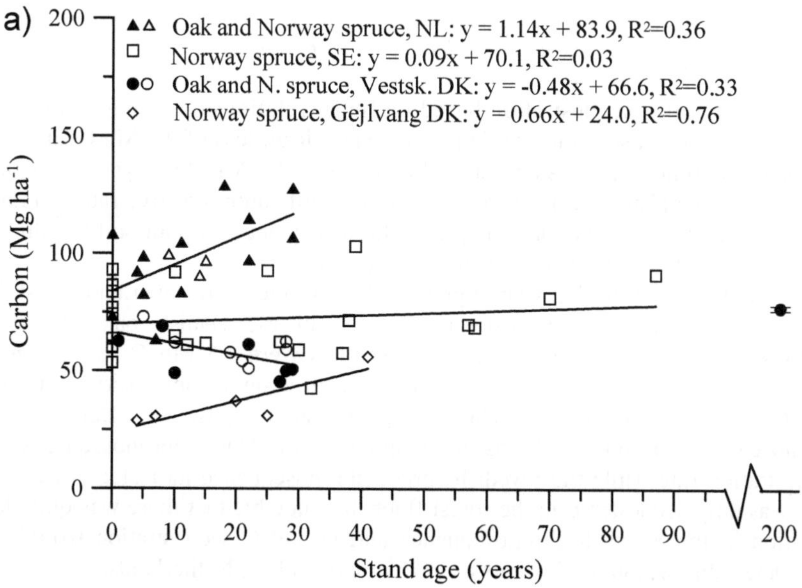

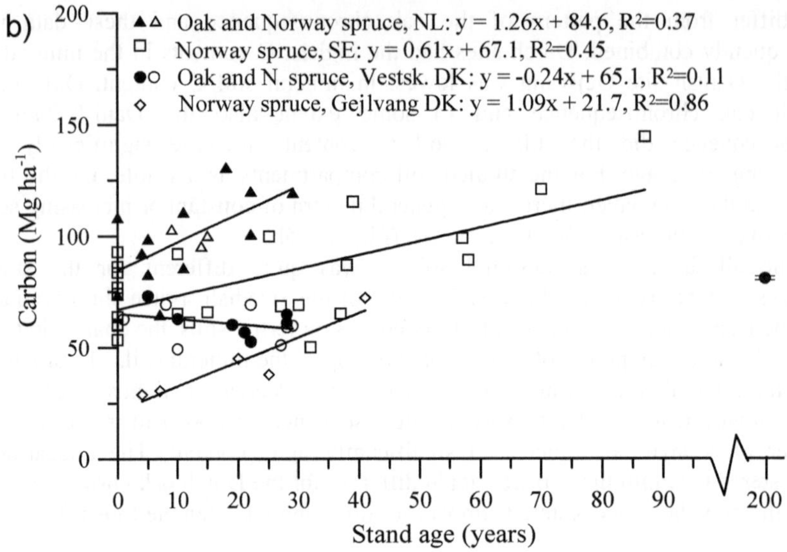

Figure 2.5. *a) Mineral soil C content in the former plow layer (0-25 cm). b) Total soil C content (forest floor + former plow layer).*

The rates of soil C sequestration based on linear relationships are given in Table 2.4. For the mineral soil, rates varied from a significant loss of 0.5 Mg C ha^{-1} yr^{-1} in the Danish chronosequences on rich soil to a gain of 1.1 Mg C ha^{-1} yr^{-1} in the Dutch chronosequence (Table 2.4). There was also a significantly positive rate of mineral soil C sequestration in the Danish spruce chronosequence on poor soil but this was not the case in the Swedish spruce chronosequence.

Rates of C sequestration were higher for both former plow layer and forest floor and were not significantly different from 0 in the Danish oak and spruce chronosequence at Vestskoven. The highest rates were 1.3 Mg C ha^{-1} yr^{-1} in the Dutch chronosequence. The chronosequences with a significant positive change in soil C were all situated on relatively poor, sandy soils. The Danish spruce chronosequence on poor soil was quite similar to the Dutch chronosequence in C sequestration rate while the Swedish spruce chronosequence indicated a lower rate as C was only sequestered in the forest floor. It is notable that there was quite large variation in the Swedish chronosequence and no net C sequestration would have been detectable within 30-40 years as in Denmark and the Netherlands.

Table 2.4. *Rates of soil C sequestration (SE of regression slope) over 30-90 years in the AFFOREST chronosequences.*

Chronosequence	Forest floor C		Mineral soil C*		Total soil C*	
	Rate (Mg ha^{-1} yr)	P value	Rate (Mg ha^{-1} yr)	P value	Rate (Mg ha^{-1} yr)	P value
Oak, Vestskoven, DK	0.08 (0.03)	0.032	-0.48 (0.19)	0.031	-0.24 (0.20)	0.25
Spruce, Vestskoven, DK	0.35 (0.06)	0.003				
Spruce, Gejlvang, DK	0.43 (0.05)	0.003	0.66 (0.21)	0.056	1.09 (0.25)	0.022
Spruce, SE	0.65 (0.05)	<0.001	0.09 (0.13)	0.461	0.61 (0.14)	<0.001
Oak/spruce, NL	0.34 (0.08)	0.001	1.14 (0.40)	0.014	1.26 (0.46)	0.017

**Rates of C sequestration for mineral soil and total soil C at Vestskoven are combined for oak and spruce.*

The Swedish chronosequence included nine cropland plots. Each cropland plot was adjacent to one or two afforested stands. Under the assumption that the cropland soil C stocks of today represent conditions prior to afforestation we conducted an analysis of pairwise differences between cropland plots and afforested stands. The paired plot design reduces spatial variability attributable to differences in soil properties. Such spatial variability may possibly mask the influence of changed land use when using the chronosequence approach. The differences in soil C stocks at different stand ages are shown in Figure 2.6 for mineral soils alone and for mineral soils + forest floors. Differences in soil C stocks between afforested and cropland plots were quite variable, and differences (forest minus cropland) were in some cases negative for the plow layer. In line with the similar absolute mineral soil C stocks found along the Swedish chronosequence, there was no significant relative change in mineral soil C stocks along the chronosequence (P=0.732). When forest floor C was included there was always most soil C in afforested plots after about 30 years, and differences in soil carbon significantly increased with increasing age of the afforested stand (P=0.009)

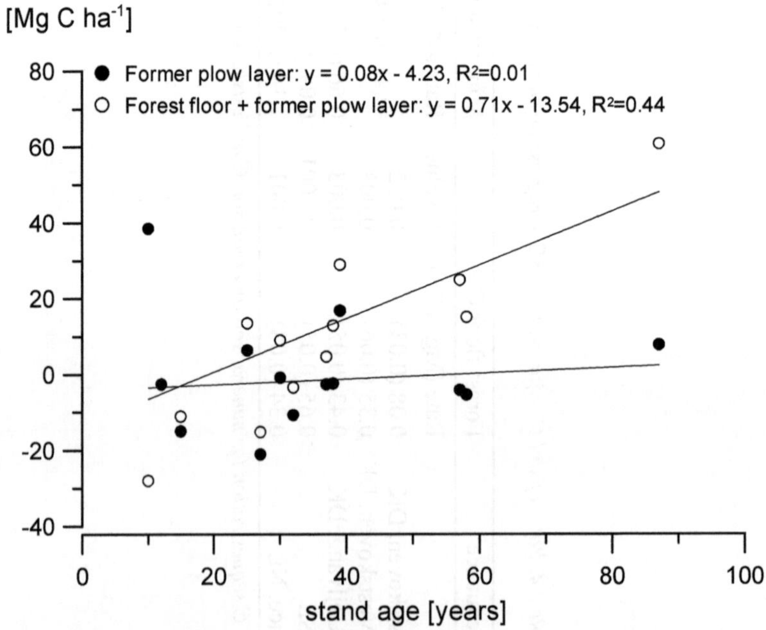

Figure 2.6. *The differences in soil C pools between paired plots of afforested stands and cropland (forest C minus cropland C) in the Swedish afforestation chronosequence.*

3.4. Carbon in the afforested ecosystem

The change in ecosystem C storage, i.e. in C contents of soil plus biomass, and the allocation of C to woody biomass and soil, respectively, is shown in Figure 2.7a for all chronosequences. It must be noted that while all Danish stands had data for both soil C and biomass C, biomass C was only assessed in the intensively studied stands in the Netherlands and in Sweden (Table 2.1). As a result, the data set for analysis of biomass and total ecosystem C sequestration did not cover all stands in these two countries.

There was a good relationship between C storage and age across all chronosequences with a general rate of C sequestration of about 2.8 Mg ha^{-1} yr^{-1} (Figure 2.7a) The relationship with age indicated sequestration of 250 Mg C ha^{-1} over the chronosequence of 90 years as a general result for all tree species and sites. The rate of C sequestration is expected to level off as stands mature as also indicated by the 200-year-old oak stand. Nevertheless, there was no basis for non-linear relationships. The age distribution of stands was skewed towards the age 0-30 years, so there is a better picture of the variability in C content within this age span.

The change in ecosystem C content was primarily caused by the growth of trees, but soils also contributed significantly. A general assessment based on regression slopes indicates that about 70% of the total C sequestered was allocated to biomass following afforestation, and the remaining 30% was sequestered in soils. As already reported for soils (Figure 2.7) there were different levels of soil C at the chronosequence sites. Therefore the general relationship between soil C and stand age was weak (Figure 2.7a). However, when the more wet Dutch sites with relatively high C contents were excluded, the relationship between soil C and age clearly improved (Figure 2.7b). Among individual chronosequences, the contribution of soils ranged from none at Vestskoven over ca. 20% in the Netherlands and at Gejlvang to 31% in Sweden.

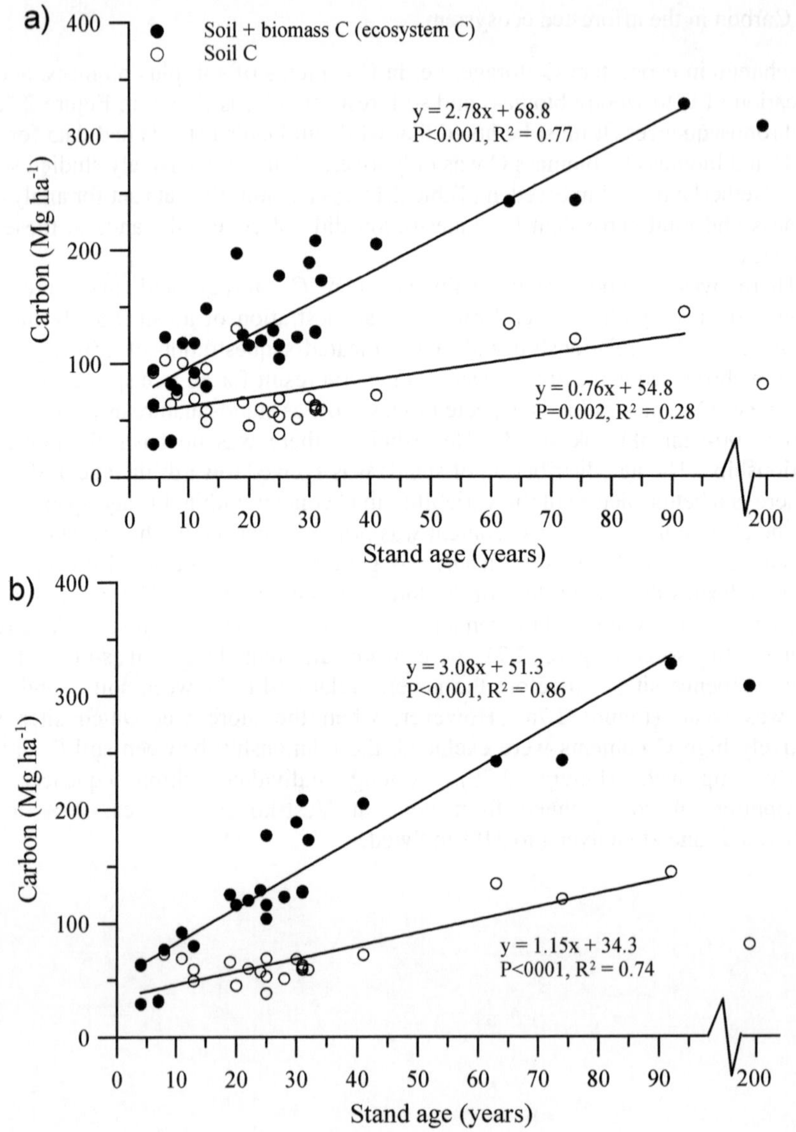

Figure 2.7. *Changes in total ecosystem C storage with time since afforestation. The graphs show the relative contribution of soils and vegetation to total ecosystem C sequestration following afforestation. The difference between the regression lines indicates the relative contribution of biomass C sequestration. a) all chronosequence sites in Denmark, the Netherlands and Sweden. b) chronosequence sites without the more wet Dutch sites.*

4. DISCUSSION

4.1. Biomass carbon

Rates of biomass carbon sequestration were estimated at 3.7 (range 2.7-4.6) Mg C ha^{-1} yr^{-1} for AFFOREST stands younger than 45 years. This range is fairly well in line with the median sequestration rate of 4.1 Mg C ha^{-1} yr^{-1} reported in a review of potential C storage in temperate regions over 25 years (Winjum & Schroeder 1997). The AFFOREST chronosequence studies only included three stands in Sweden and one stand in Denmark older than 40 years. Thus, there is little basis for conclusions regarding biomass C sequestration over a full rotation. However, from an age of 40-50 years there appears to be some differentiation between stands of different species and locations. The Swedish stands older than 45 years had sequestered less C than would be expected from the trajectories of the other chronosequences. This is mainly attributable to the fact that the Swedish chronosequence cover almost a full rotation, and current rates of increment and thus C sequestration decrease when trees mature. The sequestration rate in the Swedish chronosequence covering 90 years is well in line with the modelled sequestration rates (0.8-1.2 Mg C ha^{-1} yr^{-1}) reported by Nabuurs & Mohren (1995) for the first 100-year rotation of afforested Norway spruce in boreo-nemoral climate. Another reason for differences between chronosequences is an increasing effect of different management practices (e.g. thinning) as stands grow older. Information on thinned volumes for the older stands would have provided a more complete picture of C sequestration in biomass, but unfortunately such data were not available. Furthermore, old stands probably have a different legacy of former agriculture than the younger stands. Afforestation more than 50 years ago was carried out on soils that had not been subject to the intensive farming practices of today which include frequent liming and fertilization. The fertilized soils abandoned for afforestation today would probably also lead to higher rates of C sequestration, at least in the first decades until the organic nitrogen pool is in steady state in the forest ecosystem.

It was expected to find lower rates of C sequestration in biomass in the same tree species at poor soil compared to nutrient-rich soil. But there was little difference in the rate of C sequestration between stands of spruce on contrasting soil types within Denmark. Spruce stands at Vestskoven sequestered carbon at a slightly higher rate than at Gejlvang. Within Denmark, a difference between oak and Norway spruce was more evident with rates of C sequestration of 2.7 and 4.6 Mg C ha^{-1} yr^{-1}, respectively. This is consistent with the generally higher growth rates in spruce than in oak.

Changes in biomass C were mainly estimated for comparison with concurrent changes in soil C. As in national reporting under United Nations Framework Convention on Climate Change (e.g. Illerup et al. 2005), there are uncertainties with regard to conversion of biomass to C stock in the total biomass. These uncertainties may in part influence differences between chronosequences. However, the estimated sequestration rates are fairly conservative as we only considered on site C stocks. If thinned tree biomass and resulting storage in wood products were included, the contribution of biomass would potentially be higher in the longer chronosequences.

4.2. Forest floor carbon

When cropland is afforested, soils experience a marked reduction in management intensity. Tillage of the upper 20-30 cm ceases, as does also the application of lime and fertilizers. In addition, the vegetation changes from annual to perennial, thus accumulating a large amount of biomass C. This may change the production of dead organic matter and its quality in terms of decay. Such differences between the two types of land use would result in a phase of net accumulation of soil C until a new equilibrium level of soil C is established according to the rates of input and decomposition in a forest.

The most visible change in soil C stock at the AFFOREST sites was caused by development of a forest floor on top of the mineral soil. At a first sight, forest floors showed a quite similar development over time (Figure 2.4). Rates of C sequestration were most comparable for Danish and Dutch spruce stands and the Dutch oak stands (Table 2.4). However, as mentioned before this apparent general pattern is probably an artefact of the different age spans represented by the chronosequences. The low amount of C in the forest floor of the 200-year-old stand strongly suggests that more mature oak stands will differ from spruce stands in forest floor C storage. The Danish oak chronosequence also had a much lower rate of C sequestration (0.08 Mg C ha^{-1} yr^{-1}, Table 2.4) than that of other chronosequences. This difference in forest floor C between oak and spruce is in line with general observations in Denmark (Vesterdal & Raulund-Rasmussen 1998) and a study of litter decomposition rates in the two species (Dziadowiec 1987). In the Netherlands, however, there was no clear difference between oak and spruce, but comparison is limited by the low number of relatively young spruce stands.

The Swedish spruce stands have sequestered quite high amounts of C in forest floors, and this development may possibly be representative also for Danish and Dutch spruce stands on poor sandy soils. However, forest floor C contents rarely exceed 40 Mg ha^{-1} in Denmark (Vejre et al. 2003).

In contrast to forests with a longer history, soil type did not seem to affect forest floor development in the first decades after afforestation. Sequestration of forest floor C differed little between the Danish Gejlvang and Vestskoven sites in spite of very contrasting mineral soils. Sandy, podsolized soils usually develop thick more-like forest floors (Vesterdal & Raulund-Rasmussen 1998; Vejre et al. 2003) because of slow decomposition and presence of few macrofauna species. The similar rates of C sequestration at the two Danish sites may be attributed to the high pH and nutrient availability in recently abandoned cropland soils. The legacy of cropland management in terms of liming and fertilization can probably maintain an active microflora and fauna in both soil types during the first decades. For instance, Jussy et al. (2002) and Compton & Boone (2000) found high rates of N mineralization several decades after agricultural abandonment and afforestation. This legacy of agriculture probably offsets the differences in forest floor accumulation that would otherwise have developed at sites with such difference in parent materials (Vesterdal & Raulund-Rasmussen 1998).

The afforestation literature confirms that forest floor development contributes most to sequestration of C during the first decades . Up to 38 Mg C ha^{-1} was sequestered over 40 years (0.94 Mg C ha^{-1} yr^{-1}) in *Pinus taeda* forest floors in warm-temperate climate (Richter et al. 1999). Comparable amounts of C accumulated (rates about 0.48 Mg C ha^{-1} yr^{-1}) in cooler temperate regions under Norway spruce and mixed broadleaves following 60-80 years of stand development (Hamburg 1984; Leth & Breuning-Madsen 1992).

The rate of forest floor development partly depends on aboveground litterfall. In the AFFOREST chronosequences, annual litterfall C content increased over the first decades following afforestation (Figure 2.3), and the development in litterfall C is consistent with canopy closure usually being complete after 20 years. After canopy closure, foliar litterfall tends to be constant, and foliar litterfall constitutes about 90% of total litterfall in young stands (Pedersen & Bille-Hansen 1999). Litterfall tended to be higher in oak than in spruce stands in Denmark, and Dutch oak stands had comparable litterfall C contents to Danish oak stands. However, rates of litterfall C representative of closed stands were reached at an earlier age than in Denmark. The much higher stand densities used in oak stands in the Netherlands (Fig. 1b) may explain this, as canopy closure is attained at an earlier age. In the oldest Swedish stands, litterfall C contents are relatively low which partly may be attributed to reduced production in these mature stands.

Afforested stands reached a relatively constant level of aboveground carbon input to the soil after about 20 years. However, the composition of litterfall may change with age, e.g. there may be more woody litter (branches and twigs) as stands mature. Another important source of carbon to soils is root litter. This source of carbon may be especially important for the C pool of the mineral soil. An assessment of root litter C was not included in the project, but in cool temperate forests, root litter was estimated to contribute between 20 and 77% of the total organic matter input to forest floors (Vogt et al. 1986).

4.3. Mineral soil carbon

The obvious accumulation of C in forest floors may well lead to expectations of increases in the mineral soil too. Nevertheless, there was no general pattern in change in mineral soil C stocks in the AFFOREST chronosequences (Figure 2.5a). In fact, there was a slight net C loss in mineral soils in chronosequences (30 years) at Vestskoven in Denmark, and not even the Swedish chronosequence covering 90 years exhibited a change. For the Swedish chronosequence it was also possible to do a paired plot analysis (Figure 2.6) to see if spatial variability possibly masked the influence of changed land use when using the chronosequence approach. However, there was still no net change detectable when comparing adjacent arable and forest soils along the chronosequence. This result for Sweden suggests that there is little reason to expect mineral soil C sequestration at all sites, even over a full rotation. At Vestskoven in Denmark the soil C contents in Ledøje Plantation (200-years-old) nevertheless indicate that the mineral soil may start to sequester C on a net basis in later stages following afforestation (Figure 2.5a). There was also evidence that soil C had started to build up in the 0-5 cm of the mineral soil at Vestskoven, however, not enough to compensate a decrease in soil C in 5-25 cm (Vesterdal et al. 2002).

On the other hand, mineral soil C storage increased with age in the Netherlands and at Gejlvang, Denmark already within 30-40 years. The mechanisms behind the site-dependent response to afforestation remain elusive, but high decomposition rates in the nutrient-rich soil at Vestskoven could be responsible for the small rate of C sequestration. Furthermore the allocation of the dead organic matter in soil profiles probably differ due to different activities of macrofauna species that mix forest floor material into the mineral soil. The large forest floor C sequestration rate and lack of C allocation to mineral soils in the Swedish chronosequence (Figures 5 and 6) could be a result of limited macrofauna activity.

Other chronosequence studies also show mixed evidence of mineral soil net C sequestration. In a recent review of soil C sequestration following afforestation by Post & Kwon (2000), most studies indicated net sequestration of C in mineral soils. Post & Kwon (2000) found an average rate of 0.34 Mg C ha^{-1} yr^{-1} for all studies regardless of climate, which is fairly similar to the rates estimated for chronosequences in the Netherlands and Gejlvang, Denmark (Table 2.4.). A meta-analysis by Guo & Gifford (2002) revealed that mineral soil C stocks increased by 18% following afforestation of former cropland. In other studies, however, the mineral soil did not function as a C sink, at least initially. In a review of global data on soil C sequestration after afforestation, Paul et al. (2002) found that soil C in 0-30 cm decreased during the first three decades and finally recovered to the pre-afforestation level by age 30. The dynamics in soil C at the Danish Vestskoven site might conform to a comparable pattern where it takes longer for mineral soil C to build up. Within a time perspective of a few decades losses of C derived from agriculture may possibly more than make up for the beginning accretion of forest-derived C (Bashkin & Binkley 1998; Binkley & Resh 1999; Paul et al. 2002; Vesterdal et al. 2002). In many cases there appears to be a transient period of decreasing soil C stores following afforestation. This period is characterized by redistribution of C within the soil profile rather than a net accretion in C stores. Such a pattern was found at Vestskoven, where there was a positive change in C in 0-5 cm and a negative change in 5-25 cm (Vesterdal et al. 2002). The reason for this initial development may be the placement of produced organic matter (Post & Kwon 2000). Compared to arable land use, inputs of C to the lower parts of the plow layer are reduced due to ceased tillage. At the same time the legacy of organic material from agriculture may decompose fast in the homogenized plow layer. In the very beginning of afforestation, litter input may also be quite small to the upper soil layers, thereby contributing to a temporal decline in mineral soil C (Romanya et al. 2000).

The former agricultural land use in the AFFOREST chronosequences was mainly cropland (cereals, potatoes) or rotation between cropping and pasture phases. This is important in comparisons with reviews where data were compiled from both afforested pastures and cropland. Pastures are known to have higher mineral soil C contents than cropland (Post & Kwon 2000). Consistently with this, Paul et al.

(2002) and Guo & Gifford (2002) both found in reviews that afforestation of former cropland resulted in mineral soil C sequestration while afforestation of pastures resulted in unchanged or decreasing C stocks. The mineral soil C sequestration found at Gejlvang, Denmark is in line with that for former cropland reported by Paul et al. (2002), whereas the rates for the other chronosequences are both lower (Vestskoven, DK and SE) and higher (NL). Obviously, other factors than former agricultural land-use such as soil type, climate and site preparation may influence the magnitude of change in soil C.

4.4. Total soil carbon

For the total soil compartment studied, i.e. forest floor and mineral soil 0-25 cm, we found that afforestation as a minimum resulted in unchanged soil C contents and in most cases led to net C sequestration (Figure 2.5b, Table 2.4). The review of global data by Paul et al. (2002) included 34 sites at which changes in C were measured in both forest floor and mineral soil. For these sites, the contribution of mineral soils 0-30 cm was limited (0.15 Mg C ha^{-1} yr^{-1}), but inclusion of forest floor C resulted in a mean rate of C sequestration of 0.46 Mg C ha^{-1} yr^{-1} for 0-30 cm. In the AFFOREST study from cool temperate climate all chronosequences but one indicated C sequestration rates that were 1.3 to 2.7 times higher for the total soil compartment studied.

The rate of input of tree litter and the rate of decomposition of the litter together determine the C content of the soil. The organic matter input to arable soils is often said to be lower than to forest soils because a greater amount of the produced biomass is harvested (Bouwman & Leemans 1995). However, in conventional Danish cropland (barley and wheat) the input from aboveground biomass is about 2.4 Mg C ha^{-1} yr^{-1} (Olesen et al. 2001), whereas the annual input of C to soils with litterfall approaches about 2 Mg C ha^{-1} yr^{-1} in many cool temperate forests (Bastrup-Birk et al. 2003; Vogt et al. 1986), assuming 50% C in litterfall. Comparable amounts of C were also transferred to the soil with aboveground litterfall in the AFFOREST chronosequences (Figure 2.3), although litterfall C varied quite a lot between chronosequences. Belowground litter input from root systems is an important contributor to mineral soil C pools, but little is known about general differences between cropland and forestry. Even if the input of organic matter may be comparable, rates of decomposition in arable soils can be higher due to higher soil temperatures and moisture (Bouwman & Leemans 1995), more decomposable litter types (Schimel et al. 1985), and increased accessibility of organic matter to microbial attack because of frequent soil cultivation (Voroney et al. 1981; Denef et al. 2004).

The AFFOREST chronosequence studies did not reveal large differences in total soil C sequestration between oak and Norway spruce although oak sequestered less C in forest floors than Norway spruce. This may partly be attributed to the short period where it was possible to compare species. To our knowledge no other studies addressed species effects in afforestation using a chronosequence approach. General studies of species effects on forest floors usually agree that conifers store more C in forest floors than do broadleaves (France et al. 1989; Vesterdal & Raulund-Rasmussen 1998), but results for mineral soil C are scarce and often contradictory (Binkley 1995; Hagen-Thorn et al. 2004). Paul et al. (2003) also found no clear difference in mineral soil C sequestration after 25-50 years between afforestation with deciduous and coniferous species using isotopic C dating.

4.5. Carbon sequestration of the afforested ecosystem

One of the objectives of AFFOREST was to determine the total C sequestration and the relative contributions of soil and biomass components of the new forest ecosystems. Roughly one third of the total C was sequestered in the soil and two thirds were sequestered in biomass; thus the allocation of C to woody biomass and soil occurred approximately at a ratio of 2:1 (woody biomass : soil) for all chronosequences combined (Figure 2.7). However, this ratio was largely driven by data from the long Swedish chronosequence, and in individual chronosequences, the contribution of soils ranged from 0 to 31% (Table 2.5.). The change in ecosystem C content was primarily caused by the growth of trees as supported by other studies (Post & Kwon 2000; Paul et al. 2002).

Studies of both biomass and soil C sequestration following afforestation are scarce, and no other European studies on former cropland were available for comparison with the AFFOREST chronosequences. Several studies have been conducted in the eastern United States on afforestation and natural succession following abandonment of agriculture since this land-use change was widespread after the western United States was opened up in late 19th century. Table 2.5 therefore lists rates of C sequestration in forest floors, mineral soils and biomass and the relative contribution of soils for the AFFOREST chronosequences for comparison with similar studies in the eastern United States and a single European study from the Italian Alps. As for the AFFOREST chronosequences, most of the other studies found that simple linear relationships best described the development in ecosystem C stocks. This facilitates the comparison of sequestration rates between studies.

Table 2.5. *Rates of C sequestration following afforestation in the temperate region. Rates of C sequestration may not add up to ecosystem total due to different treatment of non-significant regressions and unlisted pools (e.g. woody debris).*

Forest type	Time period (yr)	Rate of C sequestration (Mg ha^{-1} yr^{-1})				Soil contrib. to ecosystem C seq. (%)	Mineral soil contrib. to total soil C seq. (%)	Source
		Plant biomass	Forest floor	Mineral soil (sample depth in cm)	Eco-system total			
Vestskoven, DK, Oak	31	2.72	0.08	-0.47 (25)	2.36	~0	~0	This study
Sellingen, NL Oak, N. spruce	29	4.55*	0.34	1.14 (25)	7.14*	20	77	This study
Vestskoven, DK Norway spruce	32	4.61	0.35	-0.47 (25)	4.46	~0	~0	This study
Gejlvang, DK Norway spruce	41	3.76	0.43	0.66 (25)	4.84	23	61	This study
Tsjh, SE Norway spruce	92	1.20	0.65	0.09 (25)	2.36	31	14	This study
Rhode Island, US White pine	115	1.53	0.37	0.15 (70)	2.10	24	29	Hooker & Compton (2003)

Table 2.5. Continued.

New Hampshire Mixed dec./con.	65	1.67	0.40	-0.05 (50)	1.95	18	~0	Hamburg (1984) in Hooker & Compton (2003)
South Car., US Loblolly pine	35	4.26	0.95	0.04 (60)	5.24	19	4	Richter et al. (1999)
Minn., US Oak	39	0.96	0.12	0.13 (25)	1.21	21	52	Johnston et al. (1996)
Virginia, US Loblolly pine	47	3.03	0.23	0.20 (100)	3.52	12	47	Schiffman & Johnson (1989)
Michigan, US Red pine	38	1.91†	0.28	~0 (10)	2.18	13	~0	Pregitzer & Palik (1997)
Northern Italy, Norway spruce	62	2.73‡	0.36	~0§ (A horizon)	3.18	13	~0	Thuille et al. (2000)

‡ Stem biomass only.
§ Former pasture soil, no linear trend.
**Biomass and ecosystem rates apply to 18-year period only.*
†Aboveground biomass only.

Other studies largely agree with the relative contribution of soils in spite of various methods applied for estimating changes in C stocks, including the soil depth considered. A contribution of soils of between 15 and 20% of total ecosystem storage appears to be the common picture. However, several studies in eastern United States had lower relative soil contribution because there were no changes in mineral soil C stock. The Swedish chronosequence was outstanding in terms of its large soil C sequestration almost solely caused by forest floor buildup. It follows from Table 2.5 that the relative storage by mineral soils – like in the AFFOREST chronosequences – also varied tremendously among other studies. The contribution of mineral soils relative to the total soil ranged from 0 to 52% in other studies and from 0 to about 80% in the AFFOREST chronosequences. The relative contribution of soils and also the partitioning of C between forest floor and mineral soil are probably very influenced by factors like climate, soil type, tree species, former agricultural land use, and sampling methodology. These factors all vary between studies and with this in mind the relatively stable contributions of soils to ecosystem sequestration are remarkable.

The soil type of former cropland may influence the relative effect of afforestation on ecosystem C sequestration. For biomass, soils with high nutrient availability and fine texture would build up the highest C stocks, but it is less easy to predict which soil types will be most conducive to C sequestration in the long term. Mineral soils are potentially a better storage medium than forest floors or biomass with respect to ecosystem disturbance. A large amount of C in mineral soils is bound in organo-mineral complexes that protect C from microbial oxidation (Six et al. 2002; Sollins et al. 1996; Van Veen & Kuikman 1990) whereas forest floor C is more labile. Therefore it would be particularly interesting to select soil types with high probability of increasing the mineral soil C stock. Clay soils have been suggested to be more conducive to C sequestration because of formation of such stable clay-organic matter complexes (Hagedorn et al. 2001), but the synthesis of field studies by Paul et al. (2002) did not support this in general. Several authors suggested that changes in soil C stocks are affected by the organic matter input in coarse-textured soils, and by clay mineralogy in fine textured soils (Christensen 1992; Hassink 1995; Quiroga et al. 1996). If afforestation increases the input of organic matter to soils, this hypothesis suggests that coarse textured soils would experience a greater relative increase in C content in response to afforestation.

In fact, the two Norway spruce chronosequences on contrasting soils in Denmark indicated that sandy, poor soils would sequester more C. The rate of C sequestration was somewhat higher in forest floor at the sandy soil, but the difference was most marked in the mineral soil (Figure 2.5a). Also the Dutch chronosequences with sandy texture was remarkable in the relatively large C sequestration in mineral soils. This pattern suggests that sandy, poor soils with slow decomposition are more conducive to C sequestration in the north European region. The rate of C sequestration for the ecosystem was also high for these sites, as biomass C sequestration was not lower at the soils with poor parent materials for stands younger than 45 years. According to parent materials (see Appendix 1), tree growth within the Danish chronosequences would be expected to be faster on the calcareous, nutrient-rich loamy till at Vestskoven than at the nutrient-poor, acidic glaciofluvial sand at Gejlvang. The legacy of former fertilization and liming in afforested ecosystems is probably responsible for larger nutritional homogeneity than expected based on parent materials. Thus, differences in biomass production are leveled out during the first decades, but sequestration of C in soils is still higher at the poor site, keeping ecosystem C sequestration high. The starting point in soil C stocks possibly also influence the sequestration potential. Former cropland with low initial soil C content as in Gejlvang might be more prone to increase in C stock, which could explain the difference in mineral soil C sequestration to the other Danish chronosequences. The Dutch soils, however, increased in mineral soil C in spite of high initial C stocks. The Dutch sites are less well drained than other sites as the groundwater table can be within 50 cm depth during winter months. Both the current drainage regime and pre-drainage hydrological conditions may have contributed to the high mineral soil C contents and C sequestration rates encountered in the Dutch chronosequences. The more wet soil conditions in the Dutch stands may also be the reason why the general relationship between stand age and soil C improved when Dutch stands were excluded (Figures 2.7a-b).

The AFFOREST chronosequence studies have contributed to quantification of C sequestration following afforestation in north-western Europe. No other studies within the region previously provided similar comprehensive quantitative data on C sequestration following afforestation. Still, conclusions must be tempered by some facts. For instance, sampling and analysis methods differed among the three countries, however, less than seen in compilations of independent studies. Moreover, only one chronosequence included stands older than 40 years and there was one single stand representing an age of about 200 years. As discussed in Chapter 1, there are certain drawbacks to chronosequence studies. Other sources of variation between stands cannot be distinguished from stand age, e.g., changes in agricultural practices since the oldest stand was planted and spatial variability. As pointed out by Yanai et al. (2000), these other sources of variability may lead to erroneous conclusions about the influence of stand age. Keeping the shortcomings in mind, chronosequence studies as in AFFOREST provide a first valuable estimate of changes in soil C before resampling of soils is possible. Repeated sampling at least ten years after the first sampling is planned to test the results obtained in the project. This will provide real evidence of the directional change for each stand in the chronosequences and will further test the predictive value of the chronosequence approach. Chronosequence studies also reveal little about processes responsible for changes in soil C stocks. Chronosequence studies combining pool changes and assessment of isotopic C fractions (e.g. Richter et al. 1999; Del Galdo et al. 2003) enable more insight in the dynamics of old agricultural C and new forest-derived C in soils. Lastly, to address the sustainability of sequestered soil C, i.e. the quality of C pools in sense of stability, there is a need for further studies of C fractions in the mineral soil.

5. CONCLUSIONS

The rate of biomass C sequestration was relatively similar in all chronosequences. A possible effect of parent material was not evident during the first 40-50 years. Parent material effects may be masked by the soil enrichment, which is a legacy of former agriculture. Biomass C sequestration differed more after 40-50 years, probably due to different management, tree species-specific growth patterns and less influence of former fertilization.

For soils there was good evidence that afforestation of former cropland leads to constant or increasing total soil carbon storage. Forest floor build up, and the associated C sequestration, was a general feature, but the contribution of mineral soil to total soil C sequestration differed among chronosequences. In the short term (30 years), tree species had little influence on total soil C sequestration, but in stands younger than 40 years, C sequestration in the forest floor was higher under spruce than under oak. Afforestation of nutrient-poor sandy soils results in larger forest floor C sequestration and larger total soil C sequestration than afforestation of nutrient-rich, clayey soils.

For the afforested ecosystem the general contribution of soils to C sequestration was about one third of the total C sequestration in biomass and soil. The contribution of soil varied among chronosequences from none to 31%. Rates in similar studies from eastern United States were around 15-20%. Total C sequestration was higher in afforested Norway spruce than in afforested oak in the short term (30 years). Soil type did not clearly influence the rate of ecosystem C sequestration in the short term (30-40 years).

We conclude that while biomass C sequestration initially was relatively comparable, there is still little support for generalizations regarding the potential for soil C sequestration. However, based on the AFFOREST chronosequence studies and results in the literature it is safe to conclude that soils play a minor quantitative role relative to biomass of trees in the short and possibly also the long term. Still, the contribution of soils around 20% warrants quantification of this C pool.

As a contribution to mitigation of atmospheric CO_2 concentrations, countries should report changes in C stocks in living biomass, dead organic matter and mineral soils following land-use change in national greenhouse gas budgets (IPCC 2003). The work on land use change in AFFOREST primarily focused on improving the basic knowledge of contributions of afforested cropland to sequestration of C. These results will hopefully help to bridge the gap between current knowledge and policy demands.

6. REFERENCES

Bashkin, M. A., & Binkley, D. (1998). Changes in soil carbon following afforestation in Hawaii. *Ecology, 79,* 828-833.

Bastrup-Birk, A., Hansen, K., Ro-Poulsen, H., Jørgensen, B. B., Mikkelsen, T., Pilegaard, K., & Bille-Hansen, J. (2003). Biomasse og produktion. Næringsstofkredsløb i skove - Ionbalanceprojektet (Ed. K. Hansen). *Forest and Landscape Research, 33,* 69-96. Ministry of the Environment, Danish Forest and Landscape Research Institute, Hørsholm, Denmark. (In Danish).

Billings, W. D. (1938). The structure and development of shortleaf pine stands and certain associated physical properties of soil. *Ecological Monographs, 8,* 437-499.

Binkley, D. (1995). *The influence of tree species on forest soils: Processes and patterns.* Proceeding of the Trees and Soil Workshop. Lincoln University, 28 February-2 March 1994 (Eds. D. J. Mead & I. S. Cornforth), pp. 1-33. Canterbury, New Zealand: Lincoln University Press.

Binkley, D., & Resh, S. C. (1999). Rapid changes in soils following Eucalyptus afforestation in Hawaii. *Soil Science Society of America Journal, 63,* 222-225.

Bouwman, A. F., & Leemans, R. (1995). The role of forest soils in the global carbon cycle. Carbon forms and functions in forest soils (Eds. W. F. McFee & F. M. Kelly), pp. 503-525. Madison, Wisconsin, USA: Soil Science Society of America.

Brandel, G. (1990). Volume functions for individual trees. Scots pine (Pinus sylvestris), Norway spruce (Picea abies) and birch (Betula pendula & Betula pubescens). Swedish University of Agricultural Sciences, Department of Forest Yield Research, Report No. 26. 183 pp. (In Swedish with English summary).

Christensen, B. T. (1992). Physical fractionation of soil and organic matter in primary particle size and density separates. *Advances in Soil Science, 20,* 1-89.

Compton, J. E., & Boone, R. D. (2000). Long-term impacts of agriculture on soil carbon and nitrogen in New England forests. *Ecology, 81,* 2314-2330.

Del Galdo, I., Six, J., Peressotti, A., & Cotrufo, M. F. (2003). Assessing the impact of land-use change on soil C sequestration in agricultural soils by means of organic matter fractionation and stable C isotopes. *Global Change Biology, 9,* 1204-1213.

Denef, K., Six, J., Merckx, R., & Paustian, K. (2004). Carbon Sequestration in Microaggregates of No-Tillage Soils with Different Clay Mineralogy. *Soil Science Society of America Journal, 68,* 1935-1944.

Dewar, R. C., & Cannell, M. G. R. (1992). Carbon sequestration in the trees, products and soils of forest plantations: an analysis using UK examples. *Tree Physiology, 11,* 49-71.

Dick, W. A. (1983). Organic carbon, nitrogen, and phosphorus concentrations and pH in soil profiles as affected by tillage intensity. *Soil Science Society of America Journal, 47,* 102-107.

Dziadowiec, H. (1987). The decomposition of plant litter fall in an oak-linden-hornbeam forest and an oak-pine mixed forest of the Bialowieza National Park. *Acta Societatis Botanicorum Poloniae, 56,* 169-185.

France, E. A., Binkley, D., & Valentine, D. (1989). Soil chemistry changes after 27 years under four tree species in southern Ontario. *Canadian Journal of Forest Research, 19,* 1648-1650.

Garcia-Oliva, F., & Masera, O. R. (2004). Assessment and measurement issues related to soil carbon sequestration in land-use, land-use change, and forestry (LULUCF) projects under the Kyoto Protocol. *Climatic Change, 65,* 347-364.

Guo, L. B., & Gifford, R. M. (2002). Soil carbon stocks and land use change: a meta analysis. *Global Change Biology, 8,* 345-360.

Hagedorn, F., Maurer, S., Egli, P., Blaser, P., Bucher, J. B., & Siegwolf, R. (2001). Carbon sequestration in forest soils: effects of soil type, atmospheric CO2 enrichment, and N deposition. *European Journal of Soil Science, 52,* 619-628.

Hagen-Thorn, A., Callesen, I, Armolaitis, K., & Nihlgård, B. (2004). The impact of six European tree species on the chemistry of mineral topsoil in forest plantations on former agricultural land. *Forest Ecology and Management, 195,* 373-384.

Hamburg, S. P. (1984). Effects of forest growth on soil nitrogen and organic matter pools following release from subsistence agriculture. Forest Soils and Treatment Impacts (Ed. E. L. Stone), pp. 145-158. University of Tennessee, Knoxville.

Hassink, J. (1995). Decomposition rate constants of size and density fractions of soil organic matter. *Soil Science Society of America Journal, 59,* 1631-1635.

Hoekstra, C., & Poelman, J. N. B. (1992). Dichtheid van gronden gemeten aan de meest voorkomende bodemeenheden in Nederland. Stiboka Rapport 1582, Wageningen, The Netherlands. (In Dutch).

Homann, P. S., Sollins, P., Chappell, H. N., & Stangenberger, A. G. (1995). Soil organic carbon in a mountainous, forested region: Relation to site characteristics. *Soil Science Society of America Journal, 59,* 1468-1475.

Hooker, T. D., & Compton, J. E. (2003). Forest ecosystem carbon and nitrogen accumulation during the first century after agricultural abandonment. *Ecology Applications, 13,* 299-313.

Illerup, J. B, Lyck, E., Nielsen, M., Winther, M., Mikkelsen, M. H., Hoffmann, L., Gyldenkærne, S., Sørensen, P., Vesterdal, L., Fauser, P., & Thomsen, M. (2005). Denmark's National Inventory Report 2005. Submitted under the United Nations Framework Convention on Climate Change 1990-2003. Ministry of Environment, Denmark, National Environmental Research Institute. http://cdr.eionet.eu.int/dk/Air_Emission_Inventories/Submission_UNFCCC/colql9meg/envql9mfg/NIR_April_2005_DK.doc

IPCC. (2003). Good Practice Guidance for Land Use, Land-Use Change and Forestry. http://www.ipcc-nggip.iges.or.jp/public/gpglulucf/gpglulucf.htm

Jansen, J. J., Sevenster, J., & Faber, P. J. (1996). Opbrengsttabellen voor belangrijke boomsoorten in Nederland. IBN-rapport 221, Wageningen. (In Dutch).

Johannsen, V. K., Jørgensen, B. B., & Morsing, M. (2004). Skovstatistik feltinstruks 2004. Skov & Landskab, KVL. 173 pp. (In Danish).

Johnson, D. W. (1992). Effects of forest management on soil carbon storage. *Water, Air, and Soil Pollution, 64,* 83-120.

Johnston, M. H., Homann, P. S., Engstrom, J. K., & Grigal, D. F. (1996). Changes in ecosystem carbon storage over 40 years on an old-field/forest landscape in east-central Minnesota. *Forest Ecology and Management, 83,* 17-26.

Jug, A., Makeschin, F., Rehfuess, K. E., & Hofmann-Schielle, C. (1999). Short-rotation plantations of balsam poplars, aspen and willows on former arable land in the Federal Republic of Germany. III. Soil ecological effects. *Forest Ecology and Management, 121,* 85-99.

Jussy, J. H., Koerner, W., Dambrine, E., Dupouey, J. L., & Benoit, M. (2002). Influence of former agricultural land use on net nitrate production in soils. *European Journal of Soil Science, 53,* 367-374.

Karlsson, K. (1998). A strategy for reducing the number of sample trees when estimating stand volume on sample plots. Swedish University of Agricultural Sciences, Department of Forest Yield Research, report 43, 18 pp.

Krogh, L., Nørgaard, A., Hermansen, M., Greve, M. H., Balstrøm, T., & Breuning-Madsen, H. (2003). Preliminary estimates of contemporary soil organic carbon stocks in Denmark using multiple datasets and four scaling-up methods. *Agriculture, Ecosystems & Environment, 96,* 19-28.

Kurmies, B. (1949). Humusbestimmung nach dem Biochromatverfahren ohne Kaliumjodid. *Zeitschrift für Pflanzenernährung, Dungung und Bodenkunde, 44,* 121-125.

Leth, S., & Breuning-Madsen, H. (1992). Changes in soil profile development and nutrient status due to afforestation of agricultural land. *Danish Journal of Geography, 92,* 70-74.

Lettens, S., Van Orshoven, J., van Wesemael, B., & Muys, B. (2004). Soil organic and inorganic carbon contents of landscape units in Belgium derived using data from 1950 to 1970. *Soil Use and Management, 20,* 40-47.

Madsen, S. F. (1987). Vedmassefunktioner ved forskellige aflægningsgrænser og nøjagtighedskrav for nogle vigtige danske skovtræarter. *Det Forstlige Forsøgsvæsen i Danmark, 41*, 47-242. (In Danish with English summary: Volume equations for some important Danish tree species. Standard and form class equations. Total and merchantable volumes.)

Madsen, S. F., & Heusérr, M. (1993). Volume and stem-taper functions for Norway spruce in Denmark. *Forest & Landscape Research, 1,* 51-78.

Malhi, Y, Baldocchi, D. D., & Jarvis, P. G. (1999). The carbon balance of tropical, temperate and boreal forests. *Plant, Cell and Environment, 22,* 715-740.

Marklund, L. G. (1988). Biomass functions for pine, spruce and birch in Sweden. Swedish University of Agricultural Sciences, Umeå. Dept. of Forest Survey, Report 45, 73 pp. ISSN 0348-0496. (In Swedish with English summary.)

Matejovic, I. (1993). Determination of carbon, hydrogen, and nitrogen in soils by automated elemental analysis (dry combustion method). *Communications in Soil Science and Plant Analysis, 24,* 2213-2222.

McNabb, D. H., Cromack, K., & Fredriksen, R. L. (1986). Variability of nitrogen and carbon in surface soils of six forest types in the Oregon Cascades. *Soil Science Society of America Journal, 50,* 1037-1041.

Nabuurs, G.-J., & Mohren, G. M. J. (1995). Modelling analysis of potential carbon sequestration in selected forest types. *Canadian Journal of Forest Research, 25,* 1157-1172.

Näslund, M. (1936). Skogsförsöksanstaltens gallringsförsök i tallskog. *Meddelanden från Statens Skogsförsöksanstalt, 29,* 1-169. (In Swedish).

Olesen, J. E., Andersen, J. M., Jacobsen, B. H., Hvelplund, T., Jørgensen, U., Schou, J. S., Graversen, J., Dalgaard, T., & Fenhann, J. V. (2001). Kvantificering af tre tiltag til reduktion af landbrugets emission af drivhusgasser. DJF Rapport Markbrug nr. 48, Danish Institute of Agricultural Sciences, Foulum, Denmark. (In Danish with English summary).

Paul, E. A., Morris, S. J., Six, J., Paustian, K., & Gregorich, E. G. (2003). Interpretation of soil carbon and nitrogen dynamics in agricultural and afforested soils. *Soil Science Society of America Journal, 67,* 1620-1628.

Paul, K. I., Polglase, P. J., Nyakuengama, J. G., & Khanna, P. K. (2002). Change in soil carbon following afforestation. *Forest Ecology and Management, 168,* 241-257.

Pedersen, L. B., & Bille-Hansen, J. (1999). A comparison of litterfall and element fluxes in even aged Norway spruce, sitka spruce and beech stands in Denmark. *Forest Ecology and Management, 114,* 55-70.

Post, W. M., & Kwon, K. C. (2000). Soil carbon sequestration and land-use change: processes and potential. *Global Change Biology, 6,* 317-327.

Powlson, D. S., Smith, P., Coleman, K., Smith, J. U., Glendining, M. J., Körschens, M., & Franko, U. (1998). A European network of long-term sites for studies on soil organic matter. *Soil and Tillage Research, 47,* 263-274.

Pregitzer, K. S., & Palik, B. J. (1997). Change in ecosystem carbon 46 years afte establishing red pine (Pinus resinosa Ait.) on abandoned agricultural land in the Great Lakes Region. In Soil organic matter in temperate agroecosystems: long-term experiments in North America (Eds. E. A. Paul, K. Paustian, E. T. Elliott & C. V. Cole), pp. 263-270. New York: CRC Press.

Quiroga, A. R., Buschaiazzo, D. E., & Peinemann, N. (1996). Soil organic matter particle size fractions in soils of the semiarid Argentinian pampas. *Soil Science, 161,* 104-107.

Richter, D. D., Markewitz, D., Trumbore, S. E., & Wells, C. G. (1999). Rapid accumulation and turnover of soil carbon in a re-establishing forest. *Nature, 400,* 56-58.

Rodriguez-Murillo, J. C. (2001). Organic carbon content under different types of land use and soil in peninsular Spain. *Biology and Fertility of Soils, 33,* 53-61.

Romanya, J., Cortina, J., Falloon, P., Coleman, K., & Smith, P. (2000). Modelling changes in soil organic matter after planting fast-growing Pinus radiata on Mediterranean agricultural soils. *European Journal of Soil Science, 51,* 627-641.

Römkens, P. F. A. M., van der Pflicht, J., & Hassink, J. (1999). Soil organic matter dynamics after the conversion of arable land to pasture. *Biology and Fertility of Soils, 28,* 277-284.

Rosenqvist, L., & Johansson, M.-B. (2005). Carbon sequestration and fluxes of dissolved organic carbon in a chronosequence of Norway spruce forest planted on former arable land. Manuscript in preparation.

SAS Institute (1993). SAS/STAT User's Guide Rel. 6.07. Cary, North Carolina, USA: SAS Institute Inc.

Schiffman, P. M. & Johnson, W. C. (1989). Phytomass and detrital carbon storage during forest regrowth in the southeastern United States Piedmont. *Canadian Journal of Forest Research, 19,* 69-78.

Schimel, D. S., Coleman, D. C., & Horton, K. A. (1985). Soil organic matter dynamics in paired rangeland and cropland toposequences in North Dakota. *Geoderma, 36,* 201-214.

Six, J., Conant, R. T., Paul, E. A., & Paustian, K. (2002). Stabilization mechanisms of soil organic matter: Implications for C-saturation of soils. *Plant and Soil, 241,* 155-176.

Smith, K.A. (1999). After the Kyoto Protocol: Can soil scientists make a useful contribution? *Soil Use and Management, 15,* 71-75.

Soil Survey Staff (1992). Keys to Soil Taxonomy, 5th Edition. SMSS technical monograph no. 19. Blacksburg, Virginia: Pocahontas Press, Inc.

Sollins, P., Homann, P., & Caldwell, B. A. (1996). Stabilization and destabilization of soil organic matter: mechanisms and controls. *Geoderma, 74,* 65-105.

Thuille, A., Buchmann, N., & Schulze, E.-D. (2000). Carbon stocks and soil respiration rates during deforestation, grassland use and subsequent Norway spruce afforestation in the Southern Alps, Italy. *Tree Physiology, 20,* 849-857.

Van Veen, J. A., & Kuikman, P. J. (1990). Soil structural aspects of decomposition of organic matter by micro-organisms. *Biogeochemistry, 11,* 213-233.

Vejre, H., Callesen, I., Vesterdal, L., & Raulund-Rasmussen, K. (2003). Carbon and nitrogen in Danish forest soils - contents and distribution determined by soil order. *Soil Science Society of America Journal, 67,* 335-343.

Vesterdal, L., & Raulund-Rasmussen, K. (1998). Forest floor chemistry under seven tree species along a soil fertility gradient. *Canadian Journal of Forest Research, 28,* 1636-1647.

Vesterdal, L., Ritter, E., & Gundersen, P. (2002). Change in soil organic carbon following afforestation of former arable land. *Forest Ecology and Management, 169,* 137-143.

Vogt, K. A., Grier, C. C., & Vogt, D. J. (1986). Production, turnover, and nutrient dynamics of above- and belowground detritus of world forests. *Advances in Ecological Research, 15,* 303-377.

Voroney, R. P., Van Veen, J. A., & Paul, E. A. (1981). Organic C dynamics in grassland soils. 2. Model validation and simulation of the long-term effects of cultivation and rainfall erosion. *Canadian Journal of Soil Science, 61,* 211-224.

Westman, C. J. (1995). A simple device for sampling of volumetric forest soil cores. *Silva Fennica, 29,* 247-251.

Winjum, J. K., & Schroeder, P. E. (1997). Forest plantations of the world: their extent, ecological attributes, and carbon storage. *Agricultural and Forest Meteorology, 84,* 153-167.

Yanai, R. D., Arthur, M. A., Siccama, T. G., & Federer, C. A. (2000). Challenges of measuring forest floor organic matter dynamics: Repeated measures from a chronosequence. *Forest Ecology and Management, 138,* 273-283.

CHAPTER 3

INTERCEPTION AND WATER RECHARGE FOLLOWING AFFORESTATION: EXPERIENCES FROM OAK AND NORWAY SPRUCE CHRONOSEQUENCES IN DENMARK, SWEDEN AND THE NETHERLANDS

C. VAN DER SALM[1], L. ROSENQVIST[2], L. VESTERDAL[3], K. HANSEN[3], H. DENIER VAN DER GON[4], A. BLEEKER[5], R. WIEGGERS[1] and A. VAN DEN TOORN[1]

[1]Alterra Green World Research, P.O. Box 47, 6700 AA Wageningen, the Netherlands, E-mail: Caroline.vandersalm@wur.nl, [2]Swedish University of Agricultural Sciences, Box 7001, 750 07 Uppsala, Sweden, [3]Forest and Landscape Denmark, KVL, Hørsholm Kongevej 11, 2970 Hørsholm, Denmark, [4]TNO, Environment and Geosciences, P.O. Box 342, Laan van Westenenk 501, 7300 AH Apeldoorn, the Netherlands, [5]Energy research Centre of the Netherlands (ECN), P.O. Box 1, 1755 ZG Petten, the Netherlands

Abstract. The long-term effects of afforestation on hydrological fluxes were investigated using a series of forests of different age planted on comparable soils (chronosequences) in Sweden, Denmark and the Netherlands. Rainfall, throughfall, soil moisture contents and groundwater dynamics were monitored at two oak chronosequences and four spruce chronosequences during a period of one to two years. At all chronosequences, the hydrological fluxes were simulated using a hydrological simulation model. The model was validated on measured throughfall data, soil water contents and Cl fluxes. Afforestation has a clear influence on the water recharge of the considered sites. Water recharge is generally lower under spruce compared to oak. In the spruce stands 5–30% of the incoming precipitation leads to water recharge to ground and surface water, whereas water recharge in the oak stands ranges between 20–35% of the precipitation. In general, water recharge declined with an increase of the stand age. At the oak stands leaching decreased from 35 to 20% of the precipitation in the first 30 years. In the spruce stands the water recharge varied considerable between the four investigated chronosequences but in general, the decline in water recharge was approximately 100-150 mm (10-20%). In both oak and spruce stands, losses by soil evaporation slightly declined. Transpiration slightly increased in the oak stands and transpiration remained fairly stable in the spruce stands. It can be concluded that afforestation leads to a reduction in water recharge compared to agricultural use. This reduction is mainly due to an increase in interception

G.W. Heil et al. (eds.), Environmental Effects of Afforestation in North-Western Europe, 53–77.

evaporation. The strongest reduction is found when sites are afforested with dense spruce forests. The smallest impact is found in open deciduous forest, which has lower interception evaporation.

1. INTRODUCTION

Afforestation may have adverse environmental effects such as the reduction of water recharge to groundwater and surface water due to higher transpiration of forest compared to agricultural land. Several studies were carried out to investigate the effects of changes in forest cover on the water recharge (Bosch & Hewlett 1982; Blackie 1993, Sahin & Hall 1996). These studies indicate that water recharge declines with an increase in forest cover. The decline in recharge is generally larger for coniferous forest compared to deciduous forests. The reductions in recharge strongly differ from site to site due to differences in climate, site characteristics and field layer. Most of the results are based on clear cut experiments in paired catchments and strongly focus on the difference between a mature forest stand and a clear cut. Much less information is available on changes in the water balance in the years between initial afforestation and mature forest. Moreover, data on the changes in the contribution of the various hydrological fluxes to the water balance, such as interception losses, soil evaporation and transpiration are relatively scarce.

The above mentioned studies indicated that afforestation strongly influences the hydrology of the afforested area. To make decisions on afforestation of former agricultural areas in Europe more information on the impact of afforestation on hydrological fluxes is needed. In order to increase the knowledge on the long-term effects of afforestation a series of forests of different age planted on comparable soils (chronosequences) were investigated in the AFFOREST project.

This chapter describes the results of two oak chronosequences and four spruce chronosequences. At all chronosequences, the hydrological fluxes were simulated using a hydrological simulation model. Simulated throughfall fluxes, water contents and groundwater levels were compared with monthly measured values at the different sites. The specific objectives were i) to estimate interception and water recharge after afforestation of former arable land, and ii) to determine the differences in these hydrological fluxes for coniferous and deciduous tree species.

2. METHODS

2.1. General approach

Chronosequences of oak and spruce were selected in Denmark, southern Sweden and the Netherlands (Chapter 1). At each chronosequence, hydrological conditions were measured monthly during a period of one to two years. Soil hydrological fluxes were modeled using a dynamic simulation model. The model was validated on measured throughfall data and soil water contents. The simulated hydrological fluxes were used to calculate Chloride (Cl) leaching fluxes at the bottom of the root zone (section 2.5). The calculated Cl budgets were used as a final validation of the calculated fluxes.

2.2. Study sites

The hydrology of six chronosequences was studied. Two of these chronosequences were afforested with oak and four with Norway spruce. The oak forests were located in Denmark on a clay soil and in the Netherlands on a sandy soil. The spruce chronosequences were located in Sweden (sandy soil), Denmark (one on a sandy soil and one on a clay soil) and the Netherlands (sandy soil). A description of the sites is given in Chapter 1 and on the AFFOREST website (www.sl.kvl.dk/afforest).

Throughfall was measured using 10-15 collectors at each site. Precipitation was measured in an open field within a distance of 2 km from the sites. Soil water contents were measured using TDR equipment and used for validation of the hydrological model. In Sweden and Denmark, the TDR probes (PRENART) were located within fixed depth intervals of 0-20 cm, 0-50 cm, and for Denmark only, within 0-90 cm. In the Netherlands, soil moisture contents were measured from 10-100 cm depth at an interval of 10 cm using a portable TDR probe (TRIME-T3) which was lowered in to plastic access tubes (TECANAT). TDR measurements were carried out once a month at 3-9 points within each plot (see details on www.sl.kvl.dk/afforest). The height of the groundwater table was measured once a month at the Dutch sites where the groundwater table was within 3 m of the soil surface.

2.3. Model description

The water balance of a forest stand can be described as:

$$P = E + R + D + Q + \Delta S \tag{1}$$

in which P is the precipitation, E is the evapotranspiration, Q is the vertical leaching flux (recharge) to the groundwater, R is runoff, D is the lateral drainage and ΔS is the change in the amount of water in the soil profile. Evapotranspiration is divided in interception evaporation (E_i), transpiration (E_t) and soil evaporation (E_s) according to:

$$E = E_i + E_t + E_s \tag{2}$$

The various terms of the water balance were calculated using the Darcy model SWAP (Van Dam et al. 1997). This model provides a finite difference solution to the Richards equation:

$$\frac{\delta\theta}{\delta t} = \frac{\delta}{\delta z}\left(K(h)\left(\frac{\delta h}{\delta z}+1\right)\right) - S(h) \tag{3}$$

where θ ($m^3\ m^{-3}$) is the volumetric water content, t (d) is time, z (m) the vertical position in the soil, h (m) soil water pressure head, K ($m\ d^{-1}$) hydraulic conductivity and S (d^{-1}) the sink term accounting for root water uptake (actual transpiration).

The potential loss of water by evapotranspiration is calculated using the Penman-Monteith equation (Monteith 1981):

$$E_{pm} = \frac{1}{\lambda} \frac{s R_n + \rho C_p \delta q / r_a}{s + \gamma (1 + r_s / r_a)} f_s \tag{4}$$

where E_{pm} is the potential evapotranspiration (mm d^{-1}), λ is the specific heat of evaporation (J kg^{-1}), s is the slope of the saturated water vapour curve (hPa °K^{-1}), R_n is the net radiation (W m^{-2}), ρ is the density of air (kg m^{-3}), C_p is the specific heat capacity of the air (J g^{-1} °K^{-1}), δq is the water vapour deficit (hPa), r_a and r_s are the aerodynamic and the crop resistances (s m^{-1}), γ is a psychrometer coefficient (mbar °C^{-1}) and f_s is the number of seconds per day.

The potential evapotranspiration is divided into interception evaporation, potential soil evaporation and potential transpiration. When the canopy is wet, r_s is zero and the Penman-Monteith equation reduces to:

$$E_{wet} = \frac{\frac{1}{\lambda}(sR_n + \rho C_p / r_a\ \delta q)}{(s + \gamma)} \tag{5}$$

Potential transpiration is calculated from the potential evapotranspiration, given by Eq. 4, by reducing the evapotranspiration during rainfall with the calculated interception evaporation according to:

$$E_t^* = \frac{(s + \gamma)}{(s + \gamma\ (1 + r_c / r_a))} (E_{wet} - E_i) \tag{6}$$

The total amount of interception (E_i) is calculated according to Gash (1995):

$$E_i = sc\ P \tag{7}$$

when $P < P_s$, or as:

$$E_i = sc\ P_s + sc \frac{E_{avg}}{R_{avg}} (P - P_s) \tag{8}$$

when $P > P_s$, where sc is the soil cover fraction(-), R_{avg} is the average rainfall rate (mm hr^{-1}), E_{avg} the average daily evaporation rate during rainfall (mm hr^{-1}), P the daily precipitation (mm d^{-1}) and P_s (mm d^{-1}) is the amount of rainfall to saturate the canopy:

$$P_s = \frac{R_{avg}}{E_{avg}} S_{\max} \, sc \ln\left(1 - \frac{E_{avg}}{R_{avg}}\right) \tag{9}$$

where S_{max} is the storage capacity of the crown (mm). Lateral drainage to a local drainage network (ditches) was calculated by:

$$q_d = \frac{\Phi_{avg} - \Phi_d}{\gamma_d} \tag{10}$$

where q_d (m^3 m^{-2} d^{-1}) is the flux of water to the local drainage system, Φ_d (m) is hydraulic head of drainage system, and γ_d (d) is drainage resistance. In order to distribute the discharge rates over the soil layers, first a discharge layer is determined by considering a travel-time distribution. The most important assumption in this computational procedure is that lateral discharge occurs parallel to equidistant water courses (distance Lk (m)), cf. Van Dam et al. 1997). Within this discharge layer, the lateral drainage from soil layer i to the local drainage system, $R_{d,L,i}$, is calculated with the equation (see Eq. 3 for symbols):

$$R_{d,L,i} = \frac{q_d}{\Delta z_i} \frac{K_i \Delta z_i}{\sum (K_i \Delta z_i)} \tag{11}$$

2.4. Model parameters

The model SWAP needs meteorological data, abiotic characteristics of the site (soil physical characteristics, drainage characteristics) and vegetation dependant parameters (crop resistance, LAI, parameters to calculate interception, tree height, root distribution) in order to calculate the hydrological fluxes.

2.4.1. Meteorological data

The model needs input of a set of daily meteorological data on precipitation, net radiation, temperature, wind speed and relative humidity. These data were not collected on the site but taken from nearby meteorological stations (Table 3.1).

Table 3.1. *Origin of the meteorological data used in the model calculations.*

Site	Location	Distance
Tönnersjöheden, S	Torup	25 km
	Göteborg[1)]	120 km
Vestskoven, DK	Højbakkegård	4 km
	Værløse[2)]	8 km
Gejlvang, DK	Billund	11 km
Sellingen, NL	Nieuw Beerta	30 km
Drenthe, NL	Eelde[3)]	20-30 km

[1)] *All meteorological data were derived from Torup except global radiation, which was derived from Göteborg.*
[2)] *Data from the beginning of 1996 to May 2001 are from Højbakkegård. Data from May 2001 to the end of 2002 are from grid cell Værløse.*
[3)] *For the 14 yr old spruce stand at Drenthe, data from Nieuw Beerta were used.*

Precipitation may vary considerably over short distances, leading potentially to large differences in daily precipitation between data from a (nearby) meteorological station and the actual measured precipitation at the site. To obtain the best estimate of the daily precipitation at the site, daily precipitation data were corrected based on the measured precipitation at the open field close to forest sites according to:

$$P_{i,site} = P_{i,station} \cdot \frac{P_{period,site}}{P_{period,station}} \tag{12}$$

where $P_{i,site}$ is the daily precipitation at the site, $P_{i,station}$ is the daily precipitation measured at the meteorological station, $P_{period,site}$ is the measured precipitation at the site during a period of one month and $P_{period,station}$ is the precipitation measured at the meteorological station during this period.

2.4.2. Abiotic characteristics

The model SWAP uses information on the physical characteristic of the soil, data on the hydrological conditions at the bottom boundary and drainage characteristics (cf. groundwater and drainage).

Water retention characteristics and conductivity data for Drenthe, Vestskoven and Gejlvang were selected from the Staring soil series (Wösten et al. 1994). For Tönnersjöheden, data were based on the HYPRES (Hydraulic properties of European soils) database (Wösten et al. 1999). For all sites, water retention characteristics for the organic layer were based on data from a Dutch Douglas fir site (Tiktak & Bouten 1994).

At Sellingen, water retention characteristics were measured in the mineral soil at 15-20 cm, 40-45 cm and 80-85 cm depth. These data were allocated to the different layer sin the soil profile (Table 3.2). At some sites measured saturated water contents were calibrated on measured maximum water contents in the field during the winter period (cf. section 3.1). Data on saturated conductivity were based on the Staring soil series (Wösten et al. 1994).

Table 3.2. *Soil hydrological characteristics for the Dutch oak sites at Sellingen.*

Depth	θ_{res}	Θ_{sat}	α	n	K_{sat}	l
Oak, Sellingen, NL: 4 years						
0-40	0.052	0.36	0.010	2.29	32.2	-0.983
40-60	0.023	0.34	0.013	2.02	63.9	0.039
> 60	0.023	0.34	0.015	1.91	63.9	0.039
Oak, Sellingen, NL: 8 years						
0-1*	0.000	0.5	0.100	1.25	800	0.0178
1-40	0.064	0.36	0.009	2.44	32.2	-0.983
40-60	0.039	0.34	0.011	2.02	63.9	0.039
> 60	0.005	0.34	0.013	2.14	63.9	0.039
Oak, Sellingen, NL: 11 years						
0-2*	0.000	0.5	0.100	1.25	800	0.0178
2-40	0.042	0.36	0.009	2.14	32.2	-0.983
40-60	0.018	0.34	0.012	2.20	63.9	0.039
> 60	0.000	0.34	0.014	1.73	63.9	0.039
Oak, Sellingen, NL: 18 years						
0-3*	0.000	0.5	0.100	1.25	800	0.0178
3-40	0.028	0.36	0.012	2.04	32.2	-0.983
40-60	0.007	0.34	0.011	2.17	63.9	0.039
> 60	0.005	0.34	0.012	2.13	63.9	0.039

Depth = depth of soil layer, [cm]
θ_{res} = *Residual water content,* $[cm^3\ cm^{-3}]$
Θ_{sat} = *Saturated water content,* $[cm^3\ cm^{-3}]$
α = *Shape parameter of main drying curve,* $[cm^{-1}]$
n = Shape parameter, [-]
K_{sat}= *Saturated hydraulic conductivity,* $[cm\ d^{-1}]$
l = Exponent in hydraulic conductivity function, -]
* = *Organic litter layer*

Free drainage of the soil profile was assumed as the groundwater level was deeper than 3 meters at most sites except from the Dutch oak sites at Sellingen and the two youngest Dutch spruce sites. At these sites with shallow groundwater tables, the fluxes at the bottom of the soil profile were calculated as a function of the groundwater level (Ernst & Feddes 1979):

$$Q_{bot} = a\,e^{b\phi_{avg}} \tag{13}$$

with Q_{bot} is the flux at the bottom boundary, *a* (cm d^{-1}) and *b* (cm^{-1}) are parameters. Values for the parameters *a* and *b* (Table 3.3) are based on literature data (De Visser & De Vries 1989).

Table 3.3. *Bottom boundary conditions, parameters a = [cm d^{-1}] and b = [cm^{-1}].*

Site	a	b
Sellingen	-0.7	-0.03
Drenthe (8 yr old)	-0.2	-0.01
Drenthe (13 yr old)	-0.4	-0.02

The Dutch oak stands were drained by ditches. Lateral drainage to the ditches at these sites was calculated using data on drainage resistance, the spacing, depth and water level of the ditches (Table 3.4). Water level of ditches was measured monthly and used as input for the model. Drainage resistances were calibrated on measured groundwater levels and soil moisture contents.

Table 3.4. *Drainage characteristics for the oak stands at Sellingen, the Netherlands.*

Parameter	Unit	4 years	8 years	11 years	18 years
Drainage resistance	d	150	300	1200	800
Ditch spacing	m	300	600	1000	1000
Depth of the ditches	cm	-140	-148	-166	-182
Average water level in ditches[1]	cm	-121 (± 5)	-131 (± 5)	-146 (± 5)	-161(± 5)

[1] *Average water level over the measurement period in cm below the soil surface*

2.4.3. *Vegetation characteristics.*

The most important crop parameters used by the model are tree height, canopy resistance, leaf area index (LAI), storage capacity of the canopy (S_{max}), soil cover (sc) and root distribution. Tree height and LAI were based on measurements (Table 3.5 & Table 3.6). Root distribution data were estimated based on visual examination of the soil profile and literature data on average root distribution for spruce and oak (De Visser & De Vries 1989). Parameters describing the interception were calibrated on measured throughfall data (Table 3.5 & Table 3.6).

Table 3.5. *The most important stand characteristics for the oak stands used in the model calculations.*

Site	Age in 2002	Years after afforest.	Height (m)	LAI (m^{-2} m^{2})	Canopy resistance (s m^{-1})	Interception parameters	
						S_{max} (cm)	Sc (-)
Vestskoven, DK	9	9	2.6	2.9	60	0.05	0.68
Vestskoven, DK	14	14	6.0	4.5	60	0.08	0.75
Vestskoven, DK	23	23	11.0	5.1	60	0.14	0.77
Vestskoven, DK	25	25	10.1	4.4	60	0.09	0.81
Vestskoven, DK	32	32	13.8	5.3	60	0.14	0.79
Sellingen, NL	6	4	3.2	3.6	85	0.05	0.20
Sellingen, NL	10	8	6.3	3.6	85	0.10	0.30
Sellingen, NL	14	11	7.6	4.3	85	0.15	0.50
Sellingen, NL	21	18	10.3	3.3	85	0.15	0.60

Table 3.6. *The most important stand characteristics for the spruce stands used in the model calculations.*

Site	Age in 2002	Years after afforest.	Height (m)	LAI (m^{-2} m^{2})	Canopy resistance (s m^{-1})	Interception parameters	
						S_{max} (cm)	Sc (-)
Tönnersjöheden, S	19	19	10.1	5.6	50	0.07	0.91
Tönnersjöheden, S	30	30	15.5	5.8	50	0.32	0.95
Tönnersjöheden, S	65	65	19.3	3.4	80	0.26	0.83
Tönnersjöheden, S	74	74	24.1	3.3	80	0.20	0.71
Tönnersjöheden, S	92	92	35.7	3.4	80	0.38	0.82
Vestskoven, DK	5	5	1.9	3.5	60	0.10	0.60
Vestskoven, DK	12	12	6.1	5.7	60	0.15	0.75
Vestskoven, DK	14	14	8.4	5.4	60	0.15	0.67
Vestskoven, DK	29	29	13.3	7.1	60	0.22	0.79
Vestskoven, DK	33	33	17.8	8.5	60	0.25	0.90
Gejlvang, DK	8	8	1.8	6.7	60	0.22	0.86
Gejlvang, DK	21	21	10.0	8.8	60	0.35	0.99
Gejlvang, DK	26	26	12.9	7.2	60	0.24	0.86
Gejlvang, DK	42	42	16.5	8.1	60	0.25	0.95
Drenthe, NL	11	8	6	5.6	140	0.17	0.85
Drenthe, NL	16	13	8	4.3	140	0.23	0.97
Drenthe, NL	17	14	7	5.1	140	0.40	0.99

2.5. Calibration and validation

Interception was calibrated on measured rainfall and throughfall data by adjusting the estimated parameters for soil cover (sc) and storage capacity of the canopy (S_{max}). Initial values for soil cover were based on field estimates and data for the storage capacity of the crown were based on a literature review (De Vries et al. 2001). Normally, parameter values for the storage capacity and the soil cover are derived from analyses of data from single storms using the analysis method by Leyton (Leyton et al. 1967). Such an analysis could not be made for these study sites as only monthly throughfall data were available.

Soil water fluxes were calibrated on measured groundwater levels and water contents. The simulated soil water contents are influenced by both the hydraulic characteristics, drainage characteristics, and the crop characteristics. Parameters were adjusted when clear deviations between measured and simulated water contents did occur. For example, when measured water contents in winter were systematically overestimated, the hydraulic characteristics were adapted. When measured water contents in summer were systematically lower than simulated water contents, crop resistances were increased to reduce the transpiration. At the shallow drained Dutch sites, drainage parameters (Table 3.4) affected the simulated groundwater level and water contents most strongly. These parameters were adjusted on basis of the winter period when transpiration can be neglected. The other parameters were calibrated later on using the full simulation period.

A final check on the validity of the calculated hydrological fluxes can be made by calculating Cl balances for the considered plots. Chloride is considered to behave as a conservative element in the soil solution and the long-term output of Cl by leaching should thus be equal to the long-term input of Cl by throughfall. In practice, substantial deviations in the Cl budgets may occur due to changes in water storage and uncertainties in Cl concentrations in both input and output. This uncertainty is particularly large when short time periods (< 3 years) are considered. For example, at the Solling spruce stand in Germany the Cl budgets for a period of 13 year was 0.7 kg ha-1 yr-1, whereas budgets for individual years ranged from -20 to 30 kg ha-1 yr-1 (van der Salm et al. 2004). At the chronosequences, the monitoring period was 2 years or less. This period is too short to use the budgets for fine tuning of the hydrological model. However, the budgets may give a rough indication of the reliability of the calculated fluxes.

3. RESULTS

3.1. Simulated interception

Calibrated values for soil cover of the oak stands increased from 0.2 to 0.8 with an increase in age (Table 3.5). Values for soil cover were considerably higher for the Danish sites compared to the Dutch sites. The storage capacity of the oak trees increased from 0.05 to 0.15 cm. Calibrated values for storage capacity were generally lower in Denmark than in the Netherlands. The differences in calibrated parameter values between the Dutch and Danish sites do not correspond with differences in stem density and LAI between the Netherlands and Denmark. On average, stem density was lower in Denmark (average approx. 2900 stems/ha) compared to the Netherlands (average approx. 6000 stems/ha) and LAI was somewhat higher in Denmark (average 4.4) compared to the Netherlands (average 3.7). The different values obtained for the Netherlands and Denmark are caused by the fact that the parameters soil cover and storage capacity can not be calibrated independently on monthly measured throughfall data. Lower values for soil cover together with higher values for storage capacity may lead to comparable simulated throughfall fluxes. This is not a serious problem when studying the long-tem effects of afforestation, as long as the obtained values for soil cover and storage capacity are used in combination. For a more detailed analysis of the hydrology on a daily-weekly scale, derivation of the parameters according to the method described by Leyton et al. (1967) is advised.

In the spruce stands, the calibrated values for the soil cover were higher than in the oak stands and increased from 0.60 to 0.99. The storage capacity of the crown was also considerably higher and ranged from 0.1 in the youngest stand to 0.4 in the older spruce stands.

Yearly simulated interception fluxes were generally within 5% of the measured values in 2002 (Figure 3.1). Substantial deviations were found in the youngest spruce stand at Vestskoven where interception losses are underestimated by 50% (70 mm) and the youngest oak stand in Sellingen where interception was overestimated by 68% (60 mm).

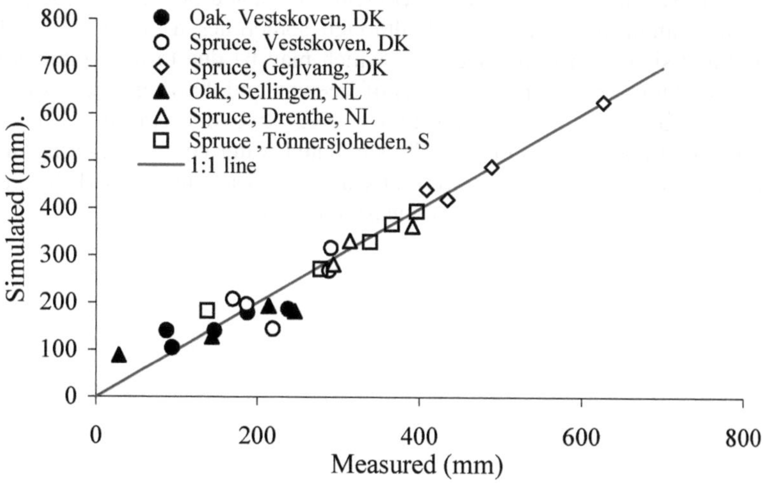

Figure 3.1. *Measured and simulated yearly interception fluxes (mm) for 2002.*

3.2. Simulated soil water contents

The model simulated measured soil water contents quite well in most stands. Examples of the simulated and the measured soil water content are shown for the 18-yr old oak stand at Sellingen and the 9-yr old oak stand at Vestskoven, the 33-yr old spruce stand at Vestskoven, the 42-yr old spruce stand at Gejlvang and the 76-yr old spruce stand at Tönnersjöheden (Figure 3.2). At the 18-yr old oak stand at Sellingen (NL), the simulated water contents at 20 cm depth were generally within the range of measured values at the 3 measurement points within the plot. However, extremely high water contents (0.42 cm^3 cm^{-3}) occurring in autumn and winter 2001/2002 were underestimated (0.36 cm^3 cm^{-3}), indicating that the simulated drainage fluxes are too high. On the other hand, in summer, simulated water contents at 20 cm depth tended to be slightly (0.05 cm^3 cm^{-3}) higher than measured values. Simulated water contents at the 9-yr old oak stand at Vestskoven were close to measured values throughout the year. However, measurements in summer are relatively sparse making it difficult to compare the simulated results with the actual situation in the summer period.

In all the spruce stands, the measured water contents were generally close to measured values. The best results were obtained for the site at Gejlvang with the exception of spring 2001, where the model underestimated the soil water contents with up to 0.07 cm^3 cm^{-3}. At Vestskoven, simulated water contents tended to be lower than measured values in winter 2000-2001, whereas water contents were slightly overestimated in winter 2002. At Tönnersjöheden the simulated seasonal variation in water contents was much lower than at the Dutch and Danish sites. In winter, measured and simulated water contents fluctuated between 0.25-0.30. This pattern was simulated by the model although sometimes a slight shift between measured and simulated peaks occurred, which may be due to uncertainties in the prediction of snow melt events or effects of freezing and thawing of the soil, which is not included in the model. In summer 2002, two dry episodes occurred at Tönnersjöheden that led to a drop in the simulated water contents. During the first dry episode, TDR measurements were absent, but during the second dry episode simulated water contents were close to measured values.

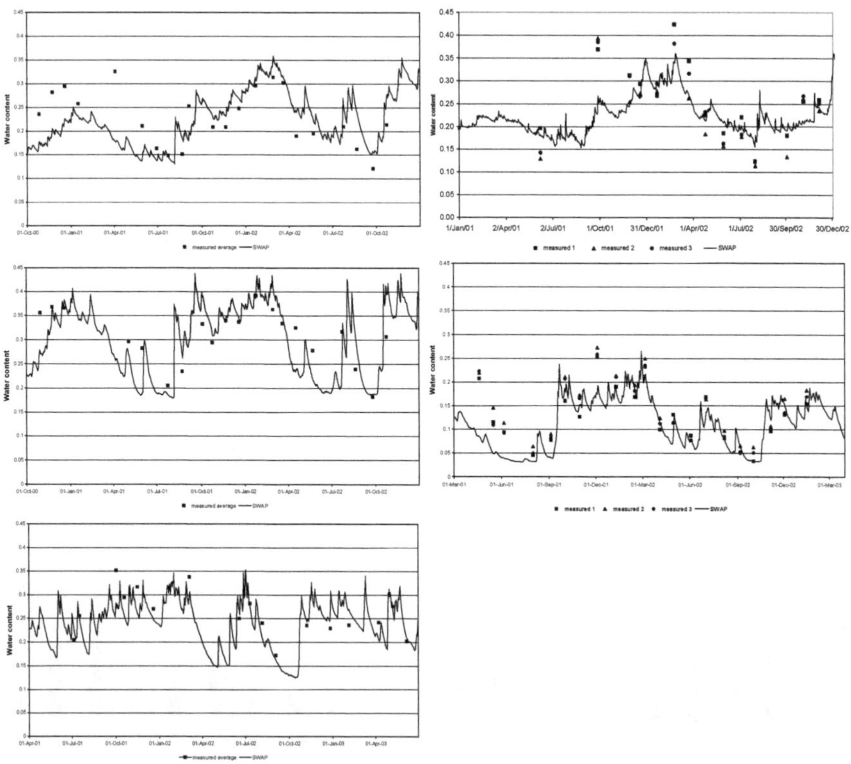

Figure 3.2. *Measured and simulated water content at 20 cm below surface for (top left) the 33-yr old spruce stand at Vestskoven, (top right) the 18-yr old oak stand at Sellingen, (middle left) the 9-yr old oak stand at Vestskoven, (middle right) the 42-yr old spruce stand at Gejlvang and (bottom left) the 73-yr old spruce stand at Tönnersjöheden.*

3.3. Chloride balances

Chloride input ranged from approximately 17 kg ha^{-1} yr^{-1}at the youngest oak stand at Vestskoven to more than 120 kg ha^{-1} yr^{-1}at the oldest spruce stand at Gejlvang. Budgets ranged from -30 to 23 kg ha^{-1} yr^{-1} (Figure 3.3). One third of the stands had a negative budget (losing more Cl than what comes in). At the remaining stands, the Cl leaching flux was lower than the throughfall flux, indicating that the hydrological model tended to underestimate the leaching fluxes. This is confirmed by the mean ratio of Cl_{out}/Cl_{in} which was 0.94.

At the oak chronosequences the budgets at 90 cm depth ranged between -11.1 to 9.1 kg ha^{-1} yr^{-1}. At most of the stands, the Cl output was within 25% of the Cl input, except for the two youngest oak stands at Vestskoven where chloride output was considerably higher than the input. The median ratio of Cl_{out}/Cl_{in} was 1.16 for the

Vestskoven oak chronosequences. The output fluxes tended to be higher than input fluxes at the young stands and the opposite at the older stands. At the Dutch oak stand at Sellingen, the median ratio of Cl_{out}/Cl_{in} was 0.99. At the 4 and the 11 years old stand the Cl leaching was within 3% of the Cl throughfall fluxes. At the other two stands, a deviation of approximately 22% was found.

Deviations in the Cl budgets were generally somewhat larger for the spruce stands and ranged from -30.3 to 23.3 kg ha^{-1} yr^{-1}. In the Netherlands and Sweden the median Cl ratio was approximately 1.1. Considerable deviations were found at the older spruce stands at Drenthe due to uncertainties in throughfall and leaching fluxes due to drought (van der Salm et al. 2005). In Sweden, Cl output fluxes were substantial higher than input fluxes for the youngest and the oldest stands. This is caused by rather high Cl concentrations in January 2002. When the period April 2001 to April 2003 was considered the Cl output was about 20 kg ha^{-1} yr^{-1} lower at both sites. At the Danish spruce stands, Cl output fluxes were consistently higher than input fluxes in the young sites (up to 14 years old), leading to a median Cl ratio of 0.8.

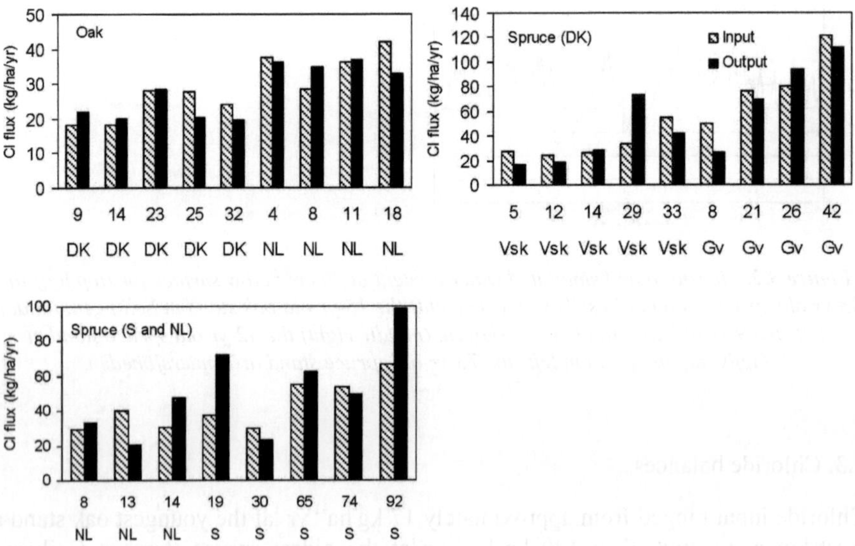

Figure 3.3. *Chloride input and output fluxes for 2002 for the six chronosequences. Vsk = Vestskoven, DK and Gv = Gejlvang, DK.*

3.4. Water balances

3.4.1. Oak stands

Water balances for the Dutch and Danish oak stands were quite comparable (Figure 3.4), despite differences in soil type, drainage and climate. The Danish oak stands received almost 900 mm of precipitation in 2002, whereas the Dutch site received a somewhat higher precipitation (1033 mm). About 10-20% (100-200 mm) of the precipitation is lost by interception (Figure 3.4). The largest amount of water is lost by transpiration (45-50%). Calculated transpiration fluxes amounted to approximately 400 mm in Denmark and 450 mm in the Netherlands. Soil evaporation fluxes ranged between 100 and 150 mm. Water recharge by drainage and leaching ranged from 200 mm in Denmark to 300 mm in the Netherlands.

a)

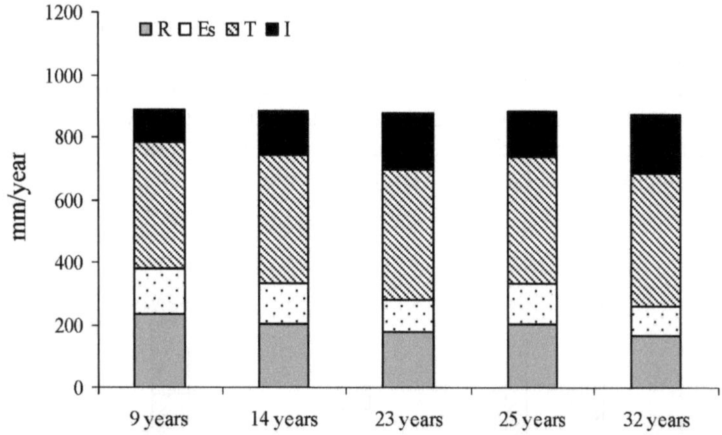

b)

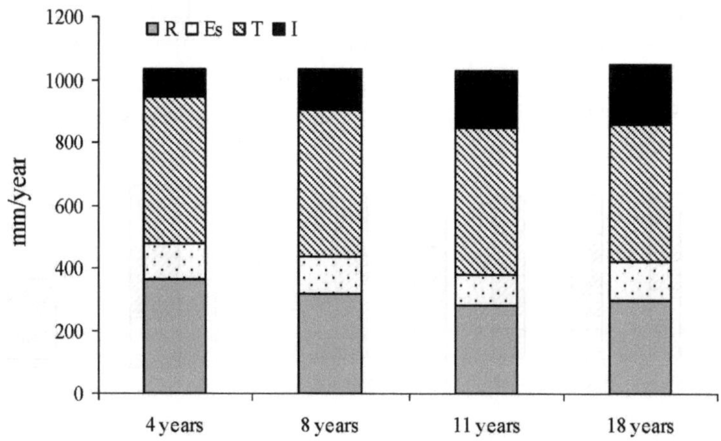

Figure 3.4. *Water balances for the oak chronosequences at a) Vestskoven and b) Sellingen (R= water recharge, E_s= soil evaporation, T= transpiration, I= interception).*

3.4.2. Spruce stands

Water balances for the four spruce stands differed considerably, mostly due to differences in climate and density of the forests (Figure 3.5). The Swedish spruce stands and the Danish spruce stands at Gejlvang received more precipitation (approximately 1100-1200 mm) than the Dutch site and the Danish site at Vestskoven (approximately 900 mm). Interception losses ranged from less than 150 mm in the youngest stand at Vestskoven to more than 600 mm in the 21-yr old spruce stand at Gejlvang. Interception accounted for 15-50% of the total water loss in the spruce stands. Losses by soil evaporation were generally low and ranged between 20 and 70 mm. Soil evaporation losses are considerably lower than in the oak stands due to the higher soil cover resulting in a lower amount of radiation reaching the forest floor. Transpiration fluxes ranged between 300 mm in the relative dry Dutch spruce stands to 450 mm in the wetter Swedish and Danish sites. Water recharge ranged from less than 50 mm in the oldest Dutch spruce stand to more than 300 mm at Gejlvang (DK) and Tönnersjöheden (S).

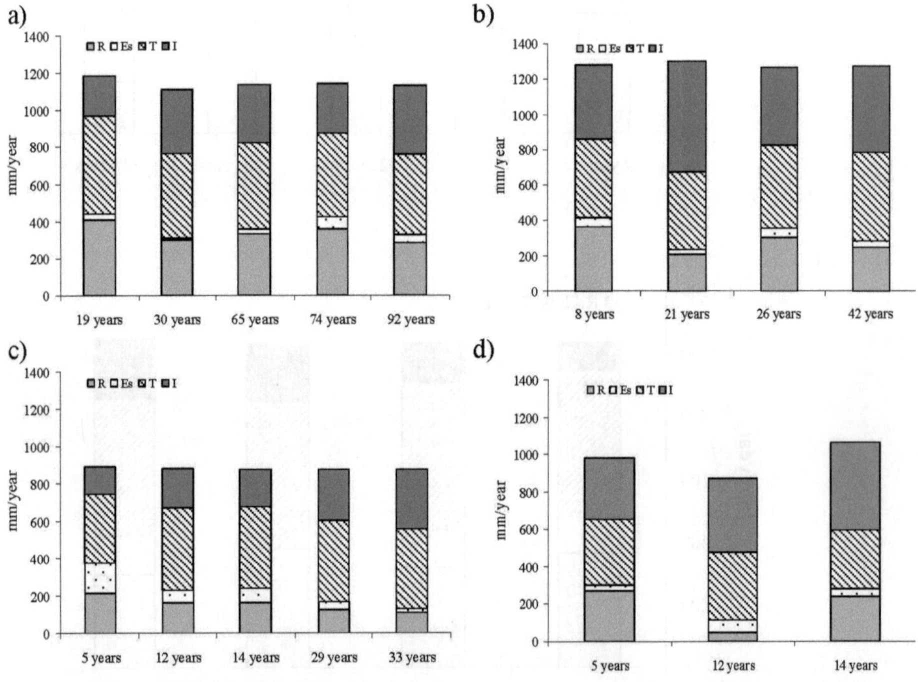

Figure 3.5. *Water balances for the spruce chronosequences at a) Tönnersjöheden, b) Gejlvang, c) Vestskoven and d) Drenthe (R= water recharge, Es= soil evaporation, T= transpiration, I= interception).*

3.5. Changes in water balances with age

3.5.1. Oak stands

The hydrological fluxes were in the same order of magnitude at the oak chronosequences in Denmark and the Netherlands,. The change in fluxes with age of the trees showed a rather coherent pattern (Figure 3.6). Interception losses increased from 90 mm in the 4-yr old stand in Sellingen to approximately 200 mm at an age of 15-20 years. The strongest increase was found in the first 15 years and leveled off afterwards. Losses by interception were less than 10% at the youngest stand and increased to approximately 20% in the oldest stands.

Soil evaporation was more or less constant in the Dutch stands (120 mm, 10% of the rainfall). In the Danish sites a slight decline was found from 145 to 95 mm (15-10%). These differences are caused by differences in management. At the Dutch sites management is absent and a dense herb layer is present, limiting soil evaporation. At the Danish sites weed control is more intensive during the first years after afforestation leading to a lower soil cover.

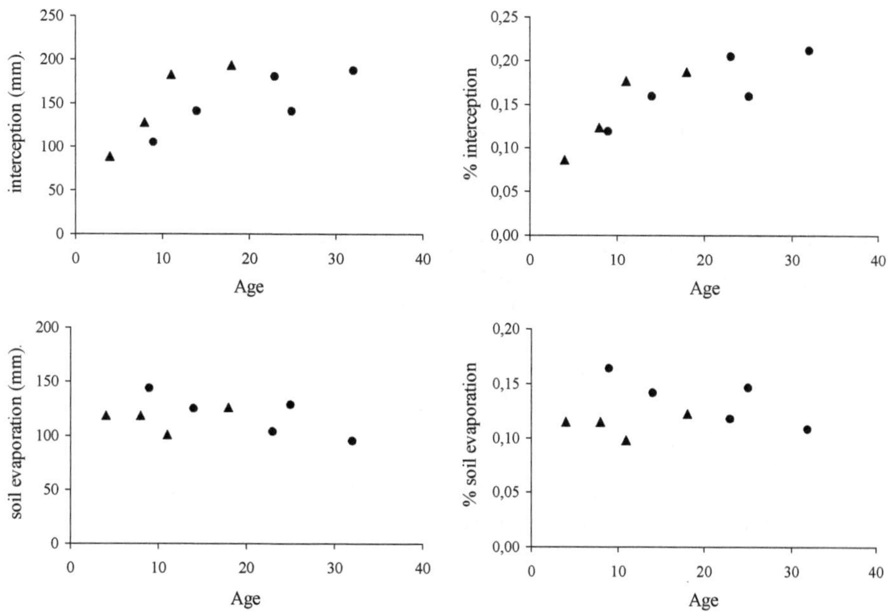

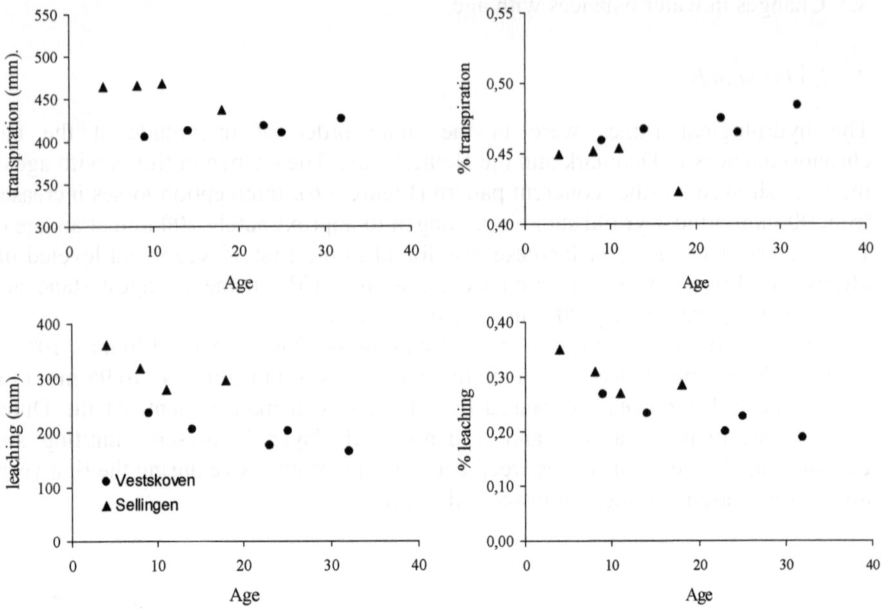

Figure 3.6. *Absolute hydrological fluxes (interception, soil evaporation, transpiration and leaching) (mm) and relative hydrological fluxes (% of precipitation) in the oak stands in 2002 as a function of age.*

Transpiration was almost constant at the Dutch sites and showed a slight increase with age at the Danish sites. Leaching clearly decreased with age from almost 400 mm in the youngest Dutch site to less than 200 mm in the older Danish sites. At the Danish sites, leaching fluxes were lower due to the lower precipitation and the decline was slightly less (from 240 mm to 170 mm). A more coherent picture of the changes in leaching with age can be obtained by expressing the leaching flux as a percentage of the precipitation. Leaching decreased from 35% in the 4-yr old oak stand to 19% in the 30-yr old stand. The fastest decline is found in the first 15-20 year and it levels off after canopy closure. Linear regression showed that the average decline is 6 mm yr^{-1} or 0.5% yr^{-1} (Table 3.7).

3.5.2. Spruce stands

The variation in hydrological fluxes between the various spruce stands was much larger than for the oak stands, due to more pronounced differences in climate, age and planting density. Interception increased with age in the first 40 years. The Swedish sites showed that interception losses remain at a fairly constant level with a further increase in age (Figure 3.7). The fraction of precipitation lost by interception at the Vestskoven chronosequence, the two oldest plots of the Gejlvang chronosequence and the youngest plots of the Tönnersjöheden chronosequences, were quite similar and increased from less than 20% at the 8-yr old stand to approximately 40% 40 years after afforestation. The interception losses at the Dutch chronosequence and the two youngest plots of the Gejlvang chronosequence also increased with age but losses were much higher and increased from 34% at the 8-yr old stand to 53% at the 20-yr old stand in Gejlvang. The high losses in the two youngest stands at Gejlvang can be explained by the high stem number (5800 stems/ha) compared to the other sites. The stem number and also the height of the Dutch chronosequence are in the same order of magnitude as most of the Vestskoven stands.

Soil evaporation generally decreased in the first 40 years after afforestation from more than 75 mm to less than 20 mm. However, differences between the individual stands were substantial. For example, simulated soil evaporation varied from 160 mm at the 8-yr old stand in Vestskoven to 28 mm at the 8-yr old stand in the Netherlands.

Table 3.7. *Change in water recharge as a function of age.*

Site	Species	Precipitation (mm)	Leaching unit	Equation	$R^2_{adj.}$
NL+DK	oak	900-1000	(mm)	345 – 6 Age	0.66
			%	$0.34 - 4.9\ 10^{-3}$ Age	0.75
S	spruce	1100	(mm)	384 – 0.9 Age	0.29
			(%)	$0.34 - 0.8\ 10^{-3}$ Age	0.29
Vestskoven, DK	spruce	900	(mm)	213 – 3.2 Age	0.91
			(%)	$0.24 - 3.2\ 10^{-3}$ Age	0.91
Gejlvang, DK	spruce	1200	(mm)	350 – 2.9 Age	0.35
			(%)	$0.29 - 2.4\ 10^{-3}$ Age	0.35
NL	spruce	1000	(mm)	324 -13.0 Age	0.27
			(%)	$0.34- 1.3\ 10^{-2}$ Age	0.27

Changes in transpiration with age differed somewhat for the four chrono-sequences. At the Dutch chronosequence, a decline in transpiration was found, probably due to drought stress caused by the high interception losses. At the Danish chronosequence at Vestskoven, transpiration was almost constant (440 mm), whereas transpiration increased from 440 to 500 mm at the Danish chronosequence at Gejlvang. At the Swedish chronosequence, a decline was found from 520 mm at the 19-yr old stand to 430 mm at the 92-yr old stand.

Leaching decreased with age at all the chronosequences (Figure 3.7, Table 3.7). The strongest decline was found at the Dutch chronosequence and at the Danish chronosequence at Vestskoven where leaching decreased with respectively 13.0 and 3.2 mm yr^{-1}. The decline in leaching flux was slightly less (2.9 mm yr^{-1}) at the other Danish chronosequence (Gejlvang) as the site is wetter and less fertile. The lowest decline in leaching rate was found at the mature forest stands in Sweden, where leaching declined with 0.9 mm yr^{-1}.

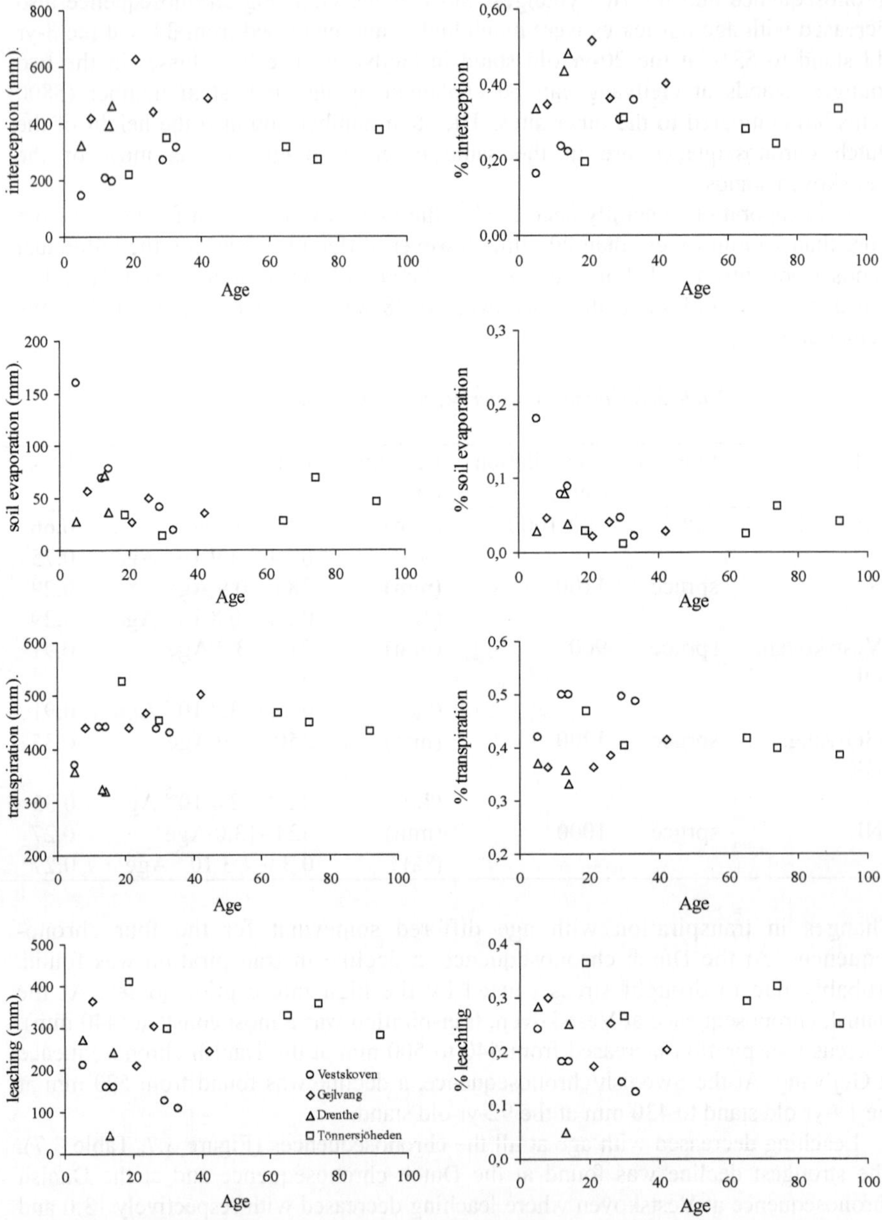

Figure 3.7. *Absolute hydrological fluxes (mm) and relative hydrological fluxes (% of precipitation) in the spruce stands in 2002 as a function of age.*

4. DISCUSSION

4.1. Validity of the simulated fluxes

The model was able to simulate measured throughfall fluxes and soil water contents quite satisfactorily in the different forest stands. However, the model was only validated for a period of one to two years. Chloride budgets were constructed to have an additional possibility for validation. Median Cl leaching fluxes were within 6% of the Cl input by throughfall. However, at two thirds of the stands the Cl output flux was lower than Cl input flux. At most stands, the deviations in the Cl budgets were due to uncertainties in the measured concentrations in throughfall and soil solution or caused by changes in the amount of water storage in the soil profile during the considered period. At the Danish spruce stands, however, Cl output fluxes were consistently underestimated in the young stands (up to 14 years old), suggesting an underestimation of the water recharge at the considered stands.

Another possibility to test the validity of the obtained results is to compare them with literature data on water recharge from various European forest stands published between 1987 and 2000 (van der Salm et al. 2006). These data originated from 34 different sites in north-western Europe, including 6 oak and 7 spruce sites. In general, leaching fluxes obtained in our study are somewhat lower compared to the literature data (Figure 3.8).

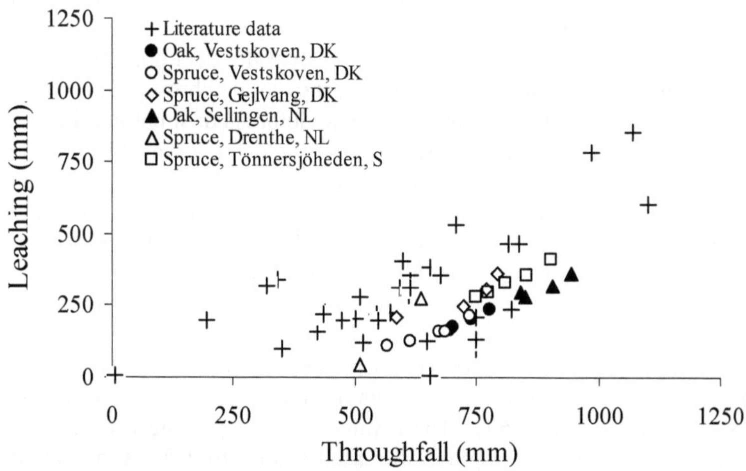

Figure 3.8. *Yearly leaching fluxes (mm) for 34 European stands (literature data) compared to the data simulated for the AFFOREST chronosequences.*

These lower fluxes are mainly caused by higher simulated transpiration fluxes. For example, simulated transpiration in the oak stands in Denmark and the Netherlands ranged from 400 to 460 mm. Literature data for oak in the Netherlands and Denmark indicated values between 260 and 300 mm (Bouten et al. 1992; Dolman 1988; Hendriks et al. 1990; Ladekarl 1998; van Grinsven et al. 1987). Transpiration fluxes for spruce in the studied area are relatively scarce. Bouten & Jansson (1995) presented values for Sollingen in Germany (323 mm), Boyle et al. (2000) found a value of 167 mm in Ireland. For Klosterhede in Denmark, transpiration was around 300 mm (Beier 1998), whereas intensive measurements in Norunda (Sweden) led to values between 181 and 243 mm during the growing season (Grelle et al. 1997; Jansson et al. 1999). All these values are considerably lower than values obtained for the chronosequences in Denmark, Sweden and the Netherlands, which ranged between 300 and 500 mm. These higher transpiration fluxes may partly be explained by the fact that the studied chronosequences are often located on fine textured soils with a good water supply (Sellingen, Vestskoven) or in high precipitation areas (Tönnersjöheden, Gejlvang).

4.2. Validity of the simulated changes in fluxes with age

Literature data indicated a clear increase in water recharge with an increase in forest area in catchments (Bosch & Hewlett 1982; Blackie 1993; Calder 1990). A statistical analysis of an international database on afforestation/deforestation experiments showed that 100% deforestation resulted in a change in water recharge of 330 mm in coniferous forests and 200 mm in deciduous forests (Sahin & Hall 1996). The data found in the chronosequences correspond quite well with these general figures. In the oak chronosequences, leaching decreased from almost 400 mm in the youngest Dutch site to less than 200 mm in the older Danish sites. In the spruce stands, the observed decline in leaching fluxes ranged between 100-230 mm. This reduction seems to be considerably lower than the data presented by Sahin & Hall (1996). However, even in the youngest stands in the chronosequence, the water recharge is strongly reduced compared to a deforested situation. For example, interception losses in the youngest stands accounted for a water loss of 150 mm. Data for the Netherlands also indicated that leaching from the 8-yr old spruce stand was 100-200 mm lower when compared to agricultural use (van der Salm et al. 2005b). When these initial losses are accounted for, the total reduction in water recharge is 200-400 mm, which is in the same order as the literature data.

The reduction in water recharge upon afforestation is mainly due to an increase in interception evaporation and transpiration. Both interception evaporation and transpiration increase with LAI (Hibbert 1967; Rudakov 1973). Reported interception losses range from 15-25% for deciduous forests and from 20-40% in coniferous forest (Hibbert 1967). These values are comparable with the data in the chronosequences, where interception decreased from 9 to 20% at the oak sites and from 16 to 52% at the spruce sites. Under comparable circumstances, the increase in interception losses with age should be faster in coniferous forest compared to oak (Aussenac & Boulangeat 1980). Similar results are found at the studied chronosequences. In the oak stands the interception increases with 4-7% in 10 years, whereas in the spruce stands the increase ranged from 7-13% in 10 years.

Transpiration is reported to increase with age by approximately 100 mm in 10 years in spruce and by approximately 50 mm in oak stands (Molchanov 1960; Aussenac 1970; Murakami et al. 2000) due to an increase in LAI. In the examined chronosequences the increase in transpiration is generally less than reported in literature. At the oak chronosequence at Sellingen, the transpiration rates remain stable and at the Vestskoven stand the increase is only 10 mm/10 year. At the spruce chronosequences in Denmark, an increase of 20 mm was found in 10 years. In the Dutch and Swedish spruce chronosequences, no increase in transpiration rate was found. The low or negligible increases in the transpiration at the chronosequences may be (partly) explained by the limited changes in LAI compared to the forest stands mentioned in literature. However, Cl budgets indicated that the transpiration fluxes tended to be somewhat overestimated in the young (spruce) stands. If the Cl budgets are correct this may also explain the limited changes in transpiration fluxes between the young and the old stands. Molchanov (1960) further reported a decline in transpiration in coniferous forest at an age of 40 years or more. These data are affirmed by the Tönnersjöheden chronosequence where transpiration declined by 35 mm between an age of 65 and 92 year.

5. CONCLUSIONS

Changes in hydrological fluxes upon afforestation were studied in six chronosequences in Sweden, Denmark and the Netherlands. Changes in interception evaporation were investigated by measuring rainfall and throughfall fluxes in the various forest stands. Changes in water recharge (leaching) were modelled using the soil hydrological model SWAP. The model was validated using measured soil water contents. The calculated fluxes were compared to Cl budgets for the various stands.

Results showed that the model was able to simulate throughfall fluxes, soil water contents and ground water levels quite well in the different forest stands. Median Cl leaching fluxes were within 6% of the chloride input by throughfall. At some sites the Cl budgets deviated substantially from zero (-30 to 23 kg ha^{-1} yr^{-1}). This is not surprising as the monitoring period is rather short and changes in soil water content and errors in individual measurements of Cl concentrations may substantially influence the budgets. Taking these uncertainties into account the results are acceptable for most of the sites. Exceptions are for the youngest spruce stands in Denmark (< 14 year old), where Cl output fluxes and thus water recharge are systematically underestimated.

Afforestation has a clear influence on the water recharge of the considered sites. Water recharge is generally lower under spruce compared to oak. In the spruce stands, 5–30% of the incoming precipitation leads to water recharge to groundwater and surface water. In the oak stands, 20–35% of the precipitation is lost as water recharge. In general, water recharge declined with increasing stand age. In the oak stands, leaching decreased from 35 to 20% of the precipitation in the first 30 years. In the spruce stands, the water recharge differed considerably for the four chronosequences that were investigated. In general, the decline in water recharge in the spruce stands was approximately 100-150 mm (10-20%). The decline in water recharge is mainly caused by an increase in interception evaporation. In the oak stands, interception losses increased by more than 100 mm (from 10% of the precipitation in the youngest stands to 20% at an age of 30 years). In the spruce stands, interception evaporation increased with 100 to 200 mm and interception was responsible for 20 to 40% of the water losses. In both the oak and spruce stands, losses by soil evaporation slightly declined. Transpiration slightly increased in the oak stands and transpiration remained fairly stable in the spruce stands.

The results found in this study support the general picture of effects of afforestation as based on literature from deforestation studies. Upon afforestation a decline in water recharge of 200-400 mm can be expected. This decline is strongly determined by the increase in interception evaporation, which is larger under spruce compared to oak. Literature data also stress the impact of increasing transpiration in the period form afforestation to canopy closure. In our study, the increase in LAI and the associated increase in transpiration was very limited (10-20 mm/10 years) compared to literature data (50-100 mm/10 years).

The above mentioned results show that afforestation leads to a reduction in water recharge compared to agricultural use. This reduction is mainly due to an increase in interception evaporation. The strongest reduction is found when sites are afforested with dense spruce forests. The smallest impact is found in open deciduous forest, which has lower interception evaporation.

6. REFERENCES

Aussenac, G. (1970). Action du couvert sur la distribution au sol des precipitations. *Annales des Sciences Forestières, 27,* 383-399.

Aussenac, G., & Boulangeat, C. (1980). Interception des précipitations et évapotranspiration réelle dans des peuplements de feuillus (Fagus Sylvatica L.) et de résineux (Pseudotsuga menziesii (Mirb) Franco.). *Annales des Sciences Forestières, 37,* 91-107.

Beier, C. (1998). Water and element fluxes calculated in a sandy forest soil taking spatial variability into account. *Forest Ecology and Management, 101,* 269-280.

Blackie, J. R. (1993). The water balance of the Balquhidder catchments. *Journal of Hydrology, 145,* 239-257.

Bosch, J. M., & Hewlett, J. D. (1982). A review of catchment experiments to determine the effect of vegetation changes on water yield and evapotranspiration. *Journal of Hydrology, 55,* 3-23.

Bouten, W., & Jansson, P. E. (1995). Water balance of the Solling spruce stand as simulated with various forest-soil-atmosphere models. *Ecological Modelling, 83,* 245-253.

Bouten, W., Schaap, M. G., Bakker, D. J., & Verstraten, J. M. (1992). Modelling soil water dyanmics in a forested ecosystem. I: a site specific evaluation. *Hydrology Processes, 6,* 435-444.

Boyle, G. M., Farrell, E. P., Cummins, T., & Nunan, N. (2000). Monitoring of forest ecosystems in Ireland. Forest Ecosystem research group Report 48, University College Dublin, Ireland.

Calder, I. R. (1990). Evaporation in the uplands. 148 pp. Chichester: Wiley.

De Visser, P., & de Vries, W. (1989). De gemidelde jaarlijkse waterbalans van bos-, heide- en graslandvegetaties. Rapport 2085, Stichting voor Bodemkartering, Wageningen, the Netherlands.

Dolman, A. J. (1988). Evaporation from an oak forest. Ph.D thesis University of Groningen.

Ernst, L. F., & Feddes, R. A. (1979). Invloed van grondwaterontrekking voor beregening en drinkwater op de grondwaterstand. Report 116, ICW, Wageningen, The Netherlands.

Gash, J. H. C., Lloyd, C. R., & Lachaud, G. (1995). Estimating sparse forest rainfall interception with an analytical model. *Journal of Hydrology, 170,* 79-86.

Grelle, A., Lundberg, A., Lindroth, A., Moren, A. S., & Cienciala, E. (1997). Evaporation components of a boreal forest: variations during the growing season. *Journal of Hydrology, 197,* 70-87.

Hendriks, M. J., Kabat, P., Homma, F., & Postma, J. (1990). Onderzoek naar de verdamping van een loofbos: meetresultaten en enkele modelberekeningen. Rapport 90. DLO-Staring Centrum, Wageningen.

Hibbert, A. R. (1967). Forest treatment effects on water yield. In W. E. Sopper & H. W. Lull (Eds.), Proceedings International Symposium on Forest Hydrology, pp. 527-544. Oxford: Pergamon Press.

Jansson, P. E., Cienciala, E., Grelle, A., Kellner, E., Lindahl, A., & Lundblad, M. (1999). Simulated evapotranspiration from the Norunda forest stand during the growing season of a dry year. *Agricultural and Forest Meteorology, 98-99,* 621-628.

Ladekarl, U. L. (1998). Estimation of the components of soil water balance in a Danish oak stand from measurements of soil moisture using TDR. *Forest Ecology and Management, 104,* 227-238.

Leyton, L., Reijnolds, E. R. C., & Thompson, F. B. (1967). Rainfall interception in forest and moorland. In W. E. Sopper & H. W. Lull (Eds.), International Symposium on forest hydrology, pp. 163-178. Oxford, U. K.: Pergamon.

Molchanov, A. A. (1960). The hydrological role of forests. Academy of Science, USSR. Institute of Forestry (translation from Russian by Israel program for scientific translations).

Monteith, J. L. (1981). Evaporation and surface temperature. *Quarterly Journal of the Royal Meteorological Society, 107,* 1-27.

Murakami, S., Tsuboyama, Y., Shimizu, T., Fujieda, M., & Noguchi, S. (2000). Variation of evapotranspiration with stand age and climate in a small forested catchment. *Journal of Hydrology, 227,* 114-127.

Rudakov, V. E. (1973). Transpiration rate and leaf mass in woody plants. *Soviet Journal of Ecology, 4,* 534-535.

Sahin, V., & Hall, M. J. (1996). The effects of afforestation and deforestation on water yields. *Journal of Hydrology, 178,* 293-309.

Tiktak, A., & Bouten, W. (1994). Soil water dynamics and long-term water balances of a Douglas fir stand in the Netherlands. *Journal of Hydrology, 156,* 265-283.

Van Dam J. C., Huygen, J., Wesseling, J. G., Feddes, R. A., Kabat, P., van Walsum, P. E. V., Groenendijk, P., & van Diepen, C. A. (1997). Simulation of water flow, solute transport and plant growth in the soil-water-atmosphere-plant environment. Theory of SWAP version 2.0. SC-DLO, Tchnical Document 45, Wageningen, The Netherlands.

Van der Salm, C., Reinds, G. J., & de Vries, W. (2004). Assesment of the water balance of european forests: a model study. *Water, Air, and Soil Pollution: Focus, 4,* 175-190.

Van der Salm, C., Denier van der Gon, H., Wieggers, R., Bleeker, A., & Van den Toorn, A. (2005). The effect of afforestation on water recharge and nitrogen leaching in an oak and spruce chronosequence in the Netherlands. *Forest Ecology and Management, 221,* 170-182.

Van der Salm, C., Reinds, G. J., & de Vries, W. (2006). Water balances in intensively monitored forest ecosystems in Europe. Submitted to Environmental Pollution.

Van Grinsven, J. J. M., van Breemen, N., & Mulder, J. (1987). Impacts of acid atmospheric deposition on woodland soils in the Netherlands. 1. Calculation of hydrologic and chemical budgets. *Soil Science Society of America Jorunal, 51,* 1629-1634.

Wösten, J. H. M., Veerman, G. J., & Stolte, J. (1994). Wateretentie- en doorlatendheidskarakteristieken van boven- en ondergronden in Nederland: de Staringreeks. Vernieuwde uitgave 1994. DLO-Staring Centrum, Wageningen, the Netherlands, Technical document 18. 66 pp.

Wösten, J. H. M., Lilly, A., Nemes, A., & Le Bas, C. (1999). Development and use of a database of hydraulic properties of European soils. *Geoderma, 90,* 169-185.

CHAPTER 4

NITROGEN DEPOSITION AND NITRATE LEACHING FOLLOWING AFFORESTATION: EXPERIENCES FROM OAK AND NORWAY SPRUCE CHRONOSEQUENCES IN DENMARK, SWEDEN AND THE NETHERLANDS

L. ROSENQVIST[1], K. HANSEN[2], L. VESTERDAL[2], H. DENIER VAN DER GON[3], C. VAN DER SALM[4], A. BLEEKER[5] and M.-B. JOHANSSON[1]

[1]Swedish University of Agricultural Sciences, Box 7001, 750 07 Uppsala, Sweden, E-mail: lars.rosenqvist@sml.slu.se, [2]Forest and Landscape Denmark, KVL, Hørsholm Kongevej 11, 2970 Hørsholm, Denmark, [3]TNO, Environment and Geosciences, P.O. Box 342, Laan van Westenenk 501, 7300 AH Apeldoorn, the Netherlands, [4]Alterra Green World Research, P.O. Box 47, 6700 AA Wageningen, the Netherlands, [5]Energy research Centre of the Netherlands (ECN), P.O. Box 1, 1755 ZG Petten, the Netherlands

Abstract. Knowledge on the impact of afforestation of arable land on N deposition and leaching of nitrate to groundwater and surface waters is limited. In the AFFOREST project we evaluated nitrogen (N) deposition and nitrate leaching following afforestation of cropland. Two oak (*Quercus robur*) and four Norway spruce (*Picea abies*) afforestation chronosequences (age range 1 to 90 years) were studied with respect to deposition and nitrate leaching in Denmark, Sweden and the Netherlands. This paper presents a synthesis of these six chronosequence experiments. Three to six forest stands of Norway spruce and/or common oak were monitored in each chronosequence for a period of two years. In each stand, throughfall and soil solutions beneath the root zone were sampled and nitrate leaching was calculated. For all sites and tree species, the throughfall deposition of N increased with stand height (and age). The forests varied substantially in their ability to retain N in the ecosystem. No consistent pattern was apparent in the three countries. However, in some chronosequences nitrate leaching was low or negligible in the early phase of afforestation and increased after canopy closure (> 15-20 years). In general, nutrient-rich clayey soils leached more nitrate than nutrient-poor sandy soils. In the first approximate 35 years after afforestation, nitrate leaching below the root zone was generally higher

G.W. Heil et al. (eds.), Environmental Effects of Afforestation in North-Western Europe, 79–108.

below oak than below Norway spruce. The presented results are compared to available studies and discussed.

1. INTRODUCTION

During the latest years incentives for afforestation include environmental concerns, e.g. protection of groundwater reserves. Compared to the former agricultural land use, planted trees influence the amount and quality of water reaching the soil by affecting the evaporation (Chapter 3) and nutrient cycling. Afforestation on former arable land also strongly influences the amount of atmospheric deposition since the surface roughness and collecting surface area is increased. Therefore, the amount of dry deposition of air pollutants can be expected to increase (Bleeker & Draaijers 2002).

Decreased nitrate leaching to water bodies is a specific expected environmental effect of afforestating arable land (Rijtema & De Vries 1994; Rowe & Pearce 1994; Jussy et al. 2000). Leaching losses of nitrogen (N) from ecosystems mainly occur as dissolved N in seepage water. Dissolved N may be in the form of nitrate (NO^{3-}), ammonium (NH^{4+}), or dissolved organic nitrogen (DON). Nitrate is highly mobile in the soil profile and easily lost from the system by leaching. Nitrate is therefore the most relevant N compound for water quality and addressed as such in specific EC directives (91/676/EEC and 98/83/EEC) concerned with the protection of waters against pollution from agricultural sources. In this line, the World Health Organisation has established a requirement of 11.3 mg N dm-3 (50 mg dm-3 NO^{3-}) for the quality of drinking water (WHO 1998).

Nitrogen leaching is at risk when ecosystems become saturated with N, i.e. when the availability of inorganic N exceeds the demand from plants and microorganisms (Aber et al. 1989; Gundersen 1991). However, even in N-limited systems substantial leaching of nitrate may occur after events of high precipitation in the winter period when evapotranspiration rates and biological uptake of N are low (Gundersen & Rasmussen 1995). Once a forest becomes N saturated, the excess nitrate is leached by percolating soil water down to the saturated zone. Leaching of nitrate may be induced by (i) increased input of N (e.g. atmospheric N deposition, fertilization, planting of N_2 fixing species), (ii) decreased biological uptake of N (e.g. following clear-cutting, thinning, weed control, site preparation, degradation of vegetation), (iii) increased net mineralization and nitrification rates (e.g. resulting from liming, site preparation, change in litter substrate quality, lowering the groundwater table), and (iv) increased water recharge (e.g. decreased canopy cover after thinning, large precipitation events).

In regions dominated by agricultural activities, N is recognised as a major pollutant of water environments. During the last 40-50 years management of farmland has intensified and caused arable soils to carry large pools of N bound in organic matter, and soils have a high nitrifying capacity (Jussy et al. 2000). Fertilizer application is the dominant source of groundwater nitrate contamination (van der Voet et al. 1996). In contrast to agricultural soils, old existing forests are characterised by a more tight N cycle. Water from old forest land is, therefore, generally of good quality with a relatively low concentration of dissolved N compared to other land uses (Thornton et al. 2000). A change in land use from agriculture to forestry may induce major changes of the N cycle, including inputs, internal cycling and losses. The external input of N to an ecosystem is dependent on the N capture efficiency of the vegetation, which again is depending on the effective surface roughness. Trees are highly efficient scavengers of pollutants (Allen & Chapman 2001) much better than the lower agricultural crops. During the years after afforestation the forest canopy structure changes continuously with increased surface roughness, which results in an increased capture of dry N deposition for each year (Hansen et al. 2006a). Net throughfall fluxes of NO^{3-} and NH^{4+} have been observed to correlate well with the roughness length of the canopy (Thom 1971; Draaijers 1993). The enhanced supply of N in combination with high mineralization may result in potential N saturation and increased nitrate leaching (Allen & Chapman 2001). At the same time, uptake in biomass is high when the trees are growing vividly building up the canopy. The demand for N decreases when the canopy has closed. It is an ongoing process where the relative contribution of the mechanisms that regulate N export is likely to change over time following the forest establishment. Gradually, N stores will move towards a new steady state.

Thus, an effective measure to reduce leaching of nitrate to the groundwater could be the conversion of agricultural land to forest. However, because of the high N status of former arable soils, retention of N in afforested ecosystems may be less efficient than in old forest land, resulting in an enhanced risk of nitrate leaching. Afforestation could also constitute a potential threat for groundwater quality since scavenging of pollutants leads to enhanced supply of N, which in combination with high mineralization may potentially result in N saturation and increased nitrate leaching (Allen & Chapman 2001). However, nitrate leaching may still be less than from the former agricultural land use.

The change in deposition following afforestation has not directly been studied, however, many studies have observed increased deposition when the height and canopy roughness increased as the trees grew (e.g. Draaijers 1993). Knowledge on nitrate leaching from forests established on former arable land is scarce. Rijtema & de Vries (1994) used a simple modelling approach to evaluate the effect of a change in land use from agriculture to forestry in the Netherlands. Afforestation led to a marked decrease in nitrate leaching regardless of the chosen tree species. In Denmark, Bastrup-Birk & Gundersen (2004) simulated N leaching in the Horndrup catchment with the INCA model and found that afforestation substantially reduced N leaching. Concentrations of nitrate ten years after afforestation at three afforested sites in Denmark were much lower than concentrations in agricultural soils before afforestation (Hansen & Vesterdal 1999). However, higher concentrations of nitrate in soils below the root zone were observed in recently (< 10 years) afforested land as compared to old forest ecosystems (Callesen et al. 1999).

This chapter evaluates the effect of afforestation of former cropland on N deposition to forests of changing height/age and on nitrate leaching. The basis for this evaluation is a synthesis of chronosequence experiments in three north-west European countries. The presented results from the AFFOREST project are compared to available studies on the topic and discussed. The specific objectives are i) to estimate the change in N deposition as the forest grows in chronosequences from recently planted forest to older forest, ii) to estimate nitrate leaching in the same chronosequences, iii) to study the possible differences between afforestation with deciduous (oak) and coniferous (Norway spruce) tree species, and iv) to assess the difference in nitrate leaching after planting on contrasting soil types, i.e. sandy or clayey soils.

2. MATERIALS AND METHODS

To study the effect of afforestation with oak and spruce on deposition and nitrate leaching, chronosequences of afforestation stands were selected in Denmark, southern Sweden and the Netherlands (Chapter 1). The study included six chronosequences of which two were differently aged oak stands and four were differently aged Norway spruce stands. In Denmark, one oak chronosequence and one Norway spruce chronosequence were chosen at the same clay-rich and nutrient-rich soil at Vestskoven close to Copenhagen. Another spruce chronosequence in a contrasting environment was established on sandy, nutrient-poor soil at Gejlvang in southern Jutland. In the Netherlands, one chronosequence of oak and one chronosequence of spruce were studied on similar sandy soil close to Sellingen. In Sweden, one chronosequence of spruce was established in the province of Halland east of Halmstad. A map of the locations is found in Chapter 1 and more site information is given in Chapter 1 and on the AFFOREST website (www.sl.kvl.dk/afforest).

2.1. Precipitation and throughfall sampling

Bulk precipitation and throughfall were sampled for two years in the Danish and the Swedish chronosequences and somewhat shorter in the Dutch chronosequences (Table 4.1). Sampling of both bulk precipitation and throughfall was performed using polyethylene funnels. Detailed information on the length of the monitoring periods and the number of funnels at each site is presented on the AFFOREST web site (www.sl.kvl.dk/afforest). In the Netherlands, the funnels were placed in a cross in each stand with definite distance between them. In Sweden, the funnels were placed randomly in each forest stand. In Denmark, an experimental design with circular subplots (10 m radius) was used. In each circular subplot, five funnels were installed, four at the cardinal points and one in the centre. Three circular subplots were placed as far from stand edges as possible. The Dutch throughfall samples from the 10 individual funnels were pooled to 2 samples, each representing an axis of the sampling cross. In Denmark, the samples were pooled in 3 sub-samples, each representing a circular plot. The Dutch collecting bottles were placed in dark PVC tubes hanging above ground, to prevent the influence of direct sunlight. In Sweden and Denmark, the collection bottles were placed in soil pits to prevent them from heat and light. The bulk precipitation and throughfall samples were collected and weighed in the field at each location. All sampling was carried out in monthly intervals.

Table 4.1. *Sampling periods for bulk precipitation, throughfall, and soil water in the chronosequences in Denmark, the Netherlands and Sweden.*

Country	Location	Species	Year	Start	End
DK	Vestskoven	Oak & Norway spruce	1	Jan-01	Dec-01
			2	Jan-02	Dec-02
	Gejlvang	Norway spruce	1	Apr-01	Mar-02
			2	Apr-02	Mar-03
NL	Sellingen	Oak	1	Apr-01	Mar-02
			2	Jan-02	Dec-02
	Drenthe[1]	Norway spruce	1	Apr-02	Mar-03
SE	Tönnersjöheden	Norway spruce	1	Apr-01	Mar-02
			2	Apr-02	Mar-03

[1] *Only soil water is dealt with at Drenthe.*

2.2. Soil and litterfall sampling

The forest floor is the layer of dead organic matter, i.e. leaves, needles, twigs, branches and fruits, that blankets the mineral soil of a forest. Soil N contents and C/N ratios were assessed for the forest floor and the upper 0-25 cm of the mineral soil as well as these two soil compartments together in all chronosequences. Nitrogen was analyzed using the same soil samples collected for assessment of C, and the sampling procedure was therefore the same. The sampling design and methods are summarized in Chapter 2. More detailed descriptions of the sampling of soils in Denmark and Sweden may be found in Vesterdal et al. (2002), Ritter et al. (2003) and Chapter 2.

Total litterfall was measured for two years in Denmark and for one year only in Sweden and the Netherlands. For information on litterfall sampling see Chapter 2.

2.3. Soil solution sampling

Soil solution in mineral soil was sampled monthly in all chronosequences using suction cup lysimeters. Detailed information on the length of monitoring periods, number of lysimeters at each site, installation depths of lysimeters and number of replicates at each depth is presented on the AFFOREST website (www.sl.kvl.dk/afforest). In Denmark and the Netherlands, concentrations measured at 0.9 m depth represent nitrate concentrations in soil water leaching from the root zone. In Sweden, the corresponding depth was 0.6 m. In the Dutch and Swedish chronosequences, lysimeters were placed at randomly selected points within each stand. In Denmark, five lysimeters were installed in each circular subplot, as described above. In all three countries, the lysimeters were installed about two to six months prior to the first sampling occasion. Soil water was extracted and discarded once or on several occasions before sampling in order to allow the lysimeters to equilibrate with the soil solution. In the Danish and Swedish chronosequences, concentrations of elements represent averages over monthly sampling periods. In the Netherlands, soil water was collected over 24 hours at each sampling occasion. Lysimeter samples at each site and depth were pooled before analysis. In Denmark, composite samples from each circular subplot and depth were analyzed separately.

2.4. Chemical analysis

In the laboratory, sample preparation was performed within two days from sampling. In Sweden, samples were stored in a freezer (-18°C) until chemical analyses. In Denmark, samples were stored at +4°C and analyzed within one month from sampling. In the Netherlands, samples were likewise stored in a refrigerator but chemical analysis was performed within two days from sampling. In Sweden and the Netherlands, nitrate in precipitation, throughfall and soil solution was analyzed by flow injection analysis (FIA), whereas nitrate was analyzed by ion chromatography in Denmark. In all countries, ammonium was analyzed by FIA. In Sweden, dissolved organic carbon (DOC) was analysed by a total organic C analyser (TOC-5000, Shimadzu).

In Denmark and Sweden, N concentrations in litterfall, forest floor and soil were determined by dry combustion (Dumas method) in a Leco CSN Analyzer (Matejovic 1993). The Dutch litterfall, forest floor and soil samples were analyzed for N by wet oxidation according to the Kjeldahl method (Hesse 1971). More detailed descriptions of the analysis of soils in Denmark and Sweden may be found in Vesterdal et al. (2002), Ritter et al. (2003) and Chapter 3, respectively.

2.5. Calculations and statistics

2.5.1. Total deposition and canopy exchange modelling

Throughfall is considered an underestimate of the total deposition of N since N is taken up in the canopy. The total N deposition to the forest ecosystem has been estimated using a canopy exchange model, the extended Ulrich model (Draaijers & Erisman 1995; Draaijers et al. 1998; de Vries et al. 2001). This model allows discrimination between canopy exchange and atmospheric deposition using long-term throughfall and precipitation fluxes. Dry deposition and canopy leaching of Ca^{2+}, Mg^{2+} and K^+ is computed by means of the so-called filtering approach, assuming a fixed relationship between wet and dry deposition of particles taking Na^+ as a tracer (Ulrich 1983). The total canopy uptake of H^+ and NH_4^+ is assumed to equal the total canopy leaching of Ca^{2+}, Mg^{2+}, and K^+ minus canopy leaching of Ca^{2+}, Mg^{2+} and K^+ associated with foliar excretion of weak acids. Based on experiments in the laboratory, Van der Maas et al. (1991) assumed that H^+ has an exchange capacity six times larger than NH_4^+. Draaijers & Erisman (1995) and

Draaijers et al. (1998) assumed canopy uptake of NO_3^- to be negligible. However, this assumption may not be true, particularly in low deposition areas (de Vries et al. 2001). In the present study, canopy exchange of NO_3^- was accounted for and calculated according to the slightly adapted canopy budget model presented by de Vries et al. (2001).

The total deposition of N (N_{TD}), i.e. the sum of wet and dry N deposition, to the forest ecosystem can thus be calculated as:

$$N_{TD} = N_{TF} + N_{SF} + N_{CE} \tag{1}$$

where N_{TF} and N_{SF} is N deposition by throughfall and stemflow, respectively, and N_{CE} is the exchange of N ($NH_4^+ + NO_3^-$) by the forest canopy.

The contribution of stemflow to the total flux to the forest floor varies with tree species and is usually less than 10% of the total flux to the surface (Ivens 1990; Draaijers et al. 1996). Stemflow was not measured at the chronosequences, and the deposition estimates from throughfall measurements are therefore underestimated. At the Dutch and Danish oak sites, the contribution of stemflow to the total N deposition has been approximated using the tree species-specific relationships between throughfall and stemflow as described by Ivens (1990) and de Vries et al. (2001). In the Danish and Swedish spruce chronosequences, stemflow was considered negligible and not accounted for. Rough-barked species, like nearly all conifers, typically have low stemflow values (Augusto et al. 2002; Pypker et al. 2005). For example, stemflow amounts of various nutrients were observed to vary from between 2 to 5% of the amounts by throughfall in a 30-year-old Norway spruce stand in south-western Sweden (Bergholm et al. 2001).

The canopy budget model requires data that meet several quality criteria, including an accurate charge balance between major cations and anions (Draaijers et al. 1996; de Vries et al. 2001). The acceptable percentage difference is less than 10% for bulk deposition and less than 20% for throughfall (WMO 1992). Annual throughfall and bulk precipitation concentration data passed the applied ion balance check at most sites, however, slight discrepancies at some sites (e.g. open field and oak sites at Vestskoven and spruce planted 1972 at Tönnersjöheden) may result in uncertain estimates of total N deposition.

A major uncertainty in the calculations involves the estimation of weak acid excretion in the forest canopies. In the Danish and Dutch oak and spruce stands weak acid exchange fluxes were calculated from measured concentrations of major cations (Ca^{2+}, Mg^{2+}, K^+, Na^+, H^+ and NH_4^+) and anions (Cl^-, SO_4^{2-} and NO_3^-) in bulk deposition and throughfall. In the Swedish spruce stands, calculation of weak acid exchange was based on DOC and pH measurements for the same compartments according to a method described by de Vries et al. (2001). In the Danish bulk precipitation samples, the sum of major anions often exceeded the sum of major cations, which resulted in unreliable estimates of weak acid exchange fluxes and thereby subsequent errors in the calculation of canopy N uptake. It was, therefore, assumed that the contribution of weak acids from atmospheric deposition (assumed to equal twice the bulk deposition in the canopy budget model) was negligible (zero), and that weak acids found in throughfall were exclusively derived by leaching from the canopy. This assumption was supported by field observations at the Swedish and Dutch sites of considerable enrichment of DOC in throughfall compared to the bulk deposition (Rosenqvist et al. 2006a). For comparative reasons this assumption was made at all chronosequence sites in Denmark, Sweden and the Netherlands. Excluding the contribution from weak acids in atmospheric deposition leads to slight overestimation of the calculated canopy leaching fluxes of weak acids, and thus to a slight underestimation of the total canopy N uptake. At the Swedish sites, this resulted in about 1-1.5 kg ha^{-1} yr^{-1} less canopy N uptake compared to if atmospheric deposition of weak acids was accounted for.

Element budgets were calculated by subtracting the leaching flux from the total deposition, in which total deposition fluxes were derived by the canopy exchange model.

2.5.2. Soil content and litterfall flux

Soil N dynamics (N contents and C/N ratios) are reported as the relationships with stand age. Forest floor N contents were calculated by multiplying N concentrations with forest floor mass. For the mineral soil N contents the fraction ≥ 2mm were neglected (McNabb et al. 1986; Homann et al. 1995), and soil N contents (N_{soil}) in [Mg ha^{-1}] for the soil layers to 25 cm depth were calculated using

$$N_{soil} = \rho_i \bullet (1 - (\delta_{i,\,2mm}/100)) \bullet d_i \bullet N_i \tag{2}$$

where ρ_i is the bulk density of the < 2 mm fraction in g cm^{-3}, $\delta_{i,\,2mm}$ is the relative volume of the fraction ≥ 2mm (%), d_i denotes the thickness of layer *i* (cm), and N_i denotes the N concentration (mg g^{-1}) in layer *i*.

The litterfall N flux was quantified by multiplying N concentrations with annual litterfall mass, and C/N ratios were based on the ratio between concentrations of C and N.

2.5.3. Hydrological modelling

Soil water fluxes were simulated using an extended version of the dynamic simulation model SWAP (Van Dam et al. 1997). Annual fluxes were calculated by summation of monthly fluxes. Detailed information on the hydrological models and their parameterisation is presented in Chapter 3 as well as in van der Salm et al. (2005) and Rosenqvist et al. (2006b). In Denmark and Sweden, the leaching of nitrate from the root zone (i.e. at 0.9 m depth in Denmark and at 0.6 m depth in Sweden) was calculated from measured monthly concentrations times estimated monthly soil water fluxes at corresponding depths. In the Netherlands, soil solution was sampled discontinuously. Here, leaching of nitrate was calculated using the concentration, linearly interpolated on a daily basis between sampling occasions, multiplied by the estimated daily flow of soil water at 0.9 m depth.

2.5.4. Statistics

Relationships between stand age and soil N contents were explored by simple linear regression. No transformations were necessary to fulfil the requirements regarding normally distributed residuals and homogeneity of variances. All statistical tests were carried out using the procedure GLM in SAS (SAS Institute 1993). The 200-yr old stand in Denmark was not included in regressions, but was included in figures for comparison.

3. RESULTS

3.1. N input to the afforested ecosystem

The total volume of bulk precipitation was presented in Chapter 3 and the bulk N deposition in the open field is presented in Table 4.2 to illustrate the difference between countries and the significance of annual variation. The volume of bulk precipitation and its chemical composition vary considerably from location to location but also from year to year. Moreover, the annual variability in volume and N deposition is not related. The lowest average precipitation was observed at Vestskoven located at Zealand in the eastern part of Denmark. The highest average precipitation was registered in the western part of Denmark at the site Gejlvang and in Sweden at Tönnersjöheden. The annual bulk N deposition was markedly lower at Vestskoven, Denmark, compared to the other sites where the annual bulk N deposition was quite high and alike. The highest bulk N deposition was observed at the Danish site Gejlvang during the first measurement year.

For all sites and tree species, the throughfall deposition of N increased with stand height (and age) (Figure 4.1, Table 4.3). This was most evident when the throughfall deposition of N was expressed relative to the bulk N deposition (Figure 4.1). The taller the stand, the higher the input by throughfall except in the very young stands where throughfall deposition sometimes was lower than in bulk deposition. Throughfall deposition of N in the oak stands did not exceed the bulk deposition in the open field until ~7 m height (Figure 4.1b, Table 4.3) but the N deposition was approximately twice as high for the tallest and oldest stands (25-32 years old) than for the younger stands (3-13 years old). In general, N throughfall deposition did not reach as high levels in the oak stands (20 kg ha^{-1} yr^{-1}) at Sellingen in the Netherlands as in the Norway spruce stands (25 kg ha^{-1} yr^{-1}) at the Danish and Swedish chronosequences (Table 4.3).

When throughfall deposition of N in the spruce chronosequences was expressed relative to the bulk deposition two clear country-specific patterns emerge. The relative N deposition is considerably higher in the Danish spruce stands than in the Swedish stands, especially for the taller trees (> 10 m) (Figure 4.1a). For oak, a consistent pattern of increasing N deposition with height was observed, independent of the site location (Figure 4.1b).

Table 4.2. *Annual total bulk deposition of N (kg N ha^{-1} yr^{-1}) to the open field at the chronosequences for two years.*

Country	Location	Total bulk deposition of N[a)] (kg N ha^{-1} yr^{-1})	
		Year 1	*Year 2*
NL	Sellingen	15	17
DK	Vestskoven	10	9
	Gejlvang	19	14
SE	Tönnersjöheden	16	13

[a)] NO_3-N + NH_4-N

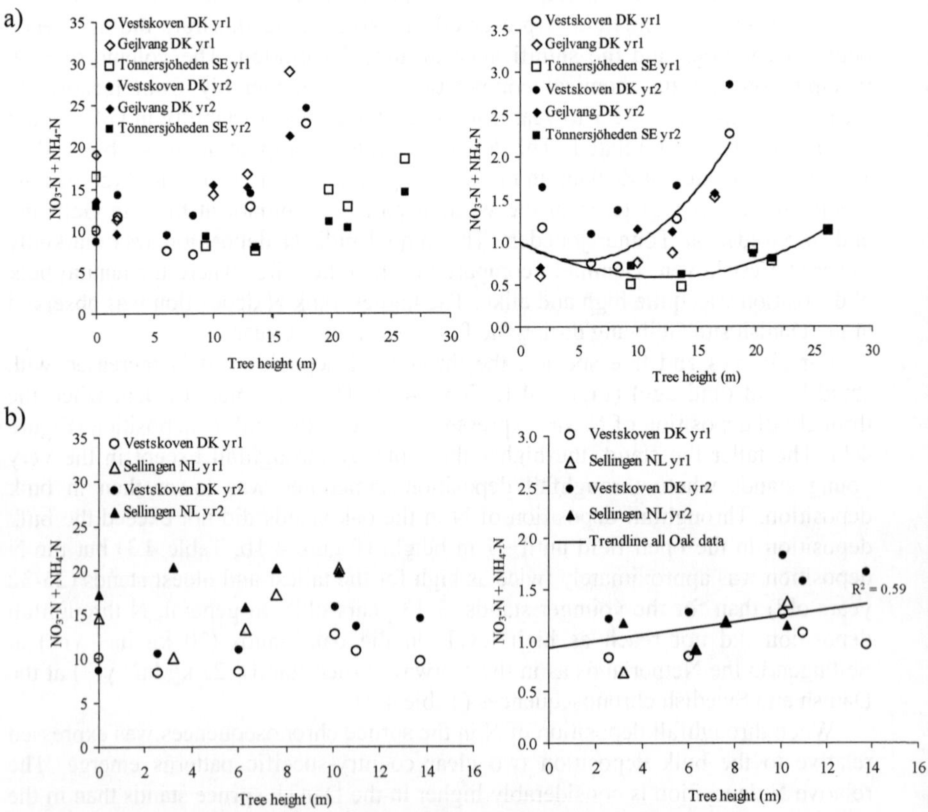

Figure 4.1. *Throughfall N (NH_4-N + NO_3-N) deposition (kg ha^{-1} yr^{-1}) (left) and throughfall N deposition relative to open field deposition of N (right) as a function of tree height (m) for the (a) Norway spruce and (b) oak chronosequences. The sampling periods (Year 1 & 2) are given in Table 4.1.*

Table 4.3. *The calculated total N deposition (N_{TD}), measured total N throughfall (N_{TF}), and calculated uptake of N (N_U) in the canopy (in kg ha^{-1} yr^{-1}) for Norway spruce and oak chronosequences in Sweden, Denmark and the Netherlands. Data are annual average values based on the following sampling periods: Tönnersjöheden (spruce), Nov 01-June 03; Vestskoven (spruce and oak), Jan 01-Dec 02; Gejlvang (spruce), Apr 01-Mar 03; Sellingen (oak), Apr 01-Dec 02. BD_N refers to the bulk N deposition at each location.*

Location	Age	Height	N_{TD}	N_{TF}	N_U	N_U in % of N_{TD}
	yr	m	kg ha^{-1} yr^{-1}	kg ha^{-1} yr^{-1}	kg ha^{-1} yr^{-1}	%
Norway Spruce						
Tönnersjöheden	BD_N		14.5			
SE	19	10.1	15.0[1]	10.1	4.3	29
	30	15.5	14.8[1]	8.4	4.3	29
	65	19.8	15.1[1]	12.6	1.2	8
	74	22.4	14.7[1]	11.6	1.6	11
	92	27.4	19.3	18.1	1.2	6
Vestskoven	BD_N		9.5			
DK	5	1.8	20.0	13.1	7.1	35
	12	6.1	11.1	8.7	2.0	18
	14	8.4	12.7	10.0	3.0	24
	29	13.9	19.8	14.1	5.9	30
	33	17.8	27.0	23.7	3.5	13
Gejlvang	BD_N		16.1			
DK	8	1.4	~16.1[1]	10.4	3.0	35
	21	10.4	24.1	14.2	9.9	41
	26	12.8	27.7	16.9	10.8	39
	42	16.1	32.8	25.2	7.7	23
Oak						
Vestskoven	BD_N		9.5			
DK	9	2.6	15.5	10.6	3.4	22
	14	6.0	14.8	10.4	3.0	20
	25	10.1	21.2	12.9	6.5	31
	23	11.0	22.7	14.7	6.0	26
	32	13.6	24.1	13.2	9.1	38
Sellingen	BD_N		15.9			
NL	4	3.2	18.2	15.1	0.9	5
	7	6.3	18.7	14.5	2.2	12
	11	7.6	27.4	18.7	6.2	23
	18	10.3	33.5	19.8	10.9	33

[1] *At these sites, N_{TD} was estimated by summation of wet and dry deposition of N (NH_4-N and NO_3-N) where dry deposition of NH_4-N was set to be zero. This approximation was made since canopy exchange modelling resulted in negative dry deposition of NH_4-N and levels of N_{TD} slightly lower than the measured bulk N deposition. As a consequence, the sum of N_{TF} and N_U (stemflow of N assumed negligible in spruce) is slightly lower than the given values of N_{TD}.*

The calculated total N deposition as well as the calculated uptake of N in the canopy is shown in Table 4.3 for both spruce and oak chronosequences in the three countries. The total N deposition in the Swedish spruce stands was only little higher than the bulk deposition whereas the total N deposition was substantially higher than the bulk N deposition at the Danish and Dutch stands. Here, also a clear increase in total N deposition with stand height was observed. The total N deposition to the spruce chronosequences was highest at Gejlvang in Denmark ranging between 16-33 kg ha^{-1} yr^{-1} while the total N deposition to the spruce stands at Tönnersjöheden only reached a maximum value of 19 kg ha^{-1} yr^{-1}. At both the Dutch and Danish oak sites, the total N deposition increased with stand height. The highest total N deposition for oak chronosequences was observed in Sellingen in the Netherlands where it reached a value of 34 kg ha^{-1} yr^{-1}. At Vestskoven in Denmark, the highest total N deposition was 24 kg ha^{-1} yr^{-1}.

The measured throughfall N fluxes differed from the total N deposition fluxes as a result of canopy exchange processes (Table 4.3). Canopy uptake of N in Norway spruce varied from 1-11 kg ha^{-1} yr^{-1} or 6-41% of the total N deposition (Table 4.3). The canopy uptake of N decreased with age for the Swedish spruce sites but no clear pattern was observed for the Danish spruce stands. The Swedish chronosequence is more complete spanning from open field to 90 years of age. It is possible that a similar pattern would be observed if older Danish spruce stands were included in the study. For Norway spruce, the largest uptake was estimated at Gejlvang in Denmark. Canopy uptake of N in oak likewise varied from 1-11 kg ha^{1} yr^{-1} or 5-38% of the total N deposition. In contrast to the spruce stands, the canopy uptake of N in the oak stands increased with age both in the Dutch and the Danish oak chronosequences.

3.2. Changes in soil N status and litterfall N flux

Immediately after afforestation there is no forest floor developed and N is only present in the mineral soil. Nitrogen accumulation in the forest floor started to develop rapidly after canopy closure (10-20 years, Figure 4.2a). At Vestskoven, oak accumulated very little N in forest floors over 30 years (85 kg ha^{-1}), whereas spruce and oak at Sellingen stored more N in forest floors within 30 years (ca. 3-400 kg ha^{-1}). Large amounts of N were accumulated in the oldest stands of the Swedish spruce chronosequence (around 1600 kg ha^{-1}). The forest floor N accumulation rate was highest for the Swedish chronosequence and lowest for oak at Vestskoven (Table 4.4).

These patterns were mainly driven by the accumulation of forest floor mass as there was little age-related change in forest floor C/N ratios in most chronosequences (Figure 4.2b). Only in the oak chronosequence at Vestskoven and spruce chronosequences at Gejlvang and in Sweden there were trends indicating increasing forest floor C/N ratios with increasing age.

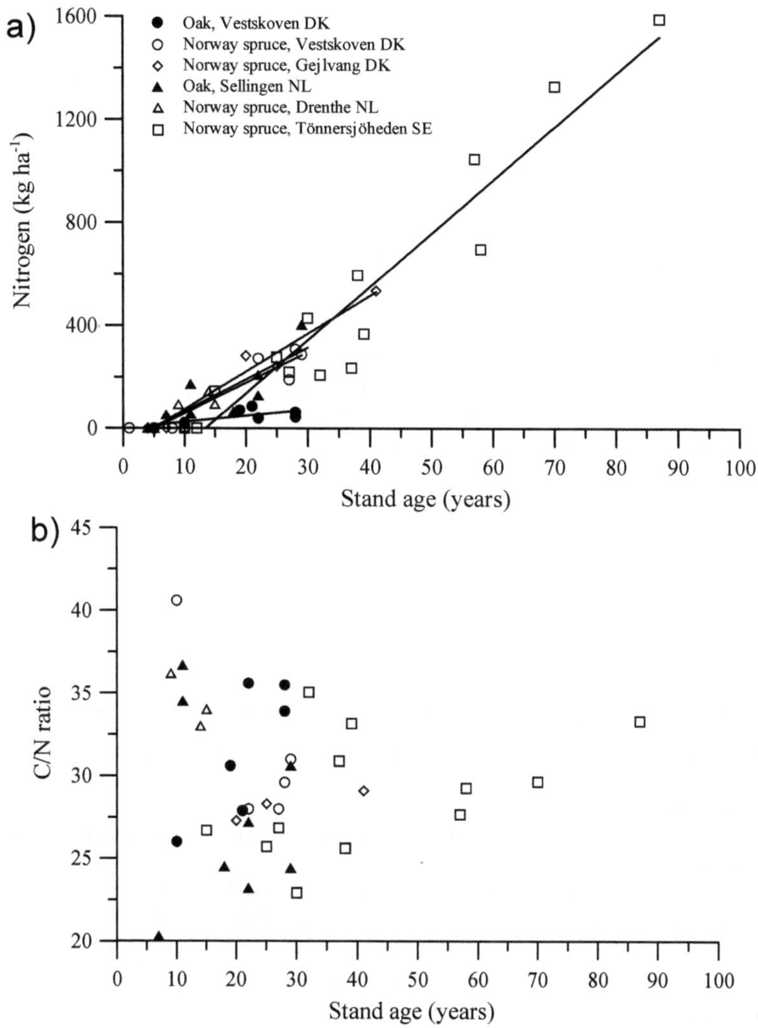

Figure 4.2. *a) Forest floor N content (kg ha^{-1}) and b) C/N ratio in forest floor in oak and Norway spruce chronosequences. Regression equations are given in Table 4.4.*

Mineral soil N contents were unchanged or tended to decrease slightly with increasing stand age in most chronosequences (Figure 4.3a and Table 4.4). Mineral soil N significantly decreased in the Swedish chronosequence. An exception to this was the significantly increasing mineral soil N content in the Dutch chronosequence. The Danish chronosequence at Gejlvang was remarkably lower in mineral soil N content regardless of stand age (around 2 Mg ha^{-1}) whereas the other Danish chronosequence at Vestskoven and the Swedish chronosequence had initial soil N contents of 5-6 Mg ha^{-1}. There was no difference between the Danish spruce and oak chronosequences at Vestskoven (Figure 4.3, Table 4.4), so all data were combined in the analysis of soil N (Ritter et al. 2003). For the studied soil compartments as a whole, i.e. the forest floor and the former plow layer (0-25 cm), there were patterns of constant, decreasing or increasing soil N content over 30-90 years after afforestation (Figure 4.3b). The only significant trend was the increasing soil N content in the Dutch chronosequence.

In the Danish Gejlvang chronosequence and in the Swedish chronosequence, mineral soil N content clearly decreased relative to C with increasing stand age whereas C/N ratios were unchanged in the Danish Vestskoven chronosequence. The chronosequence at Vestskoven also had the lowest C/N ratios regardless of stand age. The Danish and Swedish chronosequences had relatively similar C/N ratios of 10-15 for arable fields and stands younger than ten years. However, the development in C/N ratios differed strongly between the sites. The Dutch chronosequence differed from the other chronosequences by the much higher C/N ratios (Figure 4.3c). There was no significant change along this chronosequence, which covers a time span of only 14 years.

Table 4.4. *Rates of soil N sequestration (SE of regression slope) over 30-90 years in the AFFOREST chronosequences.*

Site	Forest floor N		Mineral soil N*		Total soil N*	
	Rate kg ha^{-1} yr^{-1}	P value	Rate kg ha^{-1} yr^{-1}	P value	Rate Mg ha^{-1} yr^{-1}	P value
Oak, Vestskoven DK	2.4 (1.1)	0.088	-48.2 (24.3)	0.071	-40.3 (23.4)	0.111
Spruce, Vestskoven DK	11.9 (2.2)	0.003				
Spruce, Gejlvang DK	14.7 (1.7)	0.003	-3.4 (7.6)	0.683	11.2 (8.6)	0.287
Spruce SE	20.7 (1.7)	< 0.001	-15.8 (6.3)	0.020	9.7 (7.3)	0.194
Oak/spruce NL	12.1 (2.3)	< 0.001	105.3 (33.8)	0.026	108.2 (33.2)	0.023

** Rates of soil N change for mineral soil and total soil C at Vestskoven are combined for oak and spruce.*

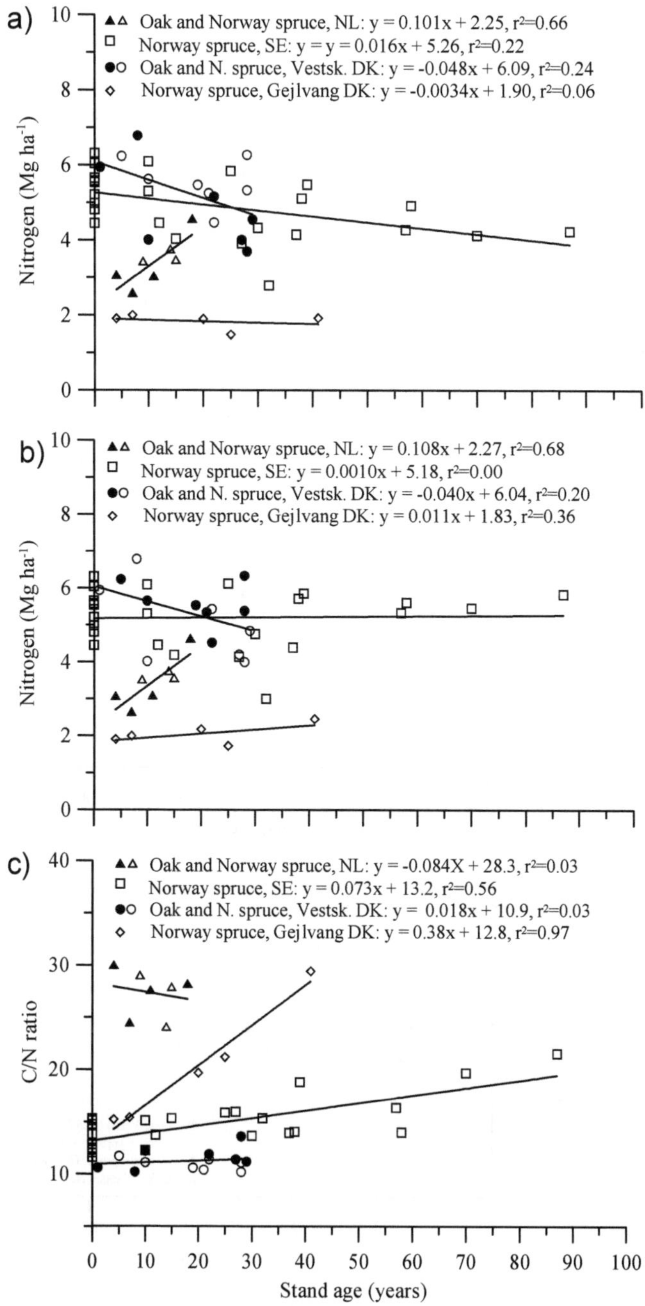

Figure 4.3. *a) N content in mineral soil, b) N content in both mineral soil and forest floor, and c) C/N ratios in mineral soil.*

There were no general trends in litterfall N flux with increasing stand age (Figure 4.4a) except that the youngest stands tended to have lower litterfall N fluxes. For oak at Vestskoven, litterfall N fluxes appeared to increase gradually with stand age. Litterfall N flux was higher in oak stands and lower in spruce stands. The few age-related dynamics were not caused by changes in litter N concentrations with stand age, as litterfall C/N ratios were quite similar along the chronosequences (Figure 4.4b).

Site- and probably also species-related differences in C/N ratio were more apparent than an effect of stand age. Oak stands had lower litterfall C/N ratios than did spruce. The oak chronosequence at Sellingen had the lowest C/N ratios (20-25) and spruce at Vestskoven had the highest C/N ratios (40-50).

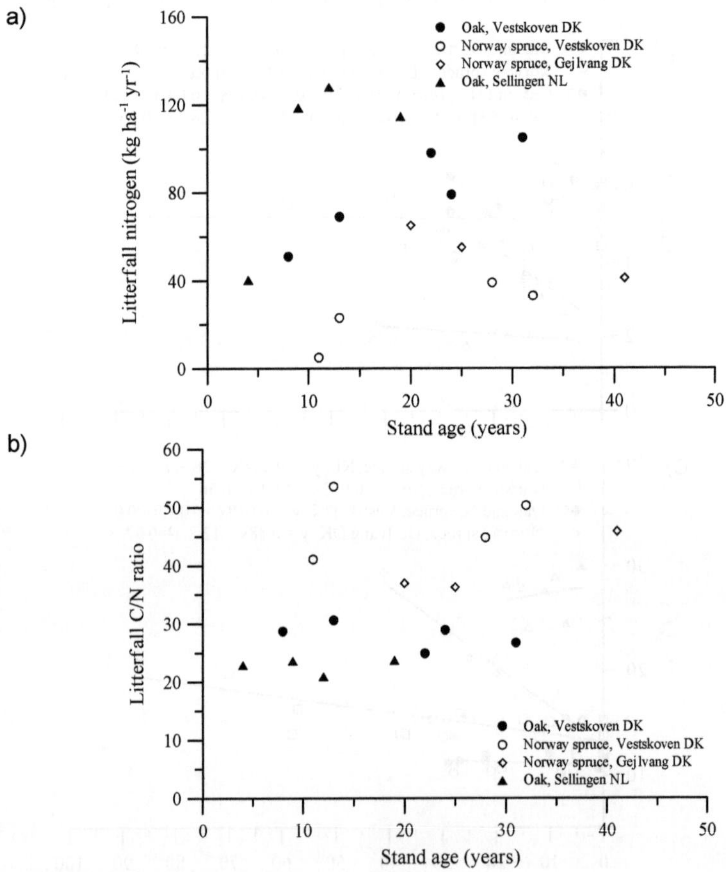

Figure 4.4. *a) Litterfall N flux (kg ha^{-1} yr^{-1}) and b) C/N ratio of litterfall in oak and Norway spruce chronosequences.*

The difference in litter C/N ratios partly contributed to the species difference in litterfall N flux, but litterfall mass and C content were also somewhat higher in oak than in spruce (Chapter 2).

3.3. Nitrate leaching following afforestation

At 90 cm depth, virtually all mineral N in soil solutions was found as nitrate. In the six chronosequences in Denmark, the Netherlands and Sweden, soil solution nitrate concentrations beneath the root zone were generally below the threshold value of 50 mg dm^{-3} NO_3^- for groundwater to be utilised as drinking water as expressed in the EU Water Framework Directive. Exceptions were consistently elevated concentrations recorded under the 33-year-old (planted 1969) spruce stand at Vestskoven in Denmark, where the concentrations frequently exceeded the drinking water quality requirements.

On an annual basis, high leaching losses of nitrate were related to high concentrations of nitrate in leaching water (Figure 4.5a, b). There was no clear relationship between nitrate leaching and stand age (Figure 4.5b) when data from all three countries were evaluated together. However, when looking at the countries separately, differences became apparent. In Denmark, leaching of nitrate from the root zone tended to be lower in the early phase of afforestation and increased after canopy closure. Here, nitrate losses by leaching only occurred in stands older than 20 years, whereas nitrate leaching always was low or negligible in stands younger than 15 years. In contrast to the Danish sites, there was a tendency of decreased nitrate leaching with forest age along the Dutch oak chronosequence at Sellingen. Furthermore, at Drenthe in the Netherlands and at Tönnersjöheden in Sweden, seepage water nitrate concentrations and leaching of nitrate from the root zone were generally negligible across the Norway spruce chronosequences, although slightly elevated levels were recorded in the 30-year-old stand (planted 1972) at Tönnersjöheden during summer periods.

During the measurement period, N budgets (i.e. total inorganic N deposition minus NO_3^--N leaching) ranged from -7 to 31 kg ha^{-1} yr^{-1} (Figure 4.6). At all except for one site (the 8-year-old oak stand at Sellingen), N leaching was lower than the estimated total N input. Nitrogen budgets tended to increase with age along the chronosequences at Gejlvang and Sellingen. In the chronosequences at Tönnersjöheden and Vestskoven, N budgets appeared relatively constant and showed no distinct trends with age. Within Denmark, higher N retention was observed in closed forest stands (> 20 years) growing on nutrient-poor sandy soil (Gejlvang) compared to stands of similar age on nutrient-rich clayey soil (Vestskoven).

Leaching of nitrate was evident in some stands at a throughfall N deposition level of approximately 8-10 kg N ha^{-1} yr^{-1}, and above a throughfall deposition level of approximately 15 kg N ha^{-1} yr^{-1} the majority of stands leached nitrate (Figure 4.7).

Over the age spans covered by the chronosequences, oak stands generally leached more nitrate than the spruce stands (Figure 4.5b). The highest leaching losses were recorded along the oak chronosequence at Sellingen in the

Netherlands, where nitrate leaching reached a level of 26 kg N ha^{-1} yr^{-1} in the 8-year-old stand. In contrast, nitrate leaching was negligible below the young Dutch spruce stands at Drenthe. At Vestskoven in Denmark, nitrate leaching appeared at an earlier age in the oak chronosequence (Figure 4.5b) and at a lower N throughfall deposition level (Figure 4.7) compared to the spruce chronosequence. Leaching was low below a throughfall deposition level of 10-12 kg N ha^{-1} yr^{-1} in the oak stands and below 14 kg N ha^{-1} yr^{-1} in the spruce stands.

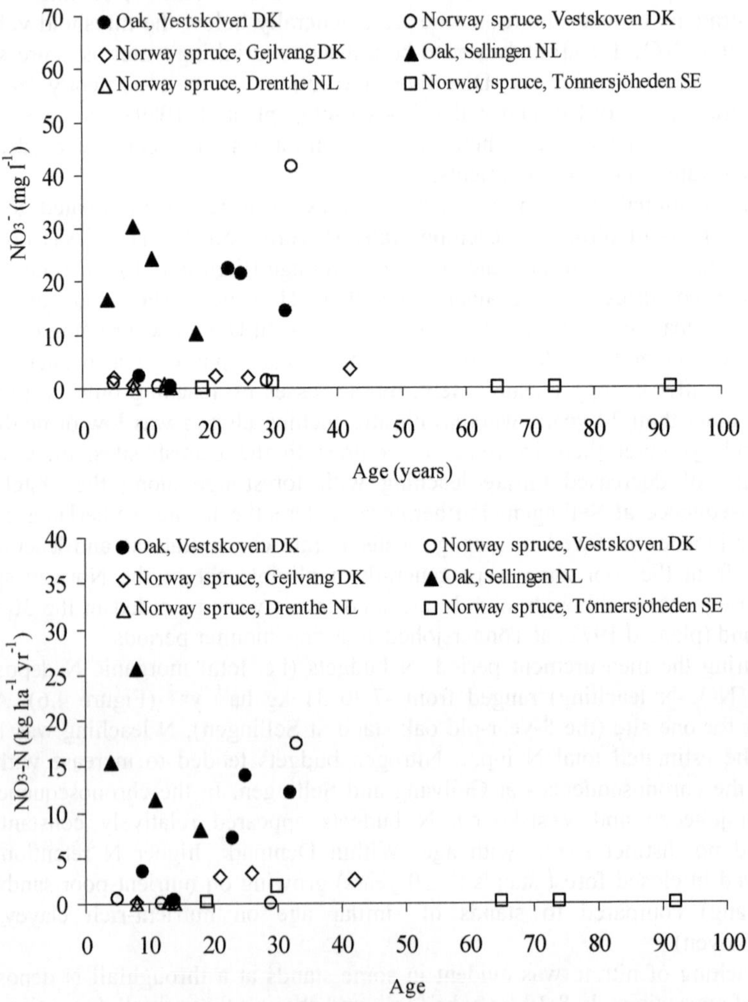

Figure 4.5. *(a) Annual average nitrate concentration (mg dm^{-3}) in soil water at 0.9 m (Tönnersjöheden, SE: 0.6 m) and (b) average nitrate leaching (kg NO_3-N ha^{-1} yr^{-1}) from the root zone in 2002 as a function of time since afforestation in the six AFFOREST chronosequences.*

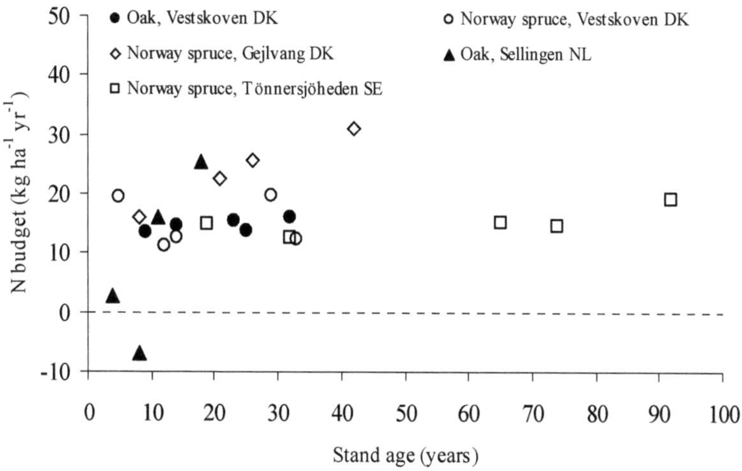

Figure 4.6. *Nitrogen budgets (kg ha^{-1} yr^{-1}), i.e. total inorganic N deposition minus NO_3^--N leaching at the bottom of the root zone, as functions of time since afforestation in five chronosequences. Positive budgets indicate that inorganic N is retained in the ecosystem, whereas negative budgets indicate a net release of inorganic N.*

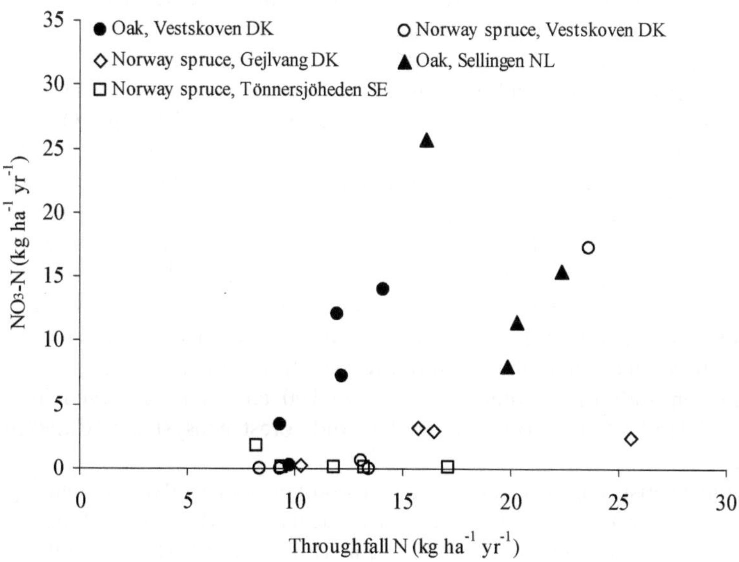

Figure 4.7. *Mean annual leaching of nitrate-N (kg ha^{-1} yr^{-1}) from the root zone in relation to throughfall N flux (kg ha^{-1} yr^{-1}) for corresponding periods in five AFFOREST chronosequences. Data are for the year 2002.*

4. DISCUSSIONS

4.1. Water quality before and after afforestation

The average nitrate concentrations below the root zone of sandy arable land in Denmark were close to the drinking water threshold value of 50 mg dm^{-3} NO_3^-, while it was estimated to be 21 mg dm^{-3} NO_3^- on clayey soils in average (Grant et al. 2004). The concentrations at the afforested stands were generally lower except for the 33-year-old spruce stand at Vestskoven. In the Netherlands, Fraters et al. (2004) showed average nitrate concentrations in groundwater beneath sandy arable land to be 75 mg dm^{-3} NO_3^- and 40 mg dm^{-3} NO_3^- beneath clayey arable land. Seepage water N concentrations reported for Swedish arable soils are similar to concentrations reported for arable soils in Denmark. Johnsson & Mårtensson (2002) estimated seepage water nitrate concentrations under arable land in southwest Sweden to be in average 57 mg dm^{-3} NO_3^- for sandy soils and 20 mg dm^3 NO_3^- for clay soils, which is substantially higher than concentrations recorded along the Swedish chronosequence (Figure 4.5a). As expected, the measured seepage water nitrate concentrations in the afforested stands were mostly lower than concentrations previously reported for arable soils in the three countries.

A high nitrate concentration in the soil solution seems to be a pre-condition for nitrate leaching (Figure 4.5a, b) and may indicate loss of nutrients from the ecosystem. In the Netherlands, nitrate leaching was estimated to be 74 kg NO_3-N ha^{-1} yr^{-1} on sandy soils under agricultural crops (Rijtema & de Vries 1994). For the whole arable area of Denmark, a leaching rate of 106 and 55 kg NO_3-N ha^{-1} yr^{-1}, respectively, were reported from sandy and clayey soils (Grant et al. 2004). Nitrate leaching below arable land in southwest Sweden (province of Halland) was estimated to be 63 kg NO_3-N ha^{-1} yr^{-1} from sandy soils and 18 kg NO_3-N ha^{-1} yr^{-1} from clayey soils, whereas an average value of 22 kg NO_3-N ha^{-1} yr^{-1} was estimated for all arable land in Sweden (Johnsson & Hoffmann 1998; Johnsson & Mårtensson 2002). In general, the simulated N leaching fluxes from the root zone in the chronosequences (Figure 4.5b) were considerably lower than the simulated N leaching from arable land in the three countries. A shift in land use from agriculture to forestry therefore presumably led to decreased nitrate leaching from the root zone. However, nitrate leaching from plantations on former arable land might still be higher than that from old forests. In Denmark, higher concentrations of nitrate in soils below the root zone (75-100 cm) were observed in recently afforested land (<10 years) compared to old forest ecosystems (Callesen et al. 1999).

In the Danish chronosequences at Vestskoven and Gejlvang, stands younger than 15 years showed practically no nitrate leaching, whereas leaching tended to increase after a stand age of around 20 years (Figure 4.5b). This may reflect higher uptake of N by trees in the early phase after afforestation due to the formation of N-rich compartments. N uptake by the canopy in the youngest stands is supported by the observation that the throughfall N deposition was lower than the bulk N deposition. When the foliar canopy is complete (about 15-20 years after planting) the trees can no longer take up the available N pool, and excess N is leached to seepage water. Increased nitrate leaching after canopy closure could also partly be attributed to increased N input with increased height (Figure 4.1, Table 4.3). Increased nitrate leaching with stand age has earlier been observed for a Sitka spruce stand in Wales (Emmet et al. 1993). The influence of stand height on N leaching has been described for Vestskoven by Hansen et al. (2006b).

At Vestskoven in Denmark, the highest concentrations of nitrate in seepage water and highest levels of nitrate leaching were encountered in the 33-year-old spruce stand (planted in 1969). Concentrations and leaching levels were negligible in the other spruce stands along the chronosequence (Figure 4.5a, b). Several factors may explain this deviant behaviour. Compared to the younger stands in the chronosequence, the annual input of N with throughfall was considerably larger in the 33-year old stand due to its larger tree height (Figure 4.1, Table 4.3) resulting in enhanced dry deposition of N. Moreover, lower uptake rates of N relative to the younger stands could have resulted in higher leaching losses of nitrate from this stand. A high rate of net N mineralizarion may indeed be expected in the nutrient-rich soil at Vestskoven because of high soil N contents and low C/N ratios (around 10-12) in the former plow layer (Figure 4.3a, c). Possibly, recent thinning (1998) in the stand may have resulted in increased amounts of mineral N relative to other spruce stands in the chronosequence. An increase in soil nitrate concentrations was visible in a thinning experiment in lodgepole pine where 60% of the trees were removed (Knight et al. 1991). Bäumler & Zech (1999) also observed a low increase in nitrate concentrations after a 40% thinning. The concentrations were back to pre-cutting conditions after one year.

Maintenance of high rates of N mineralizarion in soil in the early phase after the conversion of farmland to forest may explain the high levels of nitrate leaching beneath the oak stands at Sellingen in the Netherlands (van der Salm et al. 2005) (Figure 4.5b). In addition, throughfall N deposition was quite high to the oak stands (Table 4.3, Figure 4.7). The rates of N mineralizarion are, as discussed earlier, likely to decline following afforestation. A decreased mineralizarion rate with age may partly explain the decreasing trend in nitrate leaching along the 18-year chronosequence at Sellingen. A very possible contributing factor may be an increase in the rate of N uptake with age of the young aggrading oak forest. However, the stands in the Dutch chronosequence are all below the age of 20 years and it is possible that leaching of nitrate will increase when the stands mature and scavenging of atmospheric N increases. In addition, denitrification and the subsequent release of N_2 might be an important process that removes nitrate from the soil solution under the oak stands at Sellingen due to anaerobic conditions. The groundwater level in the forest fluctuated between -20 cm in wet periods in winter to -160 cm in summer.

No age-related trends in nitrate leaching were found along the Norway spruce chronosequences at Drenthe in the Netherlands and at Tönnersjöheden in Sweden, since N leaching was generally negligible at these sites. Here, trees and field layer were able to take up all available N. Even the old Norway spruce stands (64-92 years old) at Tönnersjöheden were able to completely retain N in the ecosystem (Figure 4.5a, b) in spite of a relatively high ambient N deposition in the region (Hallgren-Larsson 2002) and an anticipated low demand for N for growth in these mature stands. The high capacity of the old spruce stands to retain N may partly be caused by less intensive fertilisation practices during former agriculture compared to soils more recently abandoned and compared to soils abandoned in Denmark and the Netherlands. Also, the old plantations experienced a lower level of N deposition in their early growth stages than more recently planted forests (Lövblad 2000). These old stands also had abundant field layer vegetation consisting mainly of mosses, which most likely contributed to a higher uptake of N in biomass. Since the majority of Swedish forest ecosystems are N limited (Näsholm et al. 2000), low levels of nitrate leaching were anticipated. However, Andersson (2002) suggested that N saturation might be a potential problem in forests in southwest Sweden, where N deposition is relatively high (>15kg N ha^{-1} yr^{-1}). In this respect, recently afforested ecosystems would be particularly vulnerable to disturbance of the N cycle and to increased N inputs.

Several authors observed nitrate leaching to occur mainly above a threshold value of about 10 kg N ha^{-1} yr^{-1} in throughfall input (Dise et al. 1998; Gundersen et al. 1998; Kristensen et al. 2004), but at high deposition rates the leaching seemed to be largely dependent on ecosystem properties (e.g. soil type, N status etc). In old-growth forest ecosystems several indices have been used in order to predict the onset of N saturation and the potential risk of nitrate leaching. Especially, the C/N ratio of the forest floor, total throughfall N flux and foliage N concentration are well established predictors of the likeliness of N leaching from old European forest ecosystems (Dise et al. 1998; Gundersen et al. 1998; Kristensen et al. 2004). It appears as if the throughfall N flux could be used as a predictor of nitrate leaching even in recently afforested stands (Figure 4.7). However, it is not possible from our data to identify whether increased nitrate leaching after canopy closure is the result of increased throughfall N input or decreased N uptake and changing mineralizarion or a combination of these and possibly other factors. The forest floor C/N ratio was not found to be a useful predictor in young aggrading forests planted on former arable soils, since the C/N ratio of the thin forest floors to a large extent reflects the quality of recently shed litter.

4.2. Site effects on deposition and nitrate leaching

Deposition fluxes of N may show considerably local variation due to differences in e.g. climate, local N sources, stand structure, forest age, forest management and tree species (Bleeker & Draaijers 2002). In our study, the total atmospheric input of inorganic N (NO_3^- and NH_4^+) as well as the throughfall N flux to the forests tended to decrease along a south-west (the Netherlands) – north-east (Sweden) gradient (Table 4.3). Such a gradient was, however, not observed for the measured bulk deposition fluxes of N in the open field (Table 4.2 and 3).

Throughfall N deposition in Norway spruce at Tönnersjöheden, Sweden, was considerably lower than at the Danish sites (especially for the taller trees > 10 m) although the deposition of N in the open field was equal or higher at the Swedish location (Table 4.3, Figure 4.1a). The high bulk N deposition is confirmed by a number of Swedish studies (Kindbom et al. 2001; Hallgren-Larsson 2002). The area of Halland, where Tönnersjöheden is situated, is a part of Sweden that receives among the highest load of atmospheric N. The lower throughfall deposition of N in the Swedish sites relative to the Danish sites was not due to high levels of canopy exchange (Table 4.3) but rather due to a lower contribution of dry deposition of N. A different contribution of dry deposition at the Danish and Swedish sites is not unexpected since the input by dry deposition is larger in Denmark than in Sweden caused by a higher concentration of livestock. Also, the wind climate differs largely between these sites. The Danish sites are more windy than the Swedish sites, which causes dry deposition to increase at the Danish sites relative to the Swedish sites. A strong relationship between the deposition velocity of NH_3 and the wind velocity has earlier been observed (Bleeker & Draaijers 2002). Furthermore, lower fluxes of dry N deposition along the 90-year-old Swedish chronosequence may partly be caused by a strong thinning regime. These mature spruce stands have more open canopies than the younger more dense stands at the Danish chronosequences. Thinning of a forest stand will limit the rate of atmospheric deposition because of the reduction of the total surface area of the canopy and the accompanying reduction of the aerodynamic roughness of the canopy (Bleeker & Draaijers 2002).

Within Denmark, the deposition regime varied considerably between the stands at Gejlvang in the western part of the country and the stands at Vestskoven in the eastern part. A higher total N deposition at Gejlvang was attributed to higher rates of both wet and dry N deposition compared to Vestskoven (Table 4.3). Both higher wind velocity at Gejlvang, leading to higher dry deposition of N, and a higher local emission of NH_3 due to a high concentration of livestock led to these high N depositions (Hansen 2003). Our observations of N deposition (Table 4.3) in the older Danish spruce stands (> 20 years) are in good agreement with corresponding N fluxes measured in long term monitoring plots of the same age (level II) located on similar soil types not far from the studied chronosequence stands (Hansen 2003).

Dutch agriculture is more intensive than in Denmark and Sweden in sense of both livestock and the relative area covered by agriculture. As such, the throughfall N deposition to the oak stands was higher at the Dutch stands at Sellingen than at Vestskoven in Denmark. High N deposition in the Netherlands in general has been documented earlier (Draaijers 1993; Erisman et al. 2005).

Up to about 40% of the total atmospheric N deposition was taken up by the forest canopy (21 and 26 years old spruce stands at Gejlvang, Denmark) (Table 4.3). This is consistent with a study by Johnson and Lindberg (1992) conducted in several forest stands (mainly conifers) scattered over the U.S., showing that in average 40% of the total inorganic N deposition was retained by the vegetation, whereas 60% was found back in the throughfall as NO_3^- and NH_4^+. In their study, total inorganic N uptake amounted up to 850 eq. ha^{-1} yr^{-1} (~12 kg ha^{-1} yr^{-1}), which is close to the maximum calculated average rates of canopy N uptake (11 kg ha^{-1} yr^{-1}) found at Gejlvang. Even higher uptake efficiency of total inorganic N was reported by Tomaszewski et al. (2003) in a subalpine slowly aggrading 90-year-old spruce-fir-pine forest in Colorado, U.S., where approximately 85% of the total N deposition was retained by the forest canopy in the growing season. On the other hand, an investigation performed in a Douglas-fir stand at the Speulder forest in the

Netherlands showed only 9% uptake of the total NH_x deposition (Draaijers et al. 1997). However, the total N load to the forest canopies differed substantially among these studies where the Dutch study as well as our data generally showed considerably higher N deposition than the two north American studies. Often the throughfall deposition, especially in the youngest spruce stands, was lower than in bulk deposition, which also points to considerable N uptake in the canopies in the early years after afforestation.

Canopy uptake of N has earlier been estimated in the Dutch oak chronosequence at Sellingen during the same sampling period using a slightly different calculation approach (Van der Salm et al. 2005). In their study, the uptake of N was relatively constant (5.9-6.7 kg N ha^{-1} yr^{-1}) and showed no distinct trend with age. In contrast, our results suggest an increasing trend with age (and height) from 1 to 11 kg N ha^{-1} yr^{-1} between 4 and 18 years stand age (Table 4.3). The quantitative results therefore seem to be dependent on the assumptions taken during the calculation. However, in this study the calculations were performed alike on all data and comparisons between sites are justified.

Changes in soil C and N levels and N dynamics following afforestation of former arable land were very site specific. Among the studied sites, differences between nutrient-rich and nutrient-poor sites were reflected by C/N ratios and N contents in the mineral soil (former plow layer), with generally lower soil C/N ratios and higher soil N contents in the more nutrient-rich soils (Figure 4.3). The nitrogen status at the six AFFOREST chronosequences, as suggested by these indicators, increased in the approximate order: oak and spruce, Sellingen and Drenthe, the Netherlands ≤ spruce, Gejlvang, Denmark < spruce, Tönnersjöheden, Sweden < oak and spruce, Vestskoven, Denmark.

Nitrate leaching was substantially higher from the nutrient-rich clayey soils of high N status at Vestskoven than from the nutrient-poor sandy soils of lower N status at Gejlvang (Figure 4.5b). This is consistent with previous studies reporting higher nitrate leaching (Hansen 2003) and higher seepage water nitrate concentrations (Callesen at al. 1999) below forest stands on clayey soils compared to stands on sandy soils in Denmark. However, differences in N status between sites expressed as soil C/N ratios and soil N contents could not fully explain the observed patterns of nitrate leaching between all the chronosequences. For example, seepage water nitrate concentrations and nitrate leaching were high below the young oak stands at Sellingen in the Netherlands (Figure 4.5a, b), in spite of low N status (high C/N ratio) in the sandy soils at this site. In the first years following afforestation, nitrate leaching may occur as a result of high production of N by mineralizarion of N-rich organic matter inherited from the former agricultural land use. Thus, in recently afforested stands the net N mineralizarion rate or net nitrification rate would presumably be a better predictor of the risk of nitrate leaching than the soil C/N ratio.

Within Denmark, the overall N status of the site had a marked influence on the net retention of N in the afforested ecosystem. In closed stands older than about 20 years, net ecosystem retention of N was more efficient in forests growing on nutrient-poor sandy soil at Gejlvang (western Denmark) than in forests on nutrient-rich clayey soil at Vestskoven (eastern Denmark) (Figure 4.6), regardless of tree species and despite higher total N deposition at Gejlvang (Table 4.3). The most likely reason for this is higher soil supply rates of N (N mineralizarion followed by nitrification) at Vestskoven, caused by high soil N contents and pH. Soil and vegetation are unable to take up the available N (derived by mineralizarion and deposition), which is leached from the root zone.

At the Danish and Swedish chronosequences, mineral soil C/N ratios reflected the former land use within the first decade following afforestation, after which differences between soil types became apparent (Figure 4.3c). Chronosequences with soils of low N status (Gejlvang, Denmark as well as Sellingen and Drenthe, the Netherlands) were able to accumulate C in the mineral topsoil, which can be attributed largely to incorporation of forest derived organic matter with high C/N ratio and partly to reduced rates of decomposition in these nutrient-poor soils. The relatively unchanged (Gejlvang) and increasing (the Netherlands) mineral soil N contents (Figure 4.3) are in contrast to previous studies, which show considerable N depletion in the mineral topsoil after 40-115 years of afforestation with conifers or mixed broadleaves on sandy or loamy soil types in the U.S. (Hamburg 1984; Hooker & Compton 2003; Richter et al. 2000). An indication of depletion in mineral soil N contents after afforestation was observed in soils of higher N status (higher soil N content and lower C/N ratios) at Vestskoven and at Tönnersjöheden.

4.3. Tree species effects on deposition and nitrate leaching

At Vestskoven in Denmark, the oak and spruce chronosequences were established on the same soil type (loamy till) (Chapter 1). Therefore, this site offers a unique opportunity to study the influence of different tree species on deposition and nitrate leaching following afforestation of former agricultural soils. There were no consistent differences in N throughfall and total N deposition between oak and spruce stands of similar age although there was an indication of higher levels of N deposition in the 33-year-old spruce stand than in an oak stand of similar age. Also, no distinct species difference in canopy N uptake was observed between oak and spruce stands of similar age at Vestskoven, Denmark (Table 4.3). When the two species were compared along a stand height gradient, there was also no general tree species difference in N deposition for the same stand height at Vestskoven (Table 4.3). This observation contradicts earlier studies where significant higher roughness lengths have been observed for coniferous trees compared to deciduous trees (Erisman 1992; Draaijers 1993; Bleeker & Draaijers 2002), leading to higher N throughfall deposition. However, in a review Parker (1990) found that numerous studies showed higher annual throughfall deposition under hardwoods than under adjacent coniferous stands (Cronan & Reiners 1983; Henderson et al. 1977). The absence of a tree species related difference in N deposition between oak and spruce in our study may partly be attributed to the short part of the rotation period in which we could compare species (0-35 years). Differences in N deposition to the oak and spruce stands at Vestskoven may appear when the stands grow older.

Over approximately 35 years following afforestation, oak stands generally leached more nitrate than Norway spruce stands of similar age (Figure 4.5b). These results are in contrast to findings in European studies of paired mature stands of coniferous and deciduous species at the same sites, showing higher seepage water nitrate concentrations and higher leaching of nitrate below coniferous tree species than below deciduous tree species (De Schrijver et al. 2000; Gundersen et al. 2005; de Vries & Jansen 1994; Rothe et al. 2002). The higher N leaching losses under conifers were attributed to higher N deposition to coniferous than to deciduous tree species, since conifers are evergreen and have larger foliar surfaces over the year, which receive more deposition. With the exception of the 33-year old Norway spruce stand there was no general difference in N deposition levels between even-aged oak and spruce stands at Vestskoven in Denmark that explains the difference in nitrate leaching between the two tree species. With similar input of N, Kristensen et al. (2004) did find higher N leaching in deciduous stands than coniferous stands. However, in their case the difference was mainly attributed to a general soil type difference between stands of the two forest types as represented in the European intensive monitoring (level II). This was not the case in our chronosequences, where possible causes could be differences in uptake or internal cycling of N between oak and spruce.

The difference in nitrate leaching between oak and spruce reflects the difference in growth rate between the two tree species. The growth of oak and hereby the N uptake starts out slowly while spruce has a high initial growth rate and high uptake of N. Hence, leaching fluxes were always low or negligible in young spruce stands receiving low N deposition (Figure 4.5b, Figure 4.7). At Vestskoven, the spruce stands stored more C and N in forest floors than the oak stands of similar age, although the input of C and N from litterfall was higher in the oak stands (Figure 4.4). This suggests a higher rate of decomposition of the more easily decomposable oak litter and hence a higher rate of N mineralizarion in the forest floor under oak. However, there is little conclusive evidence of such a species difference for north-western Europe. While oak litter has been reported to decompose faster than spruce litter (Dziadowiec 1987), Brüggemann et al. (2005) reported higher gross N mineralizarion rates in the forest floor under spruce than under oak in a species trial on sandy soil in Denmark. The difference in nitrate leaching between the two tree species might also partly be due to higher groundwater recharge under oak, because of lower interception loss than in the spruce stands (Chapter 3). The difference in annual water recharge between oak and spruce was most pronounced in the Dutch stands on sandy soil types, whereas tree species related differences in deep seepage was small in the less permeable clay-rich soils at Vestskoven in Denmark (Rosenqvist et al. 2006b). Similar water yield for oak and spruce has previously been reported in areas with low precipitation and less permeable soils (Augusto et al. 2002).

5. CONCLUSION

Forests on former arable land vary substantially in their ability to retain N in the ecosystem. No consistent pattern was apparent in the three countries. Differences in N retention and nitrate leaching depend on a combination of local and/or regional factors such as soil N status, soil type, N deposition, forest age, tree species and hydrology. However, an effect of the former land-use is evident and the N dynamics in the newly planted forests on former arable land differ from those in old forests. The afforested sites are not (yet) in steady state and changes in the processes are ongoing. The interpretation of results from the studied chronosequences is therefore complicated by the transient nature of the factors that regulate leaching losses of N from the system. In the course of time, the environment in the new forests will evolve towards the environment in old forests. However, the present data do not allow determination of the length of such a transition period.

Compared to nutrient-poor sandy soils, nutrient-rich clayey soils appear more vulnerable to disturbance of the N cycle and to increased N deposition, possibly leading to N saturation and enhanced nitrate leaching. Furthermore, tree species seem to influence nitrate leaching and N retention. Our results suggest that the effect of different tree species (Norway spruce vs. oak) on N leaching after afforestation must be regarded in a time perspective. In the short-term (approximately 0-35 years) ecosystem N retention is generally less efficient in oak than in spruce, leading to higher N leaching losses from the oak stands. Over larger time scales (forest harvest rotation) the higher N deposition to conifers may result in higher leaching from coniferous than from deciduous stands, as often reported in literature.

6. REFERENCES

Aber, J. D., Nadelhoffer, K. J., Steudler, P., & Melillo, J. M. (1989). Nitrogen saturation in northern forest ecosystems. *BioScience, 39,* 378-386.

Allen, A., & Chapman, D. (2001). Impacts of afforestation on groundwater resources and quality. *Hydrogeology Journal, 9,* 390-400.

Andersson, P. (2002). Nitrogen turnover in Swedish spruce forest ecosystems - effects of nitrogen deposition. Acta Universitatis Agriculturae Sueciae, Agraria 342. Doctoral thesis. Swedish University of Agricultural Sciences.

Augusto, L., Ranger, J, Binkley, D., & Rothe, A. (2002). Impact of several common tree species of European temperate forests on soil fertility. *Annals of Forest Science, 59,* 233-253.

Bastrup-Birk, A., & Gundersen, P. (2004). Water quality improvements from afforestation in an agricultural catchment in Denmark illustrated with the INCA model. *Hydrology and Earth System Sciences, 8*(4), 764-777.

Bäumler, R., & Zech, W. (1999). Effects of forest thinning on the streamwater chemistry of two forest watersheds in the Bavarian Alps. *Forest Ecology and Management, 116,* 119-128.

Bergholm, J., Berggren, D., & Alavi, G. (2001). Soil acidification induced by ammonium sulphate addition in a Norway spruce forest in Southwest Sweden. In J. Bergholm, 2001, Acta Universitatis Agricuturae Sueciae, Agraria 215. Doctoral thesis. Swedish University of Agricultural Sciences.

Bleeker, A., & Draaijers, G. P. J. (2002). Impact of tree species, stand structure, closeness to forest edge, and management options on atmospheric deposition. In K. Hansen (Ed.), Literature review for AFFOREST, -2002, Planning afforestation on previously managed arable land - influence on deposition, nitrate leaching, precipitation surplus, and carbon sequestration. http://www.sl.kvl.dk/afforest

Brüggemann, N., Rosenkranz, P., Papen, H., Pilegaard, K., & Butterbach-Bahl, K. (2005). Pure stands of temperate forest tree species modify soil respiration and N turnover. *Biogeosciences Discussions, 2,* 303-331.

Callesen, I., Rauland-Rasmussen, K., Gunderssen, P., & Stryhn, H. (1999). Nitrate concentrations in soil solutions below Danish forests. *Forest Ecology and Management, 114,* 71-82.

Cronan, C. S., & Reiners, W. A. (1983). Canopy processing of acidic precipitation by coniferous and hardwood forests in New England. *Oecologia (Berlin), 59,* 216-223.

De Schrijver, A., Van Hoydonck, G., Nachtergale, L., De Keersmaeker, L., Mussche, S., & Lust, N. (2000). Comparison of nitrate leaching under Silver Birch (Betula Pendula) and Corsican Pine (Pinus Nigra ssp. Larico) in Flanders (Belgium). *Water, Air, and Soil Pollution, 122,* 77-91.

De Vries, W., & Jansen, P. C. (1994). Effects of acid deposition on 150 forest stands in the Netherlands. 3. Input output budgets for sulphur, nitrogen, base cations and aluminium. Wageningen, the Netherlands, DLO-Winand Staring Centre for Integrated Land, Soil and Water Research, Report 69.3, 60 pp.

De Vries, W., Reinds, G. J., Van der Salm, C., Draaijers, G. P. J., Bleeker, A., Auee, J., Gundersen, P., Kristensen, H. L. Van Dobben, H., De Zwart, D., Derome, J., Voogd, J. C. H., & Vel, E.M. (2001). Intensive monitoring of forest ecosystem in Europe. Technical Report 2001. EC-UN/ECE 2001, Brussels, Geneva, 161 pp.

Dise, N. B., Matzner, E., & Gundersen, P. (1998). Synthesis of nitrogen pools and fluxes from European forest ecosystems. *Water, Air, and Soil Pollution, 105,* 143-154.

Draaijers, G. (1993). The variability of atmospheric deposition to forests. The effect of canopy structure and forest edges. Faculty of Geographical Sciences, University of Utrecht, ISBN 90-6809-166-2, 200 pp.

Draaijers, G. P. J., & Erisman, J. W. (1995). A canopy budget model to estimate atmospheric deposition from throughfall measurements. *Water, Air, and Soil Pollution, 85,* 2253-2258.

Draaijers, G. P. J., Erisman, J. W., Spranger, T., & Wyers, G. P. (1996). The application of throughfall measurements for atmospheric deposition monitoring. *Atmospheric Environment, 30,* 3349-3361.

Draaijers, G. P. J., Erisman, J. W., van Leeuwen, N. F. M., Römer, F. G., te Winkel, B. H., Veltkamp, A. C., Vermeulen, A. T., & Wyers, G. P. (1997). The impact of canopy exchange on differences observed between atmospheric deposition and throughfall fluxes. *Atmospheric Environment, 31,* 387-397.

Draaijers, G. P. J., Erisman, J. W. Lövblad, G., Spranger, T., & Vel, E. (1998). Quality and uncertainty aspects of forest deposition estimation using throughfall, stemflow and precipitation measurements. Report no. R98/093, TNO-MEP, Apeldoorn, The Netherlands.

Dziadowiec, H. (1987). The decomposition of plant litter fall in an oak-linden-hornbeam forest and an oak-pine mixed forest of the Bialowieza National Park. *Acta Societatis Botanicorum Poloniae, 56,* 169-185.

Emmett, B. A., Reynolds, B., Stevens, P. A., Norris, D. A., Hughes, S., Görres, J., & Lubrecht, I. (1993). Nitrate leaching from afforested Welsh catchments - Interactions between stand age and nitrogen deposition. *Ambio, 22,* 386-394.

Erisman, J. W. (1992). Atmospheric deposition of acidifying compounds in the Netherlands. Ph.D. thesis, University of Utrecht, the Netherlands.

Erisman, J. W., Vermeulen, A., Hensen, A., Flechard, C., Dämmgen, U., Fowler, D., Sutton, M., Grünhage, L., & Tuovinen, J-.P. (2005). Monitoring and modelling of biosphere/atmosphere exchange of gases and aerosols in Europe. *Environmental Pollution, 133,* 403-413.

Fraters, B., Hotsma, P. H., Langenberg, V. T., Leeuwen, T. C. van, Mol, A. P. A., Olsthoorn, C. S. M., Schotten, C. G. J., & Willems, W. J. (2004). Agricultural practice and water quality in the Netherlands in the 1992-2002 period. Background information for the third EU Nitrate Directive Member States report. RIVM Rapport 500003002. National Institute of Public Health and the Environment (RIVM), Bilthoven, the Netherlands, pp. 1-178.

Grant, R., Blicher-Mathiesen, G., Jensen, P. G., Pedersen, M. L., Clausen, B., & Rasmussen, P. (2004). Landovervågningsoplande 2003. NOVA 2003. Danmarks Miljøundersøgelser. 118 s. http://faglige-rapporter.dmu.dk

Gundersen, P. (1991). Nitrogen deposition and the forest nitrogen cycle: role of denitrification. *Forest Ecology and Management, 44,* 15-28.

Gundersen, P., & Rasmussen, L. (1995). Nitrogen mobility in a nitrogen limited forest at Klosterhede, Denmark, examined by NH_4NO_3 addition. *Forest Ecology and Management, 71,* 75-88.

Gundersen, P., Callesen, I., & de Vries, W. (1998). Nitrate leaching in forest ecosystems is related to forest floor C/N-ratios. *Environmental Pollution, 102,* 403-407.

Gundersen P., Schmidt I. K., and Raulund-Rasmussen K. (2005). Leaching of nitrogen from temperate forests - effects of air pollution and forest management. *Environmental Reviews, 14,* 1-57.

Hallgren-Larsson, E. (Ed.) (2002). Övervakning av luftföroreningar i Hallands län. Resultat till och med september 2001. (in Swedish) IVL-rapport B1468. 28 pp.

Hamburg, S. P. (1984). Effects of forest growth on soil nitrogen and organic matter pools following release from subsidence agriculture. In E. L. Stone (Ed.), Forest soils and treatment impacts, pp. 145-158. Knoxwille, Tennessee, USA: University of Tennessee.

Hansen, K. (Ed.) (2003). Næringsstofkredsløb i skove – Ionbalanceprojektet (In Danish). Forest & Landscape Research nr. 33., ISBN 87-7903-156-0, Skov & Landskab, Hørsholm. 300 pp.

Hansen, K., & Vesterdal, L. (1999). Skovrejsning - nitratudvaskning, jordens pH og kulstofbinding. Videnblade Skovbrug nr. 4.6-9, Forskningscentret for Skov & Landskab, Hørsholm. 2 pp.

Hansen, K. Rosenqvist, L., Vesterdal, L., & Gundersen, P. (2006a). Throughfall and canopy exchange in four chronosequences of forests planted on former intensively managed soils in Denmark and Sweden. *Atmospheric Environment.* (In preparation).

Hansen, K. Rosenqvist, L., Vesterdal, L., & Gundersen, P. (2006b). Nitrogen leaching from three chronosequences of forests planted on former intensively managed soils in Denmark. *Forest Ecology and Management.* (In preparation).

Henderson, G. S., Harris, W. F., Todd, D. E. Jr., & Grizzard, T. (1977). Quality and chemistry of throughfall as influenced by forest type and season. *Journal of Ecology, 65,* 365-374.

Hesse, P. R. (1971). A textbook of soil chemical analysis. London, UK: John Murray.

Homann, P. S., Sollins, P., Chappell, H. N., & Stangenberger, A. G. (1995). Soil organic carbon in a mountainous, forested region: Relation to site characteristics. *Soil Science Society of America Journal, 59,* 1468-1475.

Hooker, T. D., & Compton, J. E. (2003). Forest ecosystem carbon and nitrogen accumulation during the first century after agricultural abandonment. *Ecology Applications, 13*(2), 299-313.

Ivens, W. P. M. F. (1990). Atmospheric deposition onto forests: an analysis of the deposition variability by means of throughfall measurements. Ph.D. thesis, University of Utrecht, The Netherlands, 153 pp.

Johnson, D. W., & S. E. Lindberg (1992). Atmospheric deposition and forest nutrient cycling. Ecological Studies 91. New York, USA: Springer.

Johnsson, H., & Hoffmann, M. (1998). Nitrogen leaching from agricultural land in Sweden. Standard rates and gross loads in 1985 and 1994. *Ambio, 27,* 481-488.

Johnsson, H., & Mårtensson, K. (2002). Kväveläckage från svensk åkermark: Beräkningar av normalutlakning för 1995 och 1999. Underlagsrapport till TRK. (In Swedish) Naturvårdsverket, rapport 5248. 91 pp.

Jussy, J. H., Colin-Belgrand, & M., Ranger, J. (2000). Production and root uptake of mineral nitrogen in a chronosequence of Douglas-fir (Pseudotsuga menziesii) in the Beaujolais Mounts. *Forest Ecology and Management, 128,* 197-209.

Kindbom, K., Svensson, A, Sjöberg, K., & Persson, C. (2001). Nationell miljöövervakning av luft- och nederbördskemi 1997, 1998 och 1999. IVL Rapport/report 1420. SMHI Rapport/report Nr 96 (in Swedish with English summary) 102 pp.

Knight, D. H., Yavitt, J. B., & Joyce, G. D. (1991). Water and nitrogen outflow from lodgepole pine after two levels of tree mortality. *Forest Ecology and Management, 46,* 215-225.

Kristensen, H. L., Gundersen, P., Callesen, I., & Reinds, G. (2004). Atmospheric N Deposition Influences Soil Nitrate Concentration Differently in European Coniferous and Deciduous Forests. *Ecosystems, 7,* 180-192.

Lövblad, G. (2000). Nitrogen deposition now and in the future. In U. Bertils & T. Näsholm (Eds.). Effects of Nitrogen Deposition on Forest Ecosystems. Swedish Environmental Protection Agency. Report 5067.

Matejovic, I. (1993). Determination of carbon, hydrogen, and nitrogen in soils by automated elemental analysis (dry combustion method). *Communications in Soil Science and Plant Analysis, 24,* 2213-2222.

McNabb, D. H., Cromack, K., & Fredriksen, R. L. (1986). Variability of nitrogen and carbon in surface soils of six forest types in the Oregon Cascades. *Soil Science Society of America Journal, 50,* 1037-1041.

Näsholm, T., Nohrstedt, H-Ö., Kårén, O., Kytö, M., & Björkman, C. (2000) How are forest trees affected? In U. Bertils & T. Näsholm (Eds.), Effects of Nitrogen Deposition on Forest Ecosystems. Swedish Environmental Protection Agency. Report 5067.

Parker, G. G (1990). Evaluation of dry deposition, pollutant damage, and forest health with throughfall studies. In A. A. Lucier & S. G. Haines. Mechanisms of forest response to acidic deposition, Springer-Verlag, Chapter 2, 10-61.

Pypker, T. G., Bond, B. J., Link, T. E., Marks, D., & Unsworth, M. H. (2005). The importance of canopy structure in controlling the interception loss of rainfall: Examples from a young and an old-growth Douglas fir forest. *Agricultural and Forest Meteorology, 130,* 113-129.

Richter, D. D., Markewitz, D., Heine, P. R., Jin, V., Raikes, J., Tian, K., & Wells, C.G. (2000). Legacies of agriculture and forest regrowth in the nitrogen of old-field soils. *Forest Ecology and Management, 138,* 233-248.

Rijtema, P. E., & de Vries, W. (1994). Differences in precipitation excess and nitrogen leaching from agricultural lands and forest plantations. *Biomass and Bioenergy, 6*(1/2), 103-113.

Ritter, E., Vesterdal, L., & Gundersen, P. (2003). Changes in soil properties after afforestation of former intensively managed soils with oak and Norway spruce. *Plant and Soil, 249,* 319-330.

Rosenqvist, L., Berggren, D., & Johansson, M.-B. (2006a). Concentrations and fluxes of dissolved organic carbon and nitrogen in precipitation, canopy throughfall and soil solution along a 90-year chronosequence of Norway spruce forest planted on former arable land in southwest Sweden. *Forest Ecology and Management.* (In preparation).

Rosenqvist, L., Hansen, K., Van der Salm, C., Gundersen P., & Hansen S. (2006b). Stand water balance and water use in different ages of forests following land use change in Denmark and Sweden. *Journal of Hydrology.* (In preparation).

Rothe, A., Huber, C., Kreutzer, K., & Weis, W. (2002). Deposition and soil leaching in stands of Norway spruce and European Beech: Results from the Höglwald research in comparison with other European case studies. *Plant and Soil, 240,* 33-45.

Rowe, L. K., & Pearce, A. J. (1994) Hydrology and related changes after harvesting native forest catchments and establishing Pinus radiata plantations. 2. The native forest water balance and changes in streamflow after harvesting. *Hydrological Processes, 8,* 281-297.

SAS Institute (1993) SAS/STAT User's Guide Rel. 6.07. SAS Institute Inc., Cary, North Carolina, USA.

Thom, A. S. (1971). Momentum absorption by vegetation. *Quarternary Journal of the Royal Meteorological Society, 97,* 414-428.

Thornton, K. W., Holbrook, S. P., Stolte, K. L., & Landy, R. B. (2000). Effects of forest management pratices on Mid-Atlantic streams. *Environmental Monitoring and Assessment, 63,* 31-41.

Tomaszewski, T., Boyce, R. L., & Sievering, H. (2003). Canopy uptake of atmospheric nitrogen and new growth nitrogen requiremant at a Colorado subalpine forest. *Canadian Journal of Forest Research, 33,* 2221-2227.

Ulrich, B. (1983). Interactions of forest canopies with atmospheric constituents: SO_2, alkali and earth alkali cations and chloride. In B. Ulrich & J. Pankrath (Eds.), Effects of air pollutants in forest ecosystems, 33-45. Dordrecht, the Netherlands: Reidel Publ. Co.

Van Dam, J. C., Huygen, J., Wesseling, J. G., Feddes, R. A., Kabat, P., van Walsum, P. E. V., Groenendijk, P., & van Diepen, C. A. (1997). Simulation of water flow, solute transport and plant growth in the soil-water-atmosphere-plant environment. Theory of SWAP version 2.0. SC-DLO, Technical Document 45, Wageningen, The Netherlands.

Van der Maas, M. P., Van Breemen, N., & Van Langenvelde, I. (1991). Estimation of atmospheric deposition and canopy exchange in two douglas fir stands in the Netherlands. Internal publication, Department of soil science and geology, Agricultural University of Wageningen, The Netherlands.

Van der Salm, C., Denier van der Gon, H. A. C., Bleeker, A., & van dern Toon, A. (2005). The effect of afforestation on water recharge and nitrogen leaching in the Netherlands. *Forest Ecology and Management, 221,* 170-182.

Van der Voet, E., Klein, R. K., & Udo de Haes, H. K. (1996). Nitrogen pollution in the European Union – origins and proposed solutions. Environmental Conservation, 23, 120-132.

Vesterdal, L., Ritter, E., & Gundersen, P. (2002). Change in soil organic carbon following afforestation of former arable land. *Forest Ecology and Management, 169,* 137-143.

WHO. (1998). Guidelines for drinking-water quality, 2nd ed. Addendum to Vol. 2. Health criteria and other supporting information. Geneva, World Health Organization 1998. pp. 64-80.

WMO. (1992). Chemical analysis of precipitation for GAW: laboratory analytical methods and sample collections standards. World Meteorological Organization, Global Atmosphere Watch. Internal Document, WMO/TD-NO 550, 32 pp.

CHAPTER 5

MODELLING THE NITROGEN DEPOSITION TO AFFORESTED SYSTEMS

A. BLEEKER[1*] and W.P.A. VAN DEURSEN[2]

*[1] TNO, Environement and Geosciences, P.O. Box 342, 7300,Laan van Westerenk 501, AH Apeldoorn, the Netherlands, *Present address: Energy Centre of the Netherlands (ECN), P.O. Box, 1755 ZG Petten, the Netherlands, E-mail: a.bleeker@ecn.nl, [2] Carthago Consultancy, Oostzeedijk Beneden 23a, 3062VK Rotterdam, the Netherlands*

Abstract. Eutrophication largely results from deposition of atmospheric N. The emission of N mainly originates from agriculture (NH_3), traffic, power plants and industry (NO_x). The most important ammonia source is emission from animal manure. The extent of this emission depends on manure composition and meteorological conditions. After N is emitted (either as NH_3 or as NO_x) it will be transported over short or long distances. Deposition rates vary with the structure of the earth surface, and it is often assumed that the interceptive properties of vegetation, expressed as roughness length, are constant and the same for all wind speeds and all transferable quantities. Careful evaluation of individual components of the overall transfer resistance is crucial to an understanding of how and where N will be distributed within a particular forest canopy. In the context of local scale ammonia, like encountered in the AFFOREST project, NH_x will be mostly dry deposited to the forest. The Eutrend model was used for the calculation of N deposition, and it calculates concentrations and depositions as a function of surface characteristics. The model is able to describe both short and long-distance transport. The land use changes occurring in an afforestation process have different effects. The first effect is related to the removal of especially NH_3 emission, while the second is related to changing deposition characteristics when turning agricultural land into forest. On top of that, after afforestation the deposition of N will change due to the effect of growing forest. Growing forest will have an effect on the roughness length, which is an important factor determining the deposition velocity. Emission from a specific location is not only deposited at that same location but is also transported and deposited in surrounding areas. Because of such transport through the air, a spatial interaction between the emission source and the deposition receptor is introduced as a complicating factor. Further investigation was needed to evaluate the relevancy of spatial interaction for the AFFOREST-sDDS. After corrections, the AFFOREST-sDSS is realistically well able to describe the deposition situation after a certain amount of years after afforestation.

G.W. Heil et al. (eds.), Environmental Effects of Afforestation in North-Western Europe, 109–128.

1. INTRODUCTION

One of the most environmentally damaging air pollutants in north-western Europe is nitrogen (N). It is deposited at rates that vary with the structure of the earth surface. Eutrophication of ecosystems largely results from deposition of atmospheric N originating from liquid manure produced and applied in big quantities in the intensive farming systems in this part of Europe (e.g. Millennium Ecosystem Assessment 2005). It is often assumed that the interceptive properties of vegetation, expressed as roughness length, are constant and the same for all wind speeds and all transferable quantities. Careful evaluation of individual components of the overall transfer resistance is crucial to an understanding of how and where N will be distributed within a particular forest canopy.

Nitrogen deposition is an important input to afforested systems in north-western Europe, being of influence on both nitrate leaching and carbon sequestration in a direct or indirect way. In this chapter, the deposition of N is discussed with focus on the specific way in which N deposition is modelled within the AFFOREST-sDSS.

The next section will give a general description of the aspects involved in the source-receptor pathway with respect to N deposition. Hereafter, the general change in emission and deposition after afforestation is described. Lastly, an overview of some specific items that needed special attention when implementing N deposition in the AFFOREST-sDSS is given and further implementation issues are discussed.

2. NITROGEN DEPOSITION: FROM SOURCE TO RECEPTOR

In the context of afforestation in north-westen Europe, N deposited to the new forests is a combination of NH_x and NO_y components. Reduced N (NH_x) consists of ammonia (NH_3) and its reaction product ammonium (NH_4^+), while oxidised N (NO_y) consists of NO_x (NO_2 + NO) and its reaction product nitrate (NO_3^-). The following description of the source-receptor pathway is divided in two main parts: the emission of N and the transport and deposition of N.

2.1. Emission of nitrogen

The emission of N mainly originates from agriculture (NH_3), traffic, power plants and industry (NO_x). The most important ammonia source is emission from animal husbandry and manure. The extent of this emission depends on the composition of manure and meteorological conditions. The most important sources are emission during grazing, emission from stables, emission from manure storage and emission during manure spreading.

An important factor determining the overall ammonia emission is the total manure excretion, which is also a starting point for the calculation of ammonia emission. The N excretion is calculated as the difference between the N amount consumed (in the fodder) and the N immobilisation in animal products (e.g. milk, meat) for each animal species for an average situation during specific years.

While the ammonia emissions mainly originate from low level sources (small height), NO_x is emitted from sources at different heights. The NO_x emissions from traffic are emitted at a relatively small height (1-3 meters), while NO_x emissions from power plants and industrial sites are emitted at heights varying from 15-200 meters. This emission height causes the emissions to be transported over different distances and spread over areas differing in size. The small height sources will mainly contribute to areas at close range from the emission point, while the high sources also will contribute to larger distances from the source.

The NO_x emissions are mainly caused by processes in which fossil fuels are burnt. However, NO_x emissions can also be emitted from soils through denitrification and formed through conversion of NH_3 by OH in the atmosphere. Natural emissions of N comprise lightning and stratospheric destruction of N_2O.

2.2. Transport and deposition of nitrogen

After N is emitted it will be transported over short or long distances. The total deposition of N mainly consists of wet and dry deposition. A third form is the cloud or fog deposition, also called occult deposition. Cloud and fog water deposition is the process where cloud and fog water droplets are directly intercepted by the earth's surface. Because quantitative information on the contribution of this type of deposition in the different parts of Europe is scarce, the cloud or fog deposition is normally not taken into account in the description of the total deposition. However, especially for medium and high altitude areas, the contribution of cloud deposition to the total deposition can be in the range of 20-50% (Fowler et al. 1991; Bleeker et al. 2000). However, the study areas of AFFOREST are all situated in low altitude areas and occult deposition can be neglected.

2.2.1. Wet deposition

Wet deposition is the process by which atmospheric pollutants are delivered to the earth's surface by rain, hail or snow. It is defined as the natural process by which atmospheric pollutants are attached to and dissolved in cloud and precipitation droplets (or particles) and delivered to the earth's surface. The amount of compounds thus received per unit of surface area is defined as wet deposition. There are two main processes involved in wet deposition:

- In-cloud scavenging also called rain-out
- Below-cloud scavenging also called wash-out

In-cloud scavenging is the process in which cloud droplets are formed due to condensation of rather humid air (Hov et al. 1987). When the air contains particles (aerosols, dust), droplets will form more rapidly, because particles act as condensation nuclei.

Below-cloud scavenging is the process where uptake of gases and/or particles can occur during the downward transport of the droplet to the earth's surface. It is a very efficient removal mechanism for soluble gases (like NH_3) and aerosols with a diameter larger than 1 μm (Hicks et al. 1989).

2.2.2. Dry deposition

Dry deposition is the process where gases and particles are deposited directly from the atmosphere to vegetation, here forest. The process is governed by 1) the concentration in air, 2) turbulent transport processes in the boundary layer, 3) the chemical and physical nature of the depositing species and 4) the efficiency of the surface to capture or absorb gases and particles. The flux of a trace gas is given as:

$$F = V_d(z)c(z) \tag{1}$$

where:
$c(z)$ is the concentration at height z;
$V_d(z)$ is the dry deposition velocity at height z (Chamberlain 1966)

The parameterisation of the dry deposition velocity can be based on a description with a resistance analogy or Big Leaf Model (e.g. Thom 1975; Hicks et al. 1987; Fowler 1978; Erisman et al. 1994). In this resistance model, the most important deposition pathways by which the component is transported to and subsequently taken up at the surface are parameterised. The surface resistance is affected by meteorology, leaf area, stomatal physiology, soil and external leaf surface pH, and presence and chemistry of water drops and films. Especially, the state of the leaf and soil surface (i.e. the presence of water films and snow) is an important variable governing the deposition of soluble gases like NH_3. More information on the use of different resistances for calculating the deposition velocity can be found in Erisman et al. (1994).

In the context of local scale ammonia emission, like encountered in the AFFOREST region, dry deposition is the most important way in which NH_x will be deposited to the surface. It is known that NH_x is both a local and a long-distant pollutant. However, because of the very local character of the items discussed in this study, NH_x is mainly removed from the atmosphere in the form of dry deposition of NH_3 close to the source. At some distance from the source, it is removed in the form of wet deposition of NH_4^+, partly because NH_4^+ is a reaction product that is first formed only after some time (Asman & Van Jaarsveld 1990). Therefore, wet deposition of NH_x between the source and direct surroundings does not contribute much to the total deposition (dry + wet) of ammonia.

2.2.3. The Eutrend model

Due to the fact that it is almost impossible to compile spatial distributions of the deposition of N based on measurements only, models must be used for calculating the deposition (both wet and dry) on different scales. The Eutrend model used for the AFFOREST calculations (v 1.17) is a version of the Operational Priority Substances (OPS) model. OPS was developed by the National Institute of Public Health and Environment in the Netherlands (RIVM) for calculating transport and average concentrations and depositions of acidifying compounds (such as ammonia and N oxides) on a local to national scale (Asman & Van Jaarsveld 1992; Van Jaarsveld

1995). Emission, dispersion, advection, chemical conversion and wet and dry deposition were included in the model. Dry deposition is modelled using a resistance approach, in which the dry deposition velocity is constructed from resistances of both the atmosphere and the receiving surface. A general description of the modelled processes is given in Van Jaarsveld (1995).

Eutrend v1.17) calculates concentrations and depositions as a function of surface characteristics. The model is able to describe both short and long-distance transport. The advection of this model is based on meteorological data (6-hour time step, 1000 and 850 hPa pressure levels) obtained through the Netherlands Meteorological Institute (KNMI) from the European Centre for Medium Range Weather Forecasts (ECMWF) in Reading, England. The basic resolution of these data is 1° longitude x 0.5° latitude. Small scale processes such as dispersion, dry and wet deposition are described on the basis of surface observations of wind speed, cloud cover, temperature, humidity and precipitation. These surface observation data are obtained (mainly as 6-hourly values) from databases kept by ECMWF, the American National Centre for Atmospheric Research (NCAR) and Deutscher Wetterdienst (DWD). The spatial resolution of these small scale processes is limited by the (local) density of the meteorological stations.

Chemical reaction rates are independent of concentration levels, which means that only linear chemistry can be described. The model can be applied with a variable spatial resolution, using a fixed receptor grid or using a set of individual receptor points. In this case, each receptor point is characterised by its co-ordinates, land use class and roughness length. Some extensions were made to the original Eutrend model. The most important one is the introduction of local land use and surface roughness effects on dry deposition. These parameters now can be specified either in grid-form or as properties of the receptors.

3. CHANGING EMISSION/DEPOSITION PATTERNS DUE TO AFFORESTATION

The effects of afforestation on emission and deposition patterns can be divided in effects due to land use changes and effects due to growing trees. In Section 3.3 special attention will be paid to spatial interactions between source and receptor, while these spatial interactions are a complicating factor in implementing deposition processes in the AFFOREST-sDSS.

3.1. Effects due to land use changes

The land use changes occurring in a afforestation process have two different effects. The first effect is related to the removal of especially NH_3 emission, while the second is related to changing deposition characteristics when turning agricultural land into forests.

3.1.1. Removal of emission due to afforestation

When agricultural land is converted to forest a major change in local emissions will be the result. From the moment agricultural practices are terminated NH_3 emissions will rapidly decrease and eventually stop. The total mineral N pool in the soil will determine the period during which emission is still possible after ending the agricultural practices. However, the largest difference in NH_3 emission will occur during the first year. The removal of NH_3 emission will have an effect on the surrounding areas and therefore introduce spatial interaction into the AFFOREST-sDSS (section 3.3).

3.1.2. Changing deposition characteristics

When land use is changed from arable or grassland to forest N deposition will change. When calculating the dry deposition using the Eutrend model the dry deposition velocity is described by means of a resistance analogy in which:

$$v_d = (r_a + r_b + r_c)^{-1} \tag{2}$$

Where:
r_a is the aerodynamic resistance,
r_b is the boundary layer resistance and
r_c is the surface or canopy resistance.

These three resistances represent the three stages of transport from the air to uptake at the surface. The aerodynamic resistance (r_a) represents the resistance against turbulent transport of the component close to the surface (Garland 1978), the quasi-laminar sub layer resistance (r_b) accounts for the transport of the component by molecular diffusion through a laminar layer adjacent to the surface (Hicks et al. 1987), and the surface resistance (r_c) accounts for the uptake at the surface. Changing from agriculture to forest will have an effect on all these resistances. The changes in r_a and r_b are mainly related to changing roughness lengths, which will be discussed in the next section.
The effect of changing land use on the canopy resistance (r_c) is related to changing uptake processes at the surface. This will have an effect on the deposition velocity, independent of the corresponding roughness length. Calculations showed that the deposition velocity decreases when changing from agricultural land to forest, only taking the effect on r_c into account (Table 5.1). For the calculations, specific roughness lengths were used of 0.03 and 0.075 m for pasture and other arable land, respectively.

Table 5.1. *Changing deposition velocities (in m.s^{-1}) for N components for different land use, only taking the effect on r_c into account.*

Component	Former land use	Forested land use	dep. vel. (m.s^{-1})	ratio vd for/vd agr
NH_3	pasture		0.631	
		coniferous	0.533	0.845
		deciduous	0.541	0.857
	arable		0.760	
		coniferous	0.654	0.902
		deciduous	0.668	0.921
NH_4	pasture		0.059	
		coniferous	0.061	1.034
		deciduous	0.061	1.034
	arable		0.063	
		coniferous	0.065	1.032
		deciduous	0.065	1.032
NO_x	pasture		0.129	
		coniferous	0.073	0.566
		deciduous	0.107	0.829
	arable		0.135	
		coniferous	0.076	0.563
		deciduous	0.112	0.830
NO_3	pasture		0.425	
		coniferous	0.425	1.000
		deciduous	0.425	1.000
	arable		0.506	
		coniferous	0.506	1.000
		deciduous	0.506	1.000

3.2. Changing deposition patterns due to growing trees

After afforestation, the deposition of N will change due to the effect of growing trees. Growing trees will have an effect on the roughness length, which is an important factor determining the deposition velocity. A change in roughness length will result in a change of mainly r_a and r_b.

In literature, relations between tree height and roughness length are mentioned. A commonly known rule of thumb is

$$z_0 = 0.1 \cdot h \tag{3}$$

Where
h is tree height (in m)
z_0 is roughness length (in m)

Thom (1971) suggested the following relation:

$$z_0 = \lambda \cdot (h - d) \tag{4}$$

Where:
d is the displacement height (appr. 0.7 x h) and
λ is a parameter ought to be insensitive to vegetation height.

Thom gave a value for λ of 0.36, while Moore (1974) determined a value of 0.26 by analysing 105 published estimates of d, z_0 and h for several crops ranging from smooth grass to forest. When using 0.26 as a value for λ, z_0 will be 1.56 m at a tree height (h) of 20 m. In general, this approach gives reasonable results.

After establishing a relation between tree height and roughness length, the next step involves the relation between roughness length and deposition velocity. This relationship was established using Eutrend (Figure 5.1). As the trees are growing after afforestation, height and roughness length is increasing and along with this the deposition velocity is increased. The increase in deposition velocity is most important for NH_3 than for NO_x .

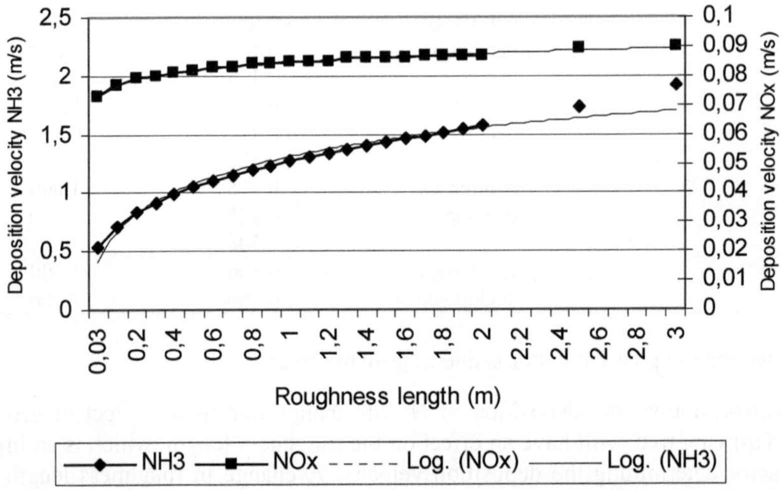

Figure 5.1. *Relation between roughness length (m) and deposition velocity (m.s^{-1}) for NH_3 and NO_x. The logarithmic trend line is shown for these two relations. Calculated using the Eutrend model.*

The respective trend lines for NH_3, NO_x and NO_3 can be represented by the following equations:

$$v_d = 0.3764 \cdot \ln(z_0) + 0.4088 \quad \text{for } NH_3\ (R^2 = 0.96) \qquad (5)$$

$$v_d = 0.0046 \cdot \ln(z_0) + 0.0736 \quad \text{for } NO_x\ (R^2 = 0.99) \qquad (6)$$

$$v_d = 0.2804 \cdot \ln(z_0) + 0.3185 \quad \text{for } NO_x\ (R^2 = 0.97) \qquad (7)$$

For NH_4 a distinction has to be made for roughness lengths smaller than 0.5 m and for roughness lengths larger or equal than 0.5 m, due to the underlying processes included in the dry deposition module of the Eutrend model (Ruijgrok et al. 1994). For these two types the corresponding equations are:

$$v_d = 0.0087 \cdot \ln(z_0) + 0.061 \quad \text{for } NH_4 \text{ where z0} < 0.5 \text{ m } (R^2 = 0.99) \tag{8}$$

$$v_d = 0.0216 \cdot (z_0) + 0.1803 \quad \text{for } NH_4 \text{ where z0} \geq 0.5 \text{ m } (R^2 = 0.99) \tag{9}$$

3.3. Spatial interaction between source and receptor

Emission from a specific location can be deposited at that same location but is can also be transported through the air and deposited to surrounding areas. A spatial interaction between the emission source and the deposition receptor is introduced through this transport as a complicating factor in the modelling of N deposition. Spatial interaction will influence the grid based procedures within the AFFOREST-sDSS and further investigation was needed to evaluate the relevancy of spatial interaction for the AFFOREST-sDDS. While the AFFOREST-sDSS is considering a change from agricultural land to forest, the focus will be mainly on changing NH_3 emissions.

The OPS model was used to determine the spatial relationships in the N deposition calculations. Calculations were performed for NH_3 in meteorological regions in the Netherlands. For a hypothetical area of 25 x 25 km, both dry and wet deposition, of NH_3 was calculated for grid cells with a size of 5 x 5 km. The emission of NH_3 originates from a source in the central grid cell and amounts to 0.5 g s^{-1} (i.e. approximately 15750 kg yr^{-1}) for the total grid cell area.

Information is needed on the specific land use characteristics of the considered area in order to calculate the deposition. The calculations were performed for grassland with a roughness length of 0.05 m. In general, the total percentage of NH_3 being deposited within the study area of 25 x 25 km was on average 20% of the emitted NH_3. Consequently, 80% of the emitted ammonia was deposited outside the area and thus will contribute to the deposition of ammonia in areas at larger distances from the source.

A distinction was made between dry, wet and total NH_3 deposition. About 90% of the deposited NH_3 will be due to dry deposition. This high contribution of dry deposition in the total deposition of ammonia will especially occur on short distances from the source. At larger distances the contribution of wet deposition will increase (Figure 5.2). The total deposition is dominated by dry deposition up to a distance of about 10 km. At larger distances the contribution of wet deposition is gradually increasing.

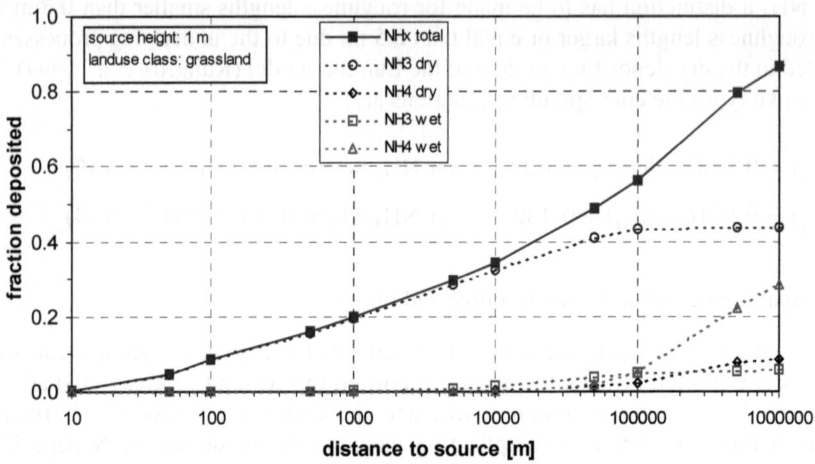

Figure 5.2. *Cumulative deposition of N compounds as a function of the downwind distance from a point source calculated using the OPS model.*

At a distance of 100 km from the source, approximately 60% of the emitted NH_3 was deposited, while at distances of about 1000 km, 10% of the emitted NH_3 is still not deposited. The contribution of emitted NH_3 to the deposition was largest in the emitting grid cell and it decreased quickly to less than 1% in the surrounding grid cells, however, neglecting these contributions would lead to a significant underestimation of the overall N deposition. This will especially be the case in areas with high NH_3 emissions. The effect of the spatial interactions on the deposition of NH_3 is shown in Table 5.2 for the nine grid cells in the centre of the area.

Table 5.2. *Calculated deposition of NH_3 (in kg yr^{-1}) in the nine central grid cells when excluding (left) and including (right) spatial interactions.*

Spatial interaction excluded		
2221	2221	2221
2221	2221	2221
2221	2221	2221

Spatial interaction included		
2426	2595	2592
2495	2781	2826
2415	2715	2805

The influence of spatial interaction on the deposition per individual grid cell varied considerably (Table 5.2), which is mainly determined by the location of a grid cell in reference to the prevailing wind direction. On a small scale of 15 x 15 km, the inclusion of spatial interaction in the calculation will lead to an NH_3 deposition, which is 18% higher compared to neglecting this interaction. In high emission areas, this will especially lead to a large underestimation of the overall NH_3 deposition.

Although the effect of spatial interaction on the deposition was only tested for NH_3, its effect is estimated to be even higher for the deposition of NO_x, since these oxidised compounds are transported over larger distances.

4. NITROGEN DEPOSITION IN THE AFFOREST-SDSS

Based on the information from the previous sections, a methodology was developed by which the effect of afforestation on N deposition can be simulated in the context of the AFFOREST-sDSS. Calculations made were compared with measurements and the uncertainties described.

4.1. Methodology of deposition calculation

The overall methodology for calculating the N deposition consists of the steps shown in Figure 5.3. These different steps will be further elaborated in this section.

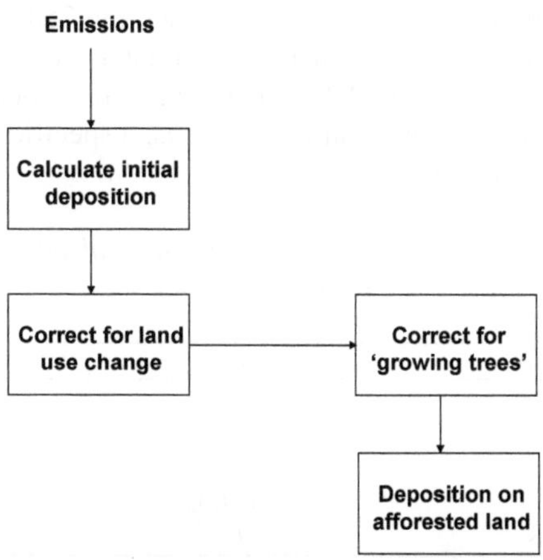

Figure 5.3. *The different steps in the procedure for calculating N deposition after afforestation.*

4.1.1. Emissions

As input to the deposition calculation procedure emissions for both NH_3 and NO_x were used. For the Belgian region Flanders, the Netherlands and Denmark, emissions from national emission databases were available at different resolutions. These emissions were redistributed over a 1x1 km grid over the respective regions on the basis of land use information. NH_3 emissions were assigned to agricultural land, while NO_x emissions were assigned to urban areas and roads. For Sweden, no specific emission data were available. Therefore, emission data from the EMEP emission database were used at a resolution of 50x50 km. These emission data were redistributed over 1x1 km grid cells using specific land use information. Besides emission data for the individual countries within the AFFOREST region also emission data for surrounding countries outside this region were required. Available emission data for NH_3 and NO_x from the EMEP database were used at a resolution of 50x50 km.

4.1.2. Calculating initial deposition

The initial deposition of N on the AFFOREST region was calculated by means of the Eutrend model and the above mentioned emission data. Information on land use and roughness length are other important input to the model. Specific roughness length and land use maps were constructed on a 1x1 km resolution. The maps were based on Corine land use information with a 250x250 m resolution (Flanders, the Netherlands and Denmark) and on Swedish land use information with a resolution of 1x1 km. When constructing the new land use database, the original Corine land use classes were reclassified. This was done by using ArcView. The result of this reclassification is shown in Color plate 1 for the AFFOREST region (excl. Sweden).

After reclassification of the initial land use data, the 250x250 m resolution data were aggregated to 1x1 km based on dominant land use by means of ArcView (Color plate 2). Based on the 250x250 m land use map a roughness map was constructed. For each of the individual land use classes, a specific roughness length was used (Table 5.3) (Color plate 3).

Table 5.3. *Land use types used in the Eutrend calculations along with assigned values for roughness length.*

Land use type	Roughness length (m)
Pasture	0.03
Arable land	0.15
Forest	1.8
Water	0.0002
Urban areas	2.0

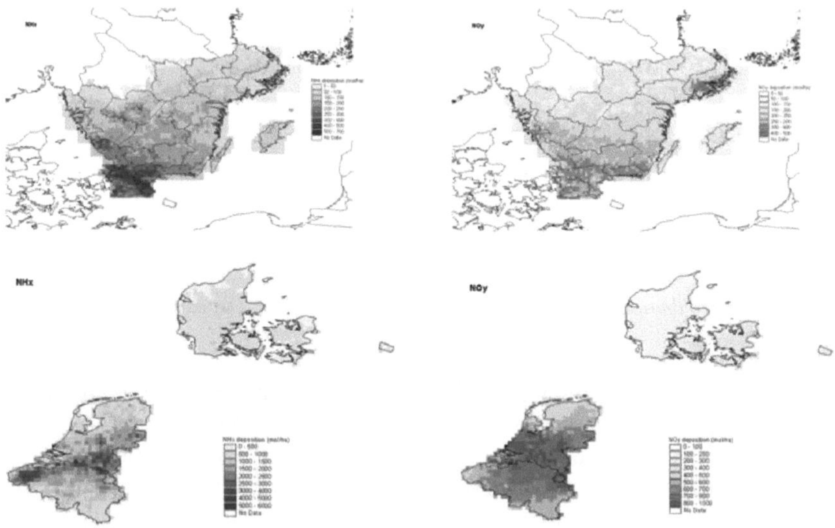

Figure 5.4. *Calculated deposition for NH_x (left) and NO_y (right) for the AFFOREST region. The results for Belgium and the Netherlands, Denmark and Sweden are shown separately to get a better overview. Note that Sweden has a different legend than the other areas.*

Using emission data, land use data and roughness length maps as an input, the Eutrend calculations were performed. The overall results of these calculations for NH_x and NO_y are shown in Figure 5.4. Apparently, the deposition was much higher for Belgium and the Netherlands than for Denmark and Sweden.

4.1.3. Correction for land use change

As described earlier, the correction for land use changes consisted of two different corrections. One correction was due to removal of emissions and the other correction was due to changing deposition characteristics because of growing forest. Therefore, the initial deposition already calculated had to be corrected for the halt in emission after change in land use. A specific correction field was calculated showing how one unit of emission was distributed over the surrounding area. The deposition of NH_x as a percentage of emitted NH_3 (1000 kg NH_3 yr^{-1}) in the grid cell from which the emission originates received about 9% of the emitted ammonia in the form of NH_x deposition (Figure 5.5).

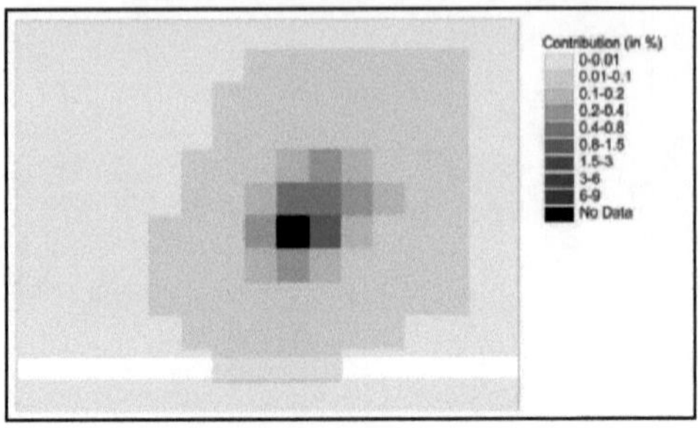

Figure 5.5. *Distribution of the emitted NH_3 to its surroundings. Deposition of NH_x as a percentage of the emitted NH_3.*

Including spatial interaction in the AFFOREST-sDSS was rather complicated because of the predefined database structure behind the sDSS. Therefore, a simplification of this spatial interaction was investigated for implementation in the AFFOREST-sDSS. It was assumed that three situations can be distinguished. These are i) removing emission only affects the emitting grid cell (no interaction), ii) removing emission has an effect on all the surrounding grid cells (complete spatial interaction) and iii) removing emission has an effect on e.g. surrounding grid cells within 5km (limited spatial interaction).

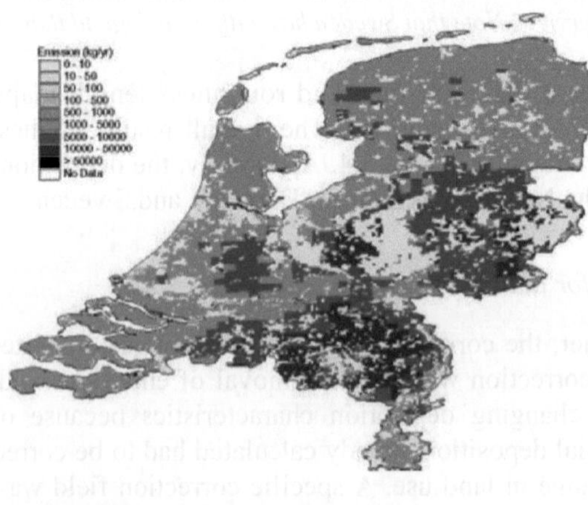

Figure 5.6. *Emission of NH_3 in the Netherlands.*

When individual cells are considered in the AFFOREST scenarios spatial interaction does not have to be taken into account. Only the contribution of the emission to that specific grid cell has to be considered. However, when grid cells in the surrounding of that specific emitting cell were also evaluated, spatial interaction played a role. On beforehand, it was not clear which situation would occur. It all depends on the question the end user of the AFFOREST-sDSS wants to have answered by the system. The different aspects involved in this correction procedure are show in

Figures 5.6 to 5.9, taking the Netherlands as an example. The first map shows the emission of NH_3, while the two following maps show the minimum (no interaction) and the maximum (complete spatial interaction) correction procedure. Figure 5.8 shows the initial total N deposition, while Figure 5.9 shows two maps with

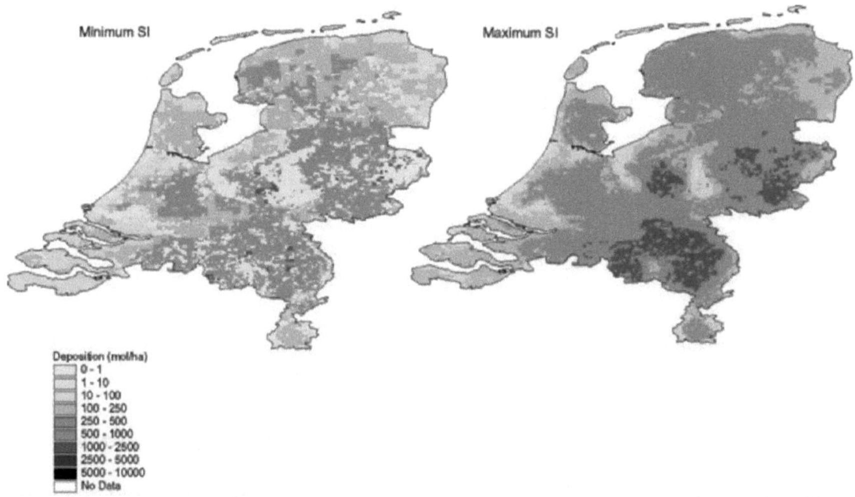

Figure 5.7. *Deposition to individual grid cells according to the minimum (no interaction) and the maximum (complete spatial interaction) scenario.*

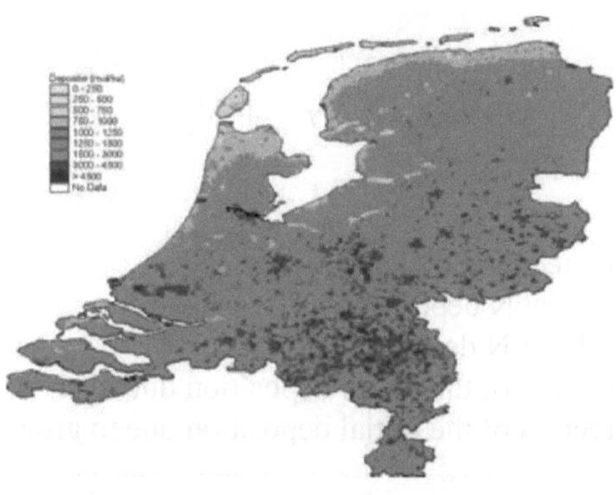

Figure 5.8. *Initial total N deposition in the Netherlands.*

the corrected N deposition according to the minimum and maximum scenario. Since it is not possible to know on forehand which scenario to choose for answering specific questions, it was decided to average the minimum and maximum scenarios as an estimate for the deposition situation after afforestation. For the different countries within the AFFOREST region these corrections were performed, both according to the minimum and the maximum scenario.

The information on the initial deposition, together with the minimum and maximum corrected deposition is inserted in the AFFOREST-sDSS database for use in the metamodel METAFORE.

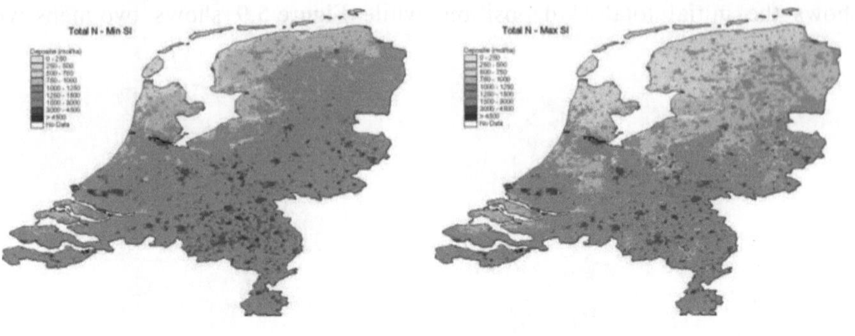

Figure 5.9. *Corrected initial N deposition in the Netherlands according to the minimum (left) and maximum (right) scenarios.*

4.1.4. Correction for growing forest and changing deposition characteristics

For the correction of the initial deposition due to the growing forest and changes in deposition characteristics, four different situations can be distinguished. These situations depend on the initial land use (pasture or arable) and the tree species used for afforestation (coniferous or deciduous). The overall correction equation used in this procedure is:

$$N_{td,cor} = N_{dd,ini} \cdot corr_{rc} \cdot corr_{z0} + N_{wd,ini} \tag{10}$$

where:

$N_{td,cor}$	corrected total N deposition
$N_{dd,ini}$	initial dry N deposition
$N_{wd,ini}$	initial wet N deposition
$corr_{rc}$	correction of the initial deposition due to changing land use
$corr_{z0}$	correction of the initial deposition due to growing forest

The general assumption is that the total N deposition is formed by 60% dry and 40% wet deposition and that only the dry deposition part changes due to changing land use and growing forest. Therefore:

$$N_{td,cor} = N_{td,ini} \cdot 0.6 \cdot corr_{rc} \cdot corr_{z0} + N_{td,ini} \cdot 0.4 \tag{11}$$

For the four situations different correction factors can be distinguished (Table 5.4).

Table 5.4. *Correction factors for different initial situations (pasture or arable land) when planting either deciduous or coniferous tree species.h is the tree height in m.*

Before-after situation		
pasture – coniferous	$corr_{rc} = 0.907$	$corr_{z0} = (0.6789\ln(h) + 0.8689)$
pasture - deciduous	$corr_{rc} = 0.9142$	$corr_{z0} = (0.6789\ln(h) + 0.8689)$
arable - coniferous	$corr_{rc} = 0.9412$	$corr_{z0} = (0.5297\ln(h) + 0.678)$
arable - deciduous	$corr_{rc} = 0.9526$	$corr_{z0} = (0.5297\ln(h) + 0.678)$

These correction factors are an average for the dry deposition of NH_x and NO_y. In general, the distribution over for NH_x and NO_y is 60%/40% for agricultural areas. For very intensive agricultural areas this distribution can change to 80%/20%, while in extensive agricultural areas in the neighbourhood of urban areas this might be 40%/60%. However, these differences will have relatively little effect on the average correction factors. Figure 5.10 shows the result for the total correction of the initial deposition as a function of tree height for the situation where pasture is converted to coniferous trees.

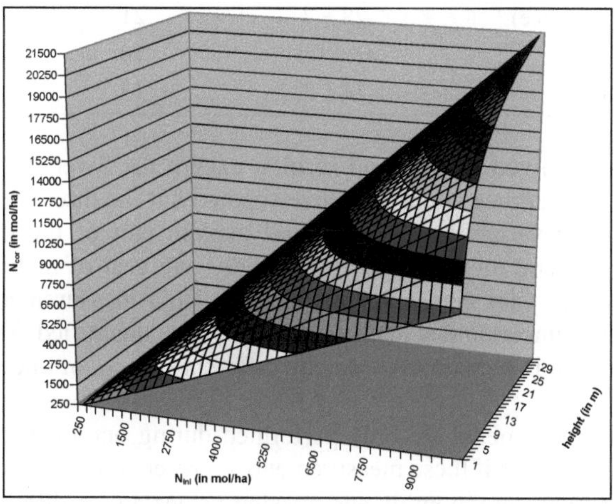

Figure 5.10. *Example of the effect the correction procedure has on the initial deposition as a function of tree height when pasture is changed to coniferous forest.*

4.1.5. Deposition to afforested land

The total procedure for calculating the deposition on afforested land can now be performed (Figure 5.3). The initial deposition was calculated for the individual AFFOREST region countries, based on country specific emissions. When certain grid cells are to be afforested, the existing deposition is corrected according to the procedure described. Hereafter, the remaining deposition is corrected due to the effect of enhanced deposition because of growing forest. The resulting deposition after these two corrections is used within the AFFOREST-sDSS to describe the deposition situation after a certain amount of years after afforestation.

4.2. Comparison with measurements and other calculation results

The calculated N deposition was compared with measured deposition on the AFFOREST chronosequence sites in Sweden, Denmark and the Netherlands. The experimental data were mapped on top of the modelled total N deposition. The data were modelled using both the Eutrend model and the EMEP model, which is used in overall European deposition studies. The comparison between measured and modelled bulk N deposition is shown in Table 5.5.

Table 5.5. *Comparison between measured and modelled N deposition (kg ha^{-1}).*

Location	N-deposition (kg N ha^{-1})		
	Measured	Eutrend	EMEP
Sweden (spruce)	18 ± 4	6	12
the Netherlands (oak)	16 ± 2	16	22
the Netherlands (spruce)	28 ± 7	21	22
Denmark 1 (spruce)	15 ± 0.3	10	15
Denmark 2 (oak)	11 ± 3	13	12

The comparison showed fairly good results, except for Sweden, which was out of range. The Dutch spruce site was underestimated but the three bulk samplers had a high site specific variability, which could explain part of the differences. The EMEP and the Eutrend models were run for only one year of meteorological data, which in reality will vary between years as well. Both measurements and calculations are subject to uncertainties that can not always be quantified. The overall comparison is influenced in a qualitative way by different sources of uncertainty:

- Field measurements are only performed during one or two years and it is not always clear if these measurements represent the average situation
- measurements are monthly and missing values due to different reasons will have a large influence on the total deposition
- deposition models have an inherent uncertainty in the way processes are modelled and in parameters used as well as input required
- emissions used have large uncertainties, especially at high resolutions.

- NH_3 emissions show a high spatial and temporal variation, which will be the cause of large uncertainties when the deposition is evaluated at specific locations
- use of different emission databases (as mentioned in previous sections) will cause additional uncertainties, because of difference in calculation approaches, input data, resolution differences, etc.
- differences in meteorological conditions from year to year can cause deposition difference of up to 30%

These topics can add up to large uncertainties. However, within the overall AFFOREST-sDSS these uncertainties have to be judged in the context of the other components of the system. Emphasis should be more on the overall trend and plausibility of the modelling results than on the absolute results for individual locations (Chapter 10).

5. DISCUSSION AND CONCLUSIONS

Since N deposition is one of the most environmentally damaging air pollutants in north-western Europe, atmospheric N is considered an important input to the AFFOREST system. The Eutrend model is used for the AFFOREST calculations Emission, dispersion, advection, chemical conversion and wet and dry deposition are included in the model.

Because of the transport of N through the air a spatial interaction between the emission source and the deposition receptor is introduced since N emissions from a specific location is transported and deposited in surrounding areas. We investigated The relevance of spatial interaction between grid cells for the AFFOREST-sDDS was investigated.

The calculated N deposition was compared with measured deposition on the AFFOREST chronosequence sites in Sweden, Denmark and the Netherlands. The comparison showed fairly good results (except for Sweden) but both measurements and calculations are subject to uncertainties that are not easily quantified.

For European scale afforestation projects there is a need for detailed information on the effect of the receiving surface on the amount of atmospheric deposition. Deposition models have an inherent uncertainty in the way processes are modelled and in parameters used as well as input required. Despite the uncertainties, the developed methodology provides a good means for producing this kind of information on a 250 x 250 m scale. This deposition modelling approach, as part of the scientific development process in the AFFOREST project, has been fairly comprehensively validated and calibrated. We conclude that the results are satisfactory to include in the METAFORE metamodel (Chapter 9).

6. REFERENCES

Asman, W. A. H., & Van Jaarsveld, J. A. (1990). A variable-resolution statistical transport model applied for ammonia and ammonium. RIVM Report no. 228471007, National Institute of Public Health and Environment, Bilthoven, The Netherlands

Asman, W. A. H., & Van Jaarsveld, J. A. (1992). A variable-resolution transport model applied for NH_x for Europe. *Atmospheric Environment, 26A,* 445-464.

Bleeker, A., Draaijers, G. P. J., Klap, J., & Van Jaarsveld, J. A. (2000). Deposition of acidifying components and base cations in Germany in the period 1987-1995. RIVM Report no. 722108027, National Institute of Public Health and Environment, Bilthoven, The Netherlands.

Chamberlain, A.C. (1966). Transport of gases to and from grass and grass-like surfaces. *Proceedings of the Royal Society Service,* A290, 236-265.

Erisman, J. W., Van Pul, A., & Wyers, P. (1994). Parameterization of dry deposition mechanisms for the quantification of atmospheric input to ecosystems. *Atmospheric Environment, 28,* 2595-2607.

Fowler, D. (1978). Dry deposition of SO_2 on agricultural crops. *Atmospheric Environment, 12,* 369-373.

Fowler, D., Duyzer, J. H., & Baldocchi, D. D. (1991). Inputs of trace gases, particles and cloud droplets to terrestrial surfaces. *Proceedings Royal Society Edinburgh, 97B,* 35-59.

Garland, J. A. (1978). Dry and wet removal of sulfur from the atmosphere. *Atmospheric Environment, 12,* 349.

Hicks, B. B., Baldocchi, D. D., Meyers, T. P., Hosker Jr., R. P., & Matt, D. R. (1987). A preliminary multiple resistance routine for deriving dry deposition velocities from measured quantities. *Water Air Soil Pollution, 36,* 311-330.

Hicks, B. B., Matt, D. R., & McMillan, R. T. (1989). A micrometeorological investigation of surface exchange of O3, SO2 and NO2: a case study. *Boundary-Layer Meteorology, 47,* 321-336.

Hov, O., Allegrini, I., Beilke, S., Cox, R. A., Eliassen, A., Elshout, A. J., Gravenhorst, G., Penkett, S. A., & Stern, R. (1987). Evaluation of atmospheric processes leading to acid deposition in Europe. Report 10, EUR 11441, CEC, Brussels.

Millennium Ecosystem Assessment, 2005. Ecosystems and Human Well-being: Biodiversity Synthesis. World Resources Institute, Washington, DC.

Thom, A.S. (1975). Momentum, mass and heat exchange of plant communities. In: J.L. Monteith (editor), Vegetation and Atmosphere. London: Academic Press, pp. 58-109.

Van Jaarsveld, J. A. (1995). Modelling the long-term atmospheric behaviour of pollutants on various spatial scales. Ph.D Thesis, University of Utrecht, The Netherlands.

CHAPTER 6

EFFECTS OF LIGHT AND N AVAILABILITY IN FORESTS ON PLANT SPECIES DIVERSITY IN THE FIELD LAYER: A PLANT ECOLOGICAL AND MODELLING APPROACH

M. ELEMANS[1,2] and G.W. HEIL[2]

[1]Karolinska Institute, Microbiology and Tumorbiology Center (MTC), Nobelsväg 16, Solna, Sweden, [2]Faculty of Science, Institute of Environmental Biology, Sorbonnelaan 16, NL 3584 CA Utrecht, the Netherlands, E-mail: g.w.heil@bio.uu.nl

Abstract. In many European countries agricultural areas are currently being converted to forest. Both spatial arrangement of new forests and habitat quality play a role in the development of the field layer in newly planted forests on former agricultural land. One important goal of afforestation is the development of a natural ecosystem with a valuable field layer.

In the course of succession, open-canopy species are replaced by climax species with a denser canopy, and the light availability on the forest floor is expected to decrease. The differences in the environmental conditions between ancient forests and afforested arable land have not been studied in a successional context. Differences between ancient forests and recent forests on former arable land can be caused by many different factors. Several studies have addressed the vegetational change in the field layer under the influence of nitrogen (N) deposition. The plant ecological approach followed in this study focused on the characteristics of individual species that determine their ecological behavior in relation to light and N availability. A greenhouse experiment was carried out in order to study the interactive effect of light and N on the performance of species of ancient and recent forests.

In addition, an field layer model has been developed of which the main growth processes are based on two compartments: a shoot compartment responsible for the acquisition of carbon (C) through photosynthesis and a root compartment involved in the uptake of N. The relative growth rate of the shoot compared to the growth rate of the root depends on the ratio between the N and C concentration in the pool.

This study showed the importance of the interactive effect of light and N on plant performance. Reduction of the light availability in the young forests will reduce the growth of fast-growing and competitive species while the growth of forest species, given that they already occur in the forest, will hardly be affected. This result stresses the importance of design (e.g. choice of tree species and density of the trees) and management of the tree layer, in controlling the development of the field layer vegetation.

G.W. Heil et al. (eds.), Environmental Effects of Afforestation in North-Western Europe, 129–148.

1. INTRODUCTION

In many European countries, agricultural areas are currently being converted to forest. Studies on the development of the forest field layer of recently planted forests and the differences with so-called ancient forest systems are scarce. Due to this scarcity of information in combination with the fact that the circumstances of the past and present situation (former land use, forest age, vicinity of seed sources, composition of the tree layer, management) are very variable, the processes affecting the development of the field layer of forests planted on former agricultural land are not yet fully understood (Honnay et al. 1999). Moreover, the forests were mostly planted on abandoned land before the significant increase in application of artificial fertilizers in agricultural practice. Still, some general trends and patterns can be derived from past research.

Practically all studies report considerable differences between the species composition of the field layer of recent and ancient forests (Petersen & Philipp 2001; Dzwonko 2001; Petersen 1994; Singleton et al. 2001). In general, species typical of the field layer of ancient forests are missing in the field layer of the recent forests, where a larger share of fast-growing, nitrophylous species is present. However, no consistent differences in species richness between ancient and recent forests were found. The difficulty for (ancient) forest species to colonize forests has been attributed to limited dispersal abilities of those species, low amounts of diaspore production and recruitment problems (Graae et al. 2003; Hermy et al. 1999; Bossuyt et al. 1999). Many forest species produce only a few large seeds and the dispersal mechanism is specialized for short-distance dispersal (Graae & Sunde 2000; Bierzychudek 1982; Hermy et al. 1999; Brunet & von Oheimb 1998a). Colonization success increases with the proximity to ancient woodland (Brunet et al. 2000; Brunet & von Oheimb 1998a; Grashof-Bokdam & Geertsema 1998).

Another factor that might play a role in the development of the forest field layer is habitat quality (Honnay et al. 1999; Peterken & Game 1984). In contrast to natural forest systems, the conditions in forests planted on former agricultural land are expected to differ largely in both light and nitrogen (N) availability. These differences will exert an influence on growth processes and the composition of the field layer.

Several studies have addressed the vegetational change in the field layer under the influence of N deposition. Most studies report an increase in the number and cover of species that are generally considered to be N indicators in the forest floor vegetation (Brunet et al. 1998; Diekmann & Dupre 1997; Diekmann et al. 1999; Lameire et al. 2000; Thimonier et al. 1992; Thimonier et al. 1994; Van Dobben et al. 1999). Species ecological preference is determined on the basis of Ellenberg's indicator values for N availability or based on literature citations on the N demand of the species. Only one study found no evidence for an effect of changing soil N levels on the vegetation composition (Kirby & Thomas 2000). Diekmann & Dupre (1997) state that eutrophication does not necessarily cause successional changes at all sites. On calcareous and N-rich, mesic soil, N is hardly a limiting factor, meaning that the relative increase in N may be negligible. This implies that the effects of atmospheric N deposition are more prominent on N-poor soils where the increase in N is comparatively large. Concerning species richness or diversity the conclusions of different studies are not consistent. Some report an increase in diversity (Thimonier et al. 1994; Thimonier et al. 1992), others a decrease (Lameire et al. 2000) while in some studies no effect of N deposition on species diversity is found (Liu & Brakenhelm 1996). Forest species are more common at low light intensity. However, several studies showed that forest species are dependent on periods of higher light availability for their survival and expansion. In general, an increased light availability seems to have a positive effect on species diversity (Halpern & Spies 1995; Willems & Boessenkool 1999; Brewer 1980). However, species diversity might not be the most appropriate measure to describe the value of the system. Halpern & Spies (1995) found that the increase in biodiversity after clear-cut logging resulted from a temporary decline in the number of forest species and a rapid accumulation of native, ruderal, non-forest species. Other studies report of a positive effect on species richness of smaller-scale recurrent periods of improved light intensity on species diversity (Brewer 1980; Willems & Boessenkool 1999). Fredericksen et al. (1999) found no effect on diversity and reported only of an effect on species composition and cover. From these studies, it can be concluded that fast-growing, weedy species tend to profit more from an improved light availability than do typical forest species, especially when nutrient conditions are favorable.

It can be concluded that both spatial arrangement of the new forests and habitat quality play a role in the development of the field layer in newly planted forests on former agricultural land. Seed introduction studies (Ehrlen & Eriksson 2000), which control for forest plant dispersal limitation, and trait based studies (Verheyen et al. 2003) suggest that dispersal limitation may be more important. Of course, the more recently and intensively used as arable land, as is the case for the areas targeted by the EU-project AFFOREST, the more habitat quality will matter. Also, when sowing practice is applied habitat quality is important.

In order to predict in which areas habitat quality is best suitable to support a successful development of a valuable field layer after afforestation, a better understanding of the growth processes, in relation to their environment, of both target species and species commonly found in young or disturbed forests, is necessary. In line with the focus of the EU-project AFFOREST, we studied the effect of light and N availability on the performance of forest field layer species and on species composition. Two important traits were studied: (Specific Leaf Area (SLA) and shoot-to-root ratio (S:R)) in a greenhouse experiment on the interactive effect of light and N on the performance of species of ancient and recent forest. Different species have specialized for survival in different environments, which generated selection on SLA and S:R. The results are used in a modeling approach to predict the performance of different species under different light and N conditions are described. Through this combination of existing knowledge, growth experiments and model simulations the effect of N availability at different light availability levels were examined on the performance of different species.

2. GREENHOUSE EXPERIMENT

The interactive effect of light and nutrients on the performance of species of ancient and recent forests was studied in a greenhouse experiment during the summer of 2001 (Figure 6.1).

2.1. Methods

The light levels represented an open forest canopy (8% of full radiation) and a very dense forest canopy (2%), the 60% light treatment is added as a control (Table 6.1). Light levels were accomplished using a plastic green film in combination with different intensities of shade cloth. The nutrient levels represented the naturally somewhat richer forest soils (30 kg N ha^{-1} y^{-1}) and the nutrient availability of agricultural land (300 kg N ha^{-1} y^{-1}). The nutrient levels were established using different levels of slow-release fertilizer (Osmocote Plus 10% 5-6 m, Grace Sierra Int. NL), providing a continued availability of nutrients during the experiment.

Figure 6.1. *The experimental design in- and outside the shadow cages.*

Table 6.1. *Overview of light-nutrient treatments in the greenhouse experiment.*

	Light		
	2%	8%	60%
High nutrient (h)	2h	8h	60h
Low nutrient (l)	2l	8l	60l

Four different species, two of which capable of growing in deep shade and the other two more commonly occurring under less shaded conditions, were grown under different combinations of light and nutrient availability. The two shade-species used were *Circaea lutetiana* and *Mercurialis perennis*. *Circaea* is a woodland pseudo-annual, a perennial plant species which behaves as vegetatively propagating annuals, that is found in the field layer of deciduous temperate woodlands of Europe, south western Asia and North America (Verburg 1998). Although it is often found on lighter spots on the forest floor, it can survive in deeper shade than most summer-flowering forest species (Weeda et al. 1987). *Mercurialis* is a spring-flowering perennial that remains in the forest field layer for a large part of the year. It is common in most of Europe except in the north-eastern part. In deep shade it can, together with *Hedera helix*, dominate the field layer vegetation (Weeda et al. 1987).

The two species of less-shaded conditions were *Aegopodium podagraria* and *Impatiens parviflora*. The perennial *Aegopodium*, belonging to the *Apiaceae*, is one of the most common plant species. It can be found on lightly shaded places and can come into prominence in places subject to increased nutrient availability (Weeda et al. 1987). *Impatiens* was the only annual species in the selection of species used in the experiment. At present the species can be found in a large part of Europe, although it originally is not indigenous (Weeda et al. 1987). It is a relatively shade-tolerant plant that is responsive to the addition of nutrients (Peace & Grubb 1982). Figure 6.2 represents the results of the experiment and the results of the statistical analysis are given in Table 6.3.

The plants were grown under these conditions for 15 weeks in total. For each species and each light and nutrient treatment the average Specific Leaf Area (SLA), and the Shoot-to-Root ratio (S:R) were determined at the end of the experiment as well as the responsiveness of these plant traits to light and nutrient supply. From these data both the effect of light and nutrient availability could be derived as well as differences between species in traits values and plasticity.

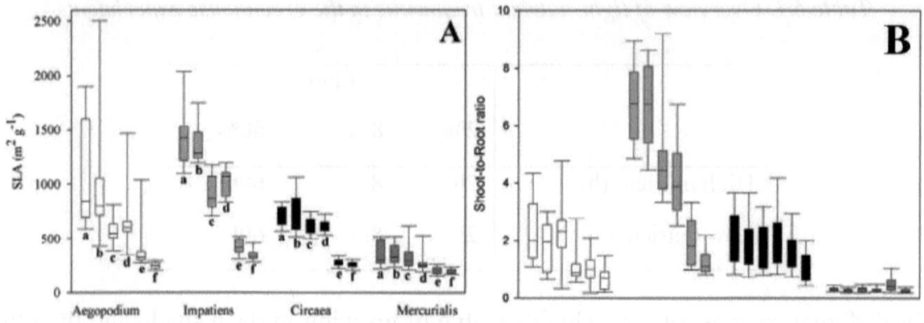

Figure 6.2. *Average values of (A) SLA and (B) S:R for the four species used in the experiment grown at a) high nutrient treatment and b) low nutrient treatment at 2% light, c) high nutrient treatment and d) low nutrient treatment at 8% light, and e) high nutrient treatment and f) low nutrient treatment at 60% light. The error bars represent the standard deviation (data taken from Elemans 2005).*

2.2. Results

A clear effect of light availability on SLA is found in all four species (Figure 6.2 A). All species showed an increase in SLA at a decrease in light availability. This is consistent with the expectations and with data from other studies (Shipley 2000).The effect of light on the SLA is found for each species and it was significantly different for each species (data not shown). The same results were found for the effect of light on the leaf mass fraction (LMF) of the different species, although the effect of light on *Mercurialis* appears to differ more than the effect on the other three species (data not shown). The plastic reaction to light was more pronounced for *Aegopodium* and *Impatiens*, the species from less shaded conditions, than for the two forest species. *Mercurialis* was least plastic in SLA, as is also illustrated by the plasticity measure in Table 6.3. Apart from a difference in plasticity, the two shade species also differed in absolute SLA from the two species adapted to less shaded conditions. Interestingly, although the plastic reaction to low light is a higher SLA, the species adapted to low light conditions maintained a lower SLA in all treatments, but especially in the low light treatments, than the species characteristic for less shaded conditions. Although the SLA was expected to increase with increasing nutrient supply (Knops & Reinhart 2000), this effect was not found (Table 6.2). This may be due to the relatively high nutrient levels as compared to naturally nutrient-poor systems.

Table 6.2. *Univariate ANOVA of the effect of light intensity (Li) and nutrient supply (Nu) on the allocation parameters of four forest species (Sp); df= degrees of freedom , MS= Mean Square, F= ratio between group variation divided by within group variation.*

Effect	df	MS	F	
SLA[a]				
Li	2	5.918	519.912	**
Nu	1	0.056	4.963	*
Sp	3	3.608	317.005	**
Li x Nu	2	0.058	5.166	**
Li x Sp	6	0.183	16.096	**
Nu x Sp	3	0.007	0.677	ns
Li x Nu x Sp	6	0.024	2.151	ns
LMF[b]				
Li	2	0.409	52.863	**
Nu	1	0.211	27.307	**
Sp	3	0.625	80.759	**
Li x Nu	2	0.023	3.074	*
Li x Sp	6	0.155	20.082	**
Nu x Sp	3	0.007	0.920	ns
Li x Nu x Sp	6	0.004	0.569	ns

[a] is log transformed
[b] is arcsin-transformed
* is significant at 0.05
** is significant at 0.01
ns is not significant

An increase in S:R is expected with a decrease in light intensity. This pattern was most manifest in *Impatiens* for which a more than five-fold increase in average S:R is found between the lowest and the highest light treatment (Figure 6.2 B, Table 6.4).

Table 6.3. *Ratio between the highest and lowest average value of yield and allocation traits as a measure of plasticity per species; SLA= Specific Leaf Area ($m^2.g^{-1}$) , LMF= Leaf Mass Fraction (g leaf.g^{-1} plant).*

	SLA	LMF
Aegopodium	4.1	3.4
Impatiens	4.2	5.4
Circaea	2.8	1.7
Mercurialis	1.8	3.0

Aegopodium also showed a clear response to light availability as well as, to a lesser extend, *Circaea.* The S:R of *Mercurialis* hardly changed between the 2 and 8% light treatment but increased significantly at the highest light level, resulting in a relatively high measure of plasticity, defined as the extent to which a plant can vary in a certain trait in response to light and nutrient availability (Table 6.3). In order to calculate plasticity, the average value of S:R was calculated in the fifth harvest for each of the six light-nutrient treatments. Plasticity was then expressed as the ratio between the highest and the lowest average value found in the six treatments. The expected decrease in S:R with increasing nutrient availability was most clearly found at the highest light availability for all species. Only *Aegopodium* clearly showed this pattern in all light treatments. Like for SLA also for absolute S:R, a clear difference was found between the species. Shoot-to-root ratio of *Aegopodium, Impatiens* and *Circaea* were mostly higher than one, indicating a preference for investment in the shoot compartment. Of these three species *Impatiens* maintained the highest S:R. *Mercurialis* showed a preference for belowground investment, possessing an average S:R smaller than one in all cases. Part of the belowground investment is probably allocated to storage organs rather than to functional root biomass. Interestingly, this species showed an increase in S:R when light availability increased.

In general, the two species *Circaea* and *Mercurialis,* capable of growing in deep shade showed a lower SLA, a lower S:R and a lower plasticity in those traits in response to light. Obviously, the species more commonly occurring in less shaded habitats thus had higher SLA and S:R and were more plastic in these traits. The prime factor affecting the studied plant traits was irradiance. The effect of nutrient availability was only evident when sufficient light was available. This might be due to the fact that the light manipulations in the experiment spanned a larger range and reached more limiting levels than the manipulations in nutrient availability.

3. MODELLING APPROACH

3.1. Advantage of a modelling approach

The greenhouse experiment was aimed at characterization of the forest species and the nitrophylic, fast-growing non-forest species. From this, in combination with patterns found in literature, a forest species and a non-forest species have been defined (Table 6.4). These are further used as model species to study the performance of the species groups over a range of light and nutrient availability. The advantage of using a mechanistic model is that different light and nutrient regimes can be applied. This results in a comparison of the performance of different species groups at different combinations of light and N availability. The model simulations also lead to a better insight in the role different plant traits and growth processes play in plant performance under different conditions.

Table 6.4. *Characteristics of the forest and forest edge species used in the model simulations, based on the results of the greenhouse experiment and patterns found in the literature (SLA = Specific Leaf Area (in $m^2.g^{-1}$), S:R = Shoot to Root ratio).*

	Forest species	Edge species
SLA	low plasticity low value	highly plastic
Leaf longevity	high	low
Photosynthetic apparatus	slow and low cost	fast and costly
S:R	low plasticity high shoot mass	highly plastic

3.2. Model description

The model plant consists of two compartments: a shoot compartment responsible for the acquisition of C through photosynthesis and a root compartment involved in the uptake of N. A schematic overview of the model plant is given in Figure 6.3.

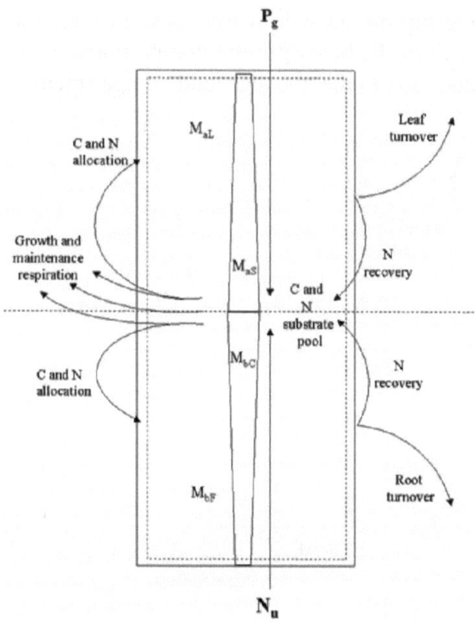

Figure 6.3. *Schematic representation of the model structure and the most important processes involved in the growth of the plant. M_{aL} is the aboveground leaf mass, M_{bF} the belowground fine root mass, M_{aS} and M_{bC} are the supportive stem and coarse root mass respectively, P_g is the gross photosynthetic C gain rate and N_u is the N uptake rate.*

Both compartments are made up of functional biomass, serving in the resource acquisition, and supportive biomass, which is not involved in acquisition processes. Both photosynthesis and N uptake are modeled realistically, in terms of an increased negative effect of biomass as the plant becomes larger (i.e. self-shading). Carbon acquired in photosynthesis and N taken up by the roots are first stored in a substrate pool. The C and N in this pool represent a stock of substrate available to invest in growth and maintenance. This substrate is freely available for investment in either above or belowground biomass. A full description of the model can be found in Elemans (2005).The rate at which substrate C and N can be invested in growth depends on the total amount of resources available in the substrate pool, i.e. if both resources are in ample supply the plant can grow at its maximum capacity, when the supply of either one or both of the resources in the substrate pool is restricted, growth rate is slowed down. A deficit of both substrates leads to a stronger growth limitation than when only one resource is in short supply. Nitrogen only has to be invested in growth, C is also required in maintenance respiration.

The relative growth rate of the shoot compared to the growth rate of the root depends on the ratio between the N and C concentration in the pool. This control mechanism of biomass allocation is based on the following line of thought. If N availability in the substrate pool is high compared to C availability, plant growth is limited by C. Accelerating the growth of the aboveground compartment, obviously at the cost of the growth of the belowground compartment, can decrease this growth limitation thus the allocation to the aboveground compartment increases.

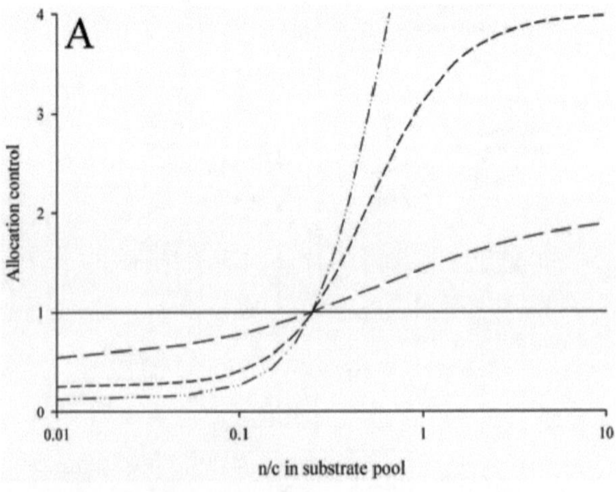

Figure 6.4. *Shape of the allocation control function. The n/c ratio on the x-axis represents the ratio at which the plant has N and C available for growth and maintenance processes; the allocation parameter on the y-axis represents the ratio between aboveground growth rate and belowground growth rate. The model plant can be set to be highly plastic or rather rigid in its allocation strategy, as is indicated by the different lines.*

This steering mechanism is shown in Figure 6.4. Differences in plasticity between species can be implemented by changing the boundaries and slope of the allocation control curve. If the C concentration in the substrate pool is high in relation to the N concentration, plant growth is limited by N, and the opposite reaction will occur to decrease N limitation.

3.3. Simulation results

First, the behavior of SLA and S:R as a function of light and N availability are presented. The SLA of the edge species reacted stronger to a decrease in light than did the forest species (Figure 6.5). Furthermore, the SLA of the forest species was below that of the edge species over the full light range, and the difference was smaller at high light than at low light. Also, the S:R of the forest species was less plastic than that of the other species (Figure 6.6).

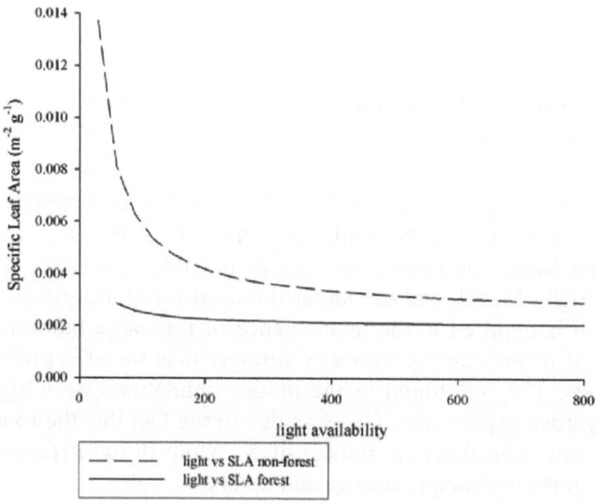

***Figure 6.5**. SLA of the model species (forest species – solid line; edge species – dashed line) as a function of light (micromole m^{-2} s^{-1}).*

The S:R of the forest species hardly reacted to either light or N. The S:R of the edge species increased with decreasing light at low N availability. At very low light levels, the opposite pattern was found.

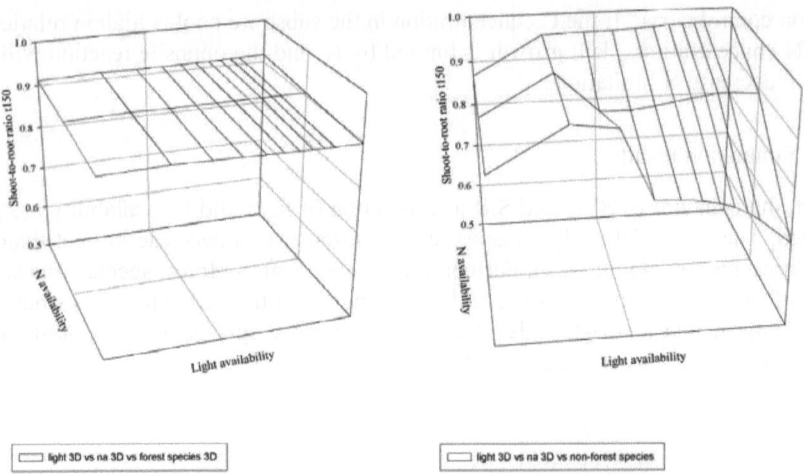

Figure 6.6. *S:R of the model species (forest species –left; non-forest species – right) after 150 days under different light and nutrient conditions.Plants started out having the same initial S:R.*

This was due to slow growth under these dark conditions, enabling only slow adjustment of the S:R. At high N availability, the S:R of the edge species increased with increasing light. The higher the light availability, the larger the plants, especially at this high N level, and the larger the need for aboveground biomass to acquire C, which is required in the maintenance of this large biomass. The effect of light on the S:R of the edge species was stronger than the effect of N, especially at low light levels. The S:R found in the model simulations were higher than those found in the garden experiment. This was due to the fact that the roots of the model plants served only one function, uptake of N, while in the experiment roots were also involved in the anchorage, storage and uptake of water.

Performance of the model plants was defined as the total plant mass, resulting from biomass production and loss, at the end of the growing season of 150 days. The light availability covered a range of 2.5 to 30% of the above-canopy light availability (Figure 6.7). The N availability was increased 10-fold in two steps. Total plant mass of the forest species increased with increasing light availability. At high light availability, N supply has a positive effect on the performance of the species. However, at low light availability hardly any effect of N on performance is found. Here, light availability is limiting growth (Figure 6.8).

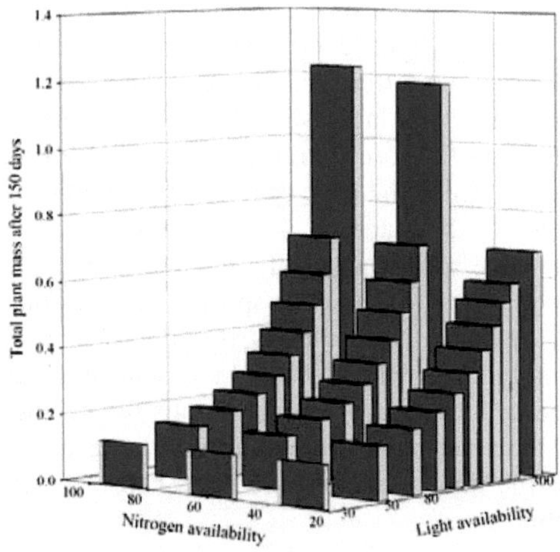

Figure 6.7. *Performance (total plant mass= g C per plant) of the forest species in an evergreen forest type under different levels of constant light (micromole m^{-2} s^{-1}) and three levels of N availability (kg N ha^{-1} yr^{-1}) representing a two-step 10-fold increase.*

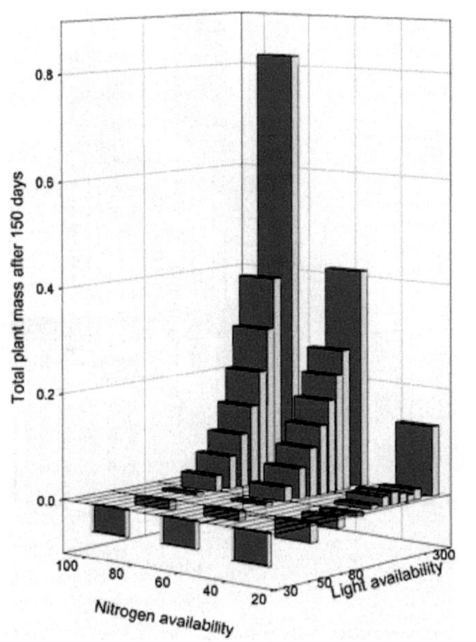

Figure 6.8. *Performance (total plant mass difference) of the edge species and the forest species at constant light availability in an evergreen forest type. Nitrogen availability was increased 10-fold in two steps. A negative value means that the forest species was performing better, a positive that the edge species had a higher biomass. The size of the bar represents the size of the difference; the higher the bar, the larger the difference.*

At low light availability, the forest species reached the highest biomass, independent of the N availability at high light levels, the edge species were performing better. When N availability was low, differences between the performances of the two species groups were small, but with increasing N supply, the differences also increased. The edge species could profit more from the increased N supply. With increasing N availability, the range at which the edge species performed better increased to lower levels. At light levels of 7.5% (approx. 50 micromole m^{-2} s^{-1}) of the above-canopy light availability, the edge species reached a higher biomass than the forest species. Owing to the higher plasticity of the edge species and thus the capability of adjusting the allocation to the higher N conditions, the edge species could invest more and faster in aboveground biomass which favored the photosynthetic C gain. According to the simulation results in an evergreen coniferous forest, edge species will be capable of improving their net plant production under the influence of increased N. However, as long as the light availability remains below a certain level, forest species will perform better.

In a light climate as in a deciduous forest, the biomass production of the forest species was higher than in an evergreen forest (Figure 6.9)

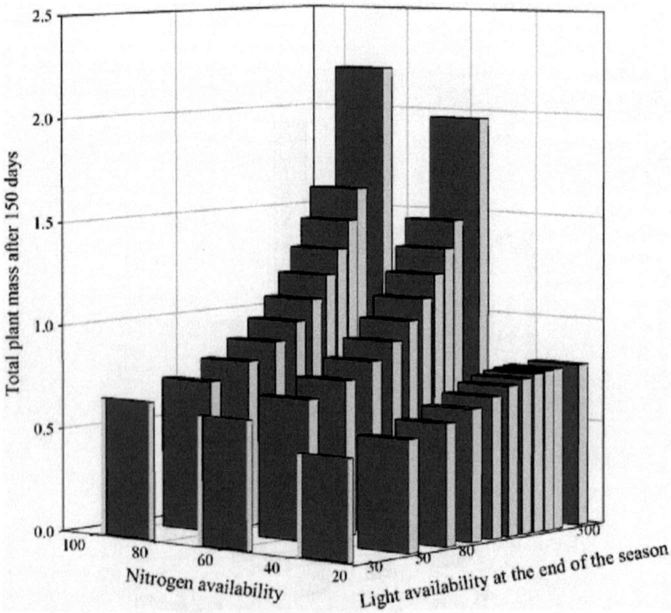

Figure 6.9. *Performance (total plant mass) of the forest species under a variable light regime in a deciduous forest type. Simulations started at the maximum light availability and decreased to the light levels indicated on the x-axis. Nitrogen availability was increased 10-fold in two steps.*

The high light intensity at the beginning of the growing season favored biomass production. The pattern with regards to light and N availability remained unchanged. Both light and N had a positive effect on total plant mass. At the deciduous light climate, the forest species could even profit from increased N at low (final) light levels.

For the edge species, the results did show some difference. At very low light levels at the end of the growing season, the edge species could not exist. For this species, the high light availability at the beginning of the season, and the accompanying high biomass production, had negative consequences at the end of the season, when light availability went down. The edge species is characterized by fast photosynthesis properties enabling fast growth under favorable conditions.

However, these traits are correlated to high costs for maintenance and high loss of biomass. The positive effect of N on the edge species was strong. The higher the N supplies the larger the range at which the edge species performed better than the forest species (Figure 6.10).

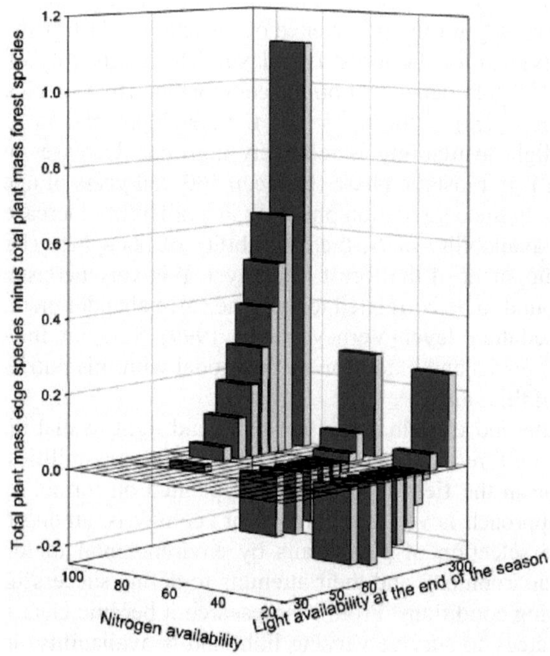

Figure 6.10. Comparison of the performance (total plant mass difference) of the edge species and the forest species at variable light availability in a deciduous forest type. Nitrogen availability was increased 10-fold in two steps. A negative value means that the forest species was performing better, a positive that the edge species had a higher biomass.

At low light conditions, however, increased N availability did not enable the plant to survive. Thus, the edge species produced more biomass in the initial high light period than the forest species, but could not keep up its biomass as light conditions deteriorated. The plant died when light levels at the end of the growing season fell below 7.5% of the above-canopy light availability (Figure 6.10). Thus, also in a deciduous light climate, the forest species performed better at the lowest light conditions although the edge species did profit more from increased N.

4. DISCUSSION AND CONCLUSION

A fast development of the forest field layer in newly planted forests on former agricultural land, is dependent on the availability of seed and habitat quality. Seed availability has been subject to many studies, but the role of habitat quality in the growth of field layer species has not been subject to many investigations. Corresponding to the focus of the AFFOREST project, we investigated the role of the availability of light and N to the field layer. The N availability in young forests on former agricultural land is higher than in natural forest ecosystems. Studies of the effect of N deposition on the species composition of existing forest showed a positive effect of N availability on the occurrence of nitrophylic, fast-growing species at the cost of species typical for the forest field layer. The availability of light to the field layer vegetation is more dependent on the composition (tree species, tree density and number of forest layers) of the forest and the developmental stage of the forest. In young forests, light availability is relatively high but decreases as the trees grow. When the forest is in biostatic phase (between 100-250 years of age), the light availability is lowest. In the degradation phase, light availability increases again.

Besides the availability of N, the availability of phosphate (P) may also play a role in the development of the forest field layer. P is very persistent in the soil and availability is found to be correlated to the time since abandonment. Only recent cultivation has raised the P level (Verheyen et al. 1999). Studying the effect of P on the growth of forest plants and extension of the model with this nutrient will be a valuable extension of this study.

To understand and explain the effect of N and light availability on the species composition of the forest field layer and to predict the possibilities for development of the vegetation in the field layer of forests planted on former agricultural land a plant ecology approach is very suitable. Plant ecology is, among other things, concerned with the selection of plant traits by environmental factors, or the 'fit' of plants to their environment, and their attempts to define successful strategies to adjust to the growing conditions. From past research it became clear that SLA and S:R are part of a strategy to survive varying light and N availability. In both traits, species exhibition phenotypic acclimation as well as genotypic adaptation to light and N. Although much research has been done, inherent shade-tolerant forest species were hardly ever included in the studies.

The greenhouse experiment conducted in this research did include typical forest species specialized to growing under very low light levels as well as species commonly occurring in less deep shade. Clear differences were found between the two species groups. The forest species generally had a lower SLA, a lower S:R and a lower plasticity in both traits than the edge species. These results agree with suggestions that species adapted to low light conditions should have a genotypically lower SLA-value (Thompson et al. 1992; Veneklaas & Poorter 1998; Walters & Reich 1996, 1999; Lusk 2002). With regards to the shade acclimation, or the level of plasticity, the results do not comply with the generally accepted idea that shade-adapted species should be more plastic. However, species adapted to severe levels of low light have not often been included in plasticity studies and the hypothesis may have to be extended. Especially because it becomes more and more clear that plasticity entails certain costs that will be a large burden to the tight C economy at low light conditions (DeWitt et al. 1998; Steinger et al. 2003).

Even without inclusion of plasticity costs, the model simulations showed that the edge species were not capable of surviving low light conditions. A positive effect was found of N availability but the edge species did not perform better than the forest species at very low light levels. In a deciduous light climate, the edge species were even not capable of surviving. The model results showed that under low light conditions, plant traits connected to conservation of existing biomass were more favorable than plant traits that enable adjustment to the growing conditions. This complies with current theory (e.g. Veneklaas & Poorter 1998, Elemans 2005).

This study showed the importance of the interactive effect of light and N on plant performance in the development of the forest field layer after afforestation of former agricultural land. Reduction of the light availability in the young forests will reduce the growth of fast-growing and competitive species while the growth of forest species, given that they already occur in the forest, will hardly be affected. This result stresses the importance of design (e.g. choice of tree species and density of the trees) and management of the tree layer, in controlling the development of the field layer vegetation. If forest management is aimed at a tight control of the light climate, measures aimed at decreasing the N availability in the soil may not be necessary.

If sowing of the desired forest species is considered, the results of this study show that there is no need to wait until N levels have decreased but sowing can be started fast provided that the light climate is controlled.

5. REFERENCES

Barkham, J. P. (1992). The effects of management on the ground flora of ancient woodland, Brigsteer park wood, Cumbria, England. *Biological Conservation 60,* 167-187.

Berendse, F., & Elberse, W. Th. (1989). Competition and nutrient losses from the plant. In H. Lambers, H. Konings, M. L. Cambridge, T. L. Pons (Eds.), Causes and consequences of variation in growth rate and productivity of higher plants, pp. 269-284. The Hague, The Netherlands: SPB Academic Publishing.

Bierzychudek, P. (1982). Life histories and demography of shade-tolerant temperate forest herbs: a review. *The New Phytologist, 90,* 757-776.

Bloom, A. J, Chapin, F. S., & Mooney, H. A. (1985). Resource limitation in plants - an economic analogy. *Annual Review of Ecology and Systematics, 16,* 363-392.

Bossuyt, B., Hermy, M., & Deckers, J. (1999). Migration of herbaceous plant species across ancient–recent forest ecotones in central Belgium. *Journal of Ecology*, 87, 629-638.

Brewer, R. (1980). A half-century of changes in the herb layer of a climax deciduous forest in Michigan. *Journal of Ecology, 68,* 823-832.

Brouwer, R. (1962a). Distribution of dry matter in the plant. *Netherlands Journal of Agricultural Science, 5,* 361-376.

Brouwer, R. (1962b). Nutritive influences on the distribution of dry matter in the plant. *Netherlands Journal of Agricultural Science, 10,* 399-408.

Brouwer, R. (1963). Some aspects of the equilibrium between overground and underground plant parts. Jaarboek van het Instituut voor Biologisch and Scheikundig onderzoek aan Landbouwgewassen 1963: 31-39.

Brunet, J., & von Oheimb, G. (1998a). Migration of vascular plants to secondary woodlands in southern Sweden. *Journal of Ecology, 86,* 429-438.

Brunet, J., & von Oheimb, G. (1998b). Colonization of secondary woodlands by Anemone nemorosa. *Nordic Journal of Botany, 18,* 369-377.

Brunet, J., Diekmann, M., & Falkengren-Grerup, U. (1998). Effects of nitrogen deposition on field layer vegetation in south Swedish oak forests. *Environmental Pollution, 102,* 35-40.

Brunet, J., von Oheimb, & G., Diekmann, M. (2000). Factors influencing vegetation gradients across ancient-recent woodland borderlines in southern Sweden. *Journal of Vegetation Science 11,* 515-524.

Bungard, R. A., Morton, J. D., McNeil, D.L., & Daly, G. T. (1998). Effects of irradiance and nitrogen on Clematis vitalba establishment in a New Zealand lowland podocarp forest remnant. *New Zealand Journal of Botany, 36,* 661-670.

DeWitt, T. J., Sih, A., & Wilson, D. S. (1998). Costs and limits of phenotypic plasticity. *Trends in Ecology & Evolution, 13,* 77-81.

Diekmann, M., & Dupre, C. (1997). Acidification and eutrophication of deciduous forests in northwestern Germany demonstrated by indicator species analysis. *Journal of Vegetation Science, 8,* 855-864.

Diekmann, M., Brunet, J., Ruhling, A., & Falkengren-Grerup, U. (1999). Effects of nitrogen deposition: Results of a temporal-spatial analysis of deciduous forests in South Sweden. *Plant Biology, 1,* 471-481.

Dzwonko, Z. (2001). Assessment of light and soil conditions in ancient and recent woodlands by Ellenberg indicator values. *Journal of Applied Ecology, 38,* 942-951.

Elemans, M. (2005). Plant traits in forest understory herbs – a modeling study. PhD thesis, Faculty of Biology, Utrecht University.

Emborg, J. (1998). Understorey light conditions and regeneration with respect to the structural dynamics of a near-natural temperate deciduous forest in Denmark. *Forest Ecology and Management, 106,* 83-95.

Falkengren-Grerup, U. (1993). Effects on beech forest species of experimental enhanced nitrogen deposition. *Flora, 188,* 85-91.

Fredericksen, T. S., Ross, B. D., Hoffman, W., Morrison, M. L., Beyea, J., Johnson, B.N., Lester, M.B., & Ross, E. (1999). Short-term understory community responses to timber-harvesting intensity on non-industrial private forestlands in Pennsylvania. *Forest Ecology and Management, 116,* 129-139.

Givnish, T. J. (1988). Adaptation to sun and shade: a whole-plant perspective. In J. R. Evans, S. Von Caemmerer, W. W. Adams (Eds.), Ecology of photosynthesis in sun and shade. Melbourne: CSIRO Editorial Services Unit, pp. 63-92.

Graae, B.J., Sunde, P.B., Fritzboger, B., (2003). Vegetation and soil differences in ancient opposed to new forests. *Forest Ecology and Management*, 177, 179-190.

Graae, B.J., & Sunde, P.B. (2000) The impact of forest continuity and management on forest floor vegetation evaluated by species traits. *Ecography*, 23, 720-731.

Grashof-Bokdam, C., & Geertsema, W. (1998) The effect of isolation and history on colonization patterns of plant species in secondary woodland. *Journal of Biogeography*, 25, 837-846.

Halpern, C. B., & Spies, T. A. (1995). Plant species diversity in natural and managed forests of the Pacific Northwest. *Ecological Applications, 5,* 913-934.

Hansen, K. (Ed.), (2002). Planning afforestation on previously managed arable land - influence on deposition, nitrate leaching, and carbon sequestration. http://www.fsl.dk/afforest.

Hermy, M., Honnay, O., Firbank, L., Grashof-Bokdam, C., & Lawesson, J. E. (1999). An ecological comparison between ancient and other forest plant species of Europe, and the implications for forest conservation. *Biological Conservation, 91,* 9-22.

Honnay, O., Hermy, M., & Coppin, P. (1999). Impact of habitat quality on forest plant species colonization. *Forest Ecology and Management, 115,* 157-170.

Kirby, K. J., & Thomas, R. C. (2000). Changes in the ground flora in Wytham Woods, southern England from 1974 to 1991 - implications for nature conservation. *Journal of Vegetation Science, 11,* 871-880.

Knops, J. M. H., & Reinhart, K. (2000). Specific leaf area along a nitrogen fertilization gradient. *American Midland Naturalist, 144,* 265-272.

Koerner, W., Dupouey, J.L., Dambrine, E., & Benoit, M. (1997). Influence of past land use on the vegetation and soils of present day forest in the Vosges mountains, France. *Journal of Ecology, 85,* 351-358.
Koike, T. (1986). Photosynthetic responses to light intensity of deciduous broad-leaved tree seedlings raised under various artificial shade. *Environmental Control in Biology, 24,* 51-58.
Lambers, H., Chapin, F. S., & Pons, T.L. (1998). Plant physiological ecology. New York: Springer-Verlag.
Lameire, S., Hermy, M., & Honnay, O. (2000). Two decades of change in the ground vegetation of a mixed deciduous forest in an agricultural landscape. *Journal of Vegetation Science, 11,* 695-704.
Leuschner, C., & Rode, M. W. (1999). The role of plant resources in forest succession: changes in radiation, water and nutrient fluxes, and plant productivity over a 300-yr-long chronosequence in NW-Germany. *Perspectives in Plant Ecology, Evolution and Systematics, 2,* 103-147.
Liu, Q. H., & Brakenhielm, S. (1996). Variability of plant species diversity in Swedish natural forest and its relation to atmospheric deposition. *Vegetatio, 125,* 63-72.
Lusk, C. H. (2002). Leaf area accumulation helps juvenile evergreen trees tolerate shade in a temperate rainforest. *Oecologia, 132,* 188-196.
Niinemets, U. (1999). Components of leaf dry mass per area - thickness and density - alter leaf photosynthetic capacity in reverse directions in woody plants. *New Phytologist, 144,* 35-47.
Nordin, A., Nasholm, T., & Ericson, L. (1998). Effects of simulated N deposition on understorey vegetation of a boreal coniferous forest. *Functional Ecology, 12,* 691-699.
Olff, H., Van Andel, J., & Bakker, J. P. (1990). Biomass and shoot root allocation of 5 species from a grassland succession series at different combinations of light and nutrient supply. *Functional Ecology, 4,* 193-200.
Onimaru, K., & Yabe, K. (1996). Comparisons of nutrient recovery and specific leaf area variation between Carex lasiocarpa var occultans and Carex thunbergii var appendiculata with reference to nutrient conditions and shading by Phragmites australis. *Ecological Research, 11,* 139-147.
Peace, W. J. H., & Grubb, P. J. (1982). Interaction of light and mineral nutrient supply in the growth of Impatiens parviflora. *New Phytologist, 90,* 127-150.
Petersen, P. M. (1994). Flora, vegetation, and soil in broadleaved ancient and planted woodland, and scrub on Rosnaes, Denmark. *Nordic Journal of Botany, 14,* 693-709.
Petersen, P. M., & Philipp, M. (2001). Implantation of forest plants in a wood on former arable land: a ten year experiment. *Flora, 196,* 286-291.
Poorter, H., & Nagel, O. (2000). The role of biomass allocation in the growth response of plants to different levels of light, CO2, nutrients and water: a quantitative review. *Australian Journal of Plant Physiology, 27,* 595-607.
Reich, P. B., Walters, M. B., & Ellsworth, D. S. (1992). Leaf life-span in relation to leaf, plant, and stand characteristics among diverse ecosystems. *Ecological Monographs, 62,* 365-392.
Reich, P. B., Walters, M. B., Ellsworth, D. S., Vose, J. M., Volin, J. C., Gresham, C., & Bowman, W. D. (1998). Relationships of leaf dark respiration to leaf nitrogen, specific leaf area and leaf life-span: a test across biomes and functional groups. *Oecologia, 114,* 471-482.
Rode, M. W. (1999). Influence of forest growth on former heathland on nutrient input and its consequences for nutrition and management of heath and forest. *Forest Ecology and Management, 114,* 31-43.
Schippers, P., & Olff, H. (2000). Biomass partitioning, architecture and turnover of six herbaceous species from habitats with different nutrient supply. *Plant Ecology, 149,* 219-231.
Shipley, B. (2000). Plasticity in relative growth rate and its components following a change in irradiance. *Plant Cell and Environment, 23,* 1207-1216.
Singleton, R., Gardescu, S., Marks, P. L., & Geber, M. A. (1901). Forest herb colonization of postagricultural forests in central New York State, USA. *Journal of Ecology, 89,* 325-338.
Steinger, T., Roy, B. A., & Stanton, M. L. (2003). Evolution in stressful environments II: adaptive value and costs of plasticity in response to low light in Sinapis arvensis. *Journal of Evolutionary Biology, 16,* 313-323.
Thimonier, A., Dupouey, J. L., & Timbal, J. (1992). Floristic changes in the herb-layer vegetation of a deciduous forest in the Lorraine Plain under the influence of atmospheric deposition. *Forest Ecology and Management, 55,* 149-167.
Thimonier, A., Dupouey, J. L., Bost, F., & Becker, M. (1994). Simultaneous eutrophication and acidification of a forest ecosystem in northeast France. *New Phytologist, 126,* 533-539.

Thompson, W. A, Kriedemann, P. E., & Craig I. E. (1992). Photosynthetic response to light and nutrients in sun-tolerant and shade tolerant rainforest trees. I. Growth, leaf anatomy and nutrient content. *Australian Journal of Plant Physiology, 19,* 1-18.

Tilman, D. (1985). The resource-ratio hypothesis of plant succession. *American Naturalist, 125,* 827-852.

Tilman, D. (1988). Plant strategies and the dynamics and structure of plant communities. Monographs in population biology 26. Princeton: Princeton University Press.

Van Dobben, H., Ter Braak, C., & Dirkse, G. M. (1999). Undergrowth as a biomonitor for deposition of nitrogen and acidity in pine forest. *Forest Ecology and Management, 114,* 83-95.

Van Duuren, L., Rider, C., Dijkstra, R., & Meijer, R. (1997). Biobase; register biodiversiteit. Van Duuren, L. Voorburg/Heerlen, Centraal buro voor de Statistiek.

Veneklaas, E. J., & Poorter, L. (1998). Growth and carbon partitioning strategies of tropical tree seedlings in contrasting light environments. In H. Lambers, H. Poorter, Van Vuuren MMI, (Eds.), Inherent variation in plant growth. Physiological mechanisms and ecological consequences. Leiden: Backhuys Publishers, 337-362.

Verburg, R. (1998). The witch at the forest floor. Reproductive allocation and clonal diversity in the pseudo-annual Circaea lutetiana L. Utrecht: Utrecht University.

Verheyen, K., Bossuyt, B., Hermy, M., & Tack, G. (1999). The land-use history (1278-1990) of a mixed hardwood forest in western Belgium and its relationship with chemical soil characteristics. *Journal of Biogeography, 26,* 1115-1128.

Verheyen, K., Honnay, O., Motzkin, G., Hermy, M., & Foster, D. R. (2003). Response of forest plant species to land-use change: a life- history trait-based approach. *Journal of Ecology, 91,* 563-577.

Walters, M. B., & Reich, P. B. (1996). Are shade tolerance, survival, and growth linked? Low light and nitrogen effects on hardwood seedlings. *Ecology, 77,* 841-853.

Walters, M. B., & Reich, P. B. (1999). Low-light carbon balance and shade tolerance in the seedlings of woody plants: do winter deciduous and broad-leaved evergreen species differ? *New Phytologist, 143,* 143-154.

Weeda, E. J., Westra, R., Westra, Ch., & Westra, T. (1987). Nederlandse Oecologische Flora. Wilde planten en hun relaties. Amsterdam: IVN.

Westoby, M., Warton, D., & Reich, P. B. (2000). The time value of leaf area. *American Naturalist, 155,* 649-656.

Westoby, M., Falster, D. S., Moles, A. T., Vesk, P. A., & Wright, I. J. (2002). Plant ecological strategies: Some leading dimensions of variation between species. *Annual Review of Ecology and Systematics, 33,* 125-159.

Willems, J. H., & Boessenkool, K. P. (1999). Coppiced woodlands and their significance for herbaceous plant species conservation. In Dong Ming & M. J. A. Werger (Eds.), A spectrum of ecological studies. In honour of Professor Zhang Zhangcheng at his 70th birthday. Southwest China Normal University Press, 188-196.

Wilson, P. J., Thompson, K., & Hodgson, J. G. (1999). Specific leaf area and leaf dry matter content as alternative predictors of plant strategies. *New Phytologist, 143,* 155-162.

Yorks, T. E., & Dabydeen, S. (1998). Seasonal and successional understory vascular plant diversity in second-growth hardwood clearcuts of western Maryland, USA. *Forest Ecology and Management, 119,* 217-230.

CHAPTER 7

MODELLING THE AFFORESTED SYSTEM: THE FOREST/TREE MODEL

G.W. HEIL[1], W. VAN DEURSEN[2], M. ELEMANS[1#], J. MOL[3] and H. KROS[3]

[1]Faculty of Science, Institute of Environmental Biology, Sorbonnelaan 16, NL 3584 CA Utrecht, the Netherlands, E-mail: g.w.heil@bio.uu.nl, [#] Current address: Karolinska Institute, Microbiology and Tumorbiology Center (MTC), P.O.Box 280, Nobelsvägen 16, 171 77 Stockholm, Sweden, [2] Carthago Consultancy, Oostzeedijk Beneden 23a, 3062 VK Rotterdam, the Netherlands, [3] Alterra Green World Research, P.O.Box 47, 6700 AA Wageningen, the Netherlands

Abstract. A forest/tree model has been developed of which the main growth processes are based on the CENW model. The model links the flows of carbon (C)), energy, nutrients and water in trees and soil organic matter. Modelled tree growth depends on physiological plant factors, the size of plant pools, such as foliage mass, environmental factors, such as temperature and rainfall, and the total amount and turn-over rates of soil organic matter, which drives mineraliZation of soil organic nitrogen (N). The forest/tree model has been developed as a generic model for coniferous trees. In addition, the model has been extended generically for deciduous trees by shedding of leaves in autumn and including growth buds in which C can be stored during winter time, and from which re-growth is initiated in spring. In spring the initial C gives the leaf activity an augmentation for photosynthesis. For the purpose of the available input parameters in the AFFOREST project it was needed to change the daily time step of the original CENW-model into a fortnightly time step. In relation to this, the original simple hump functions for N, water and temperature dependencies are replaced by smooth minimum/optimum/maximum functions. The model was validated against data obtained for Quercus robur (Oak) from the AFFOREST chronosequence in Vestskoven in Denmark. The different tree stands covered a chronosequence over a period of 35 years. The forest/tree model was successfully able to simulate all aspects of tree growth, which included water, N and C dynamics reflected in biomass production. An associated structure variable, i.e. mean tree height was also simulated successfully, which is an important input parameter for the amount of atmospheric N deposition in forests. The model includes essentially all relevant pools and processes that determine forest growth under a range of natural conditions.

G.W. Heil et al. (eds.), Environmental Effects of Afforestation in North-Western Europe, 149–174.

1. INTRODUCTION

Forest/tree models have been developed to simulate either tree growth or nutrient cycling or doing both (e.g. BIOMASS (McMurtrie et al. 1989 1992), CENW (Kirschbaum 1999), FIWALD (Schall 1995), FORGRA (Jorritsma et al 1999). However, none of these models cover the proper needs for the purpose of this research project to simulate effects of afforesting abandoned agricultural land. For this, the modelling needs were distinguished, i.e. the scale, time steps, time range, implementation of the models, what the AFFOREST-sDSS was going to look like, task sharing between the different models (atmosphere, tree, soil, hydrology), software, platforms, and data. Also the needs for end-users should be taken into account, which have implications on the model structure. For this purpose a model-based demo was developed to discuss different options of the model Graphical User Interface (GUI). This demo was developed as a visual tool to support the discussions on what an AFFOREST-sDSS was supposed to do and what integration there should be between the sDSS, the GIS and the models (see also Chapter 10). These discussions lead to the development of a list of model requirements, and what the model should be able to calculate:

- Limited amount of data requirement (requirement)
- Data should be taken from literature or from the field measurements (Chapter 2, 3 and 4) (requirement)
- Carbon (C), nitrogen (N), water dynamics (calculate)
- Tree height for N deposition (calculate)

These requirements brought us to the rough model structure as illustrated in Figure 7.1.

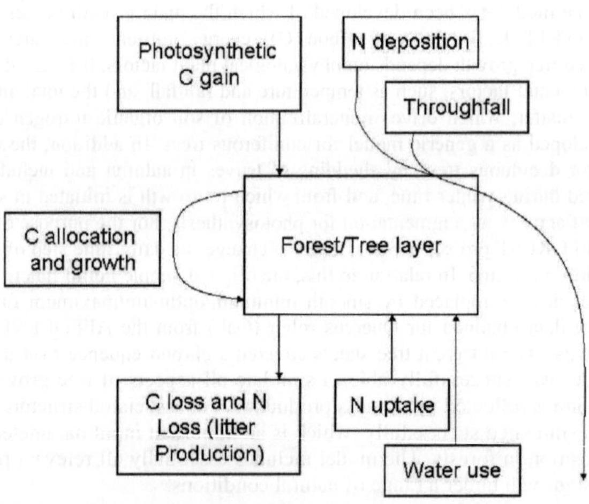

Figure 7.1. *Scheme of the forest/tree model for the AFFOREST project.*

An overview has been made of forest/tree models that could be used for the AFFOREST project at the start of the research project the following models have been reviewed: BIOMASS (McMurtrie et al 1989 1992), CENW (Kirschbaum 1999), FIWALD (Schall 1995), FORGRA (Jorritsma et al 1999), FORGRO (Mohren 1987), ForM-S (Arp & Oja 1992), FORSOL (Van Minnen et al 1995), FORSUM (Krauchi 1995), HYBRID (Friend et al 1993), JABOWA (Botkin et al 1972), NAP (Van Oene & Agren 1995), NICCCE (Van Dam & Van Breemen 1995), PnET-CN (Aber & Federer 1992), PT (Agren & Axelsson 1980), SOILN (Eckersten 1994), SOILVEG (Berdowski et al 1991), SPACE (Busing 1991), TREGRO (Weinstein & Beloin 1990), 3-PG (Landsberg & Waring 1997), (Dewar & McMurtrie 1996), (Hauhs et al 1995), (Hunt et al 1999), (Korzukhin et al 1995), and (Van Oene et al 1999).

Based on the criteria for the final metamodel of the AFFOREST research project, the CENW model, a forest growth model with linked C, energy, nutrient and water cycles, developed by Kirschbaum (1999), was preferred. The CENW model was used as basic model and the process description of the model was used for further development and prototyping in the AFFOREST project.

2. CENW

2.1. Overview CENW

The CENW (version 1.0.5) model has its name from the letters C, Energy, Nutrients and Water. Figure 7.2 gives the basic outline of the model. The model combines and links the important fluxes of C and nutrients, on the one hand, and CO_2 and water, on the other. Plants grow by fixing CO_2 from the atmosphere. However, the need to open a diffusion conduit for CO_2 uptake leads to water loss in the diffusive exchange of CO_2 and water through stomatal pores. Water can be replenished from the soil provided sufficient soil water is available. Otherwise, further water loss must be prevented by stomatal closure, which also prevents CO_2 fixation. Water use is calculated with the Penman-Monteith equation, with canopy resistance given by stomatal conductance, which, in turn, is linked to calculated C gain. Water loss by transpiration and soil evaporation and water gain by rainfall or irrigation then determine the soil water status for the following time step.

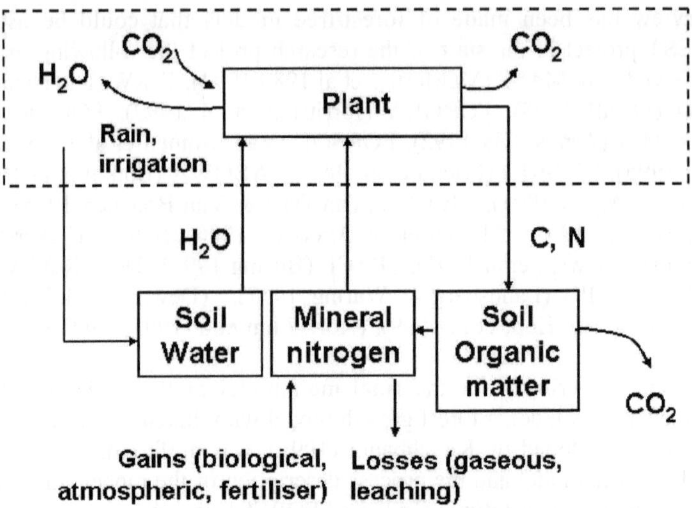

***Figure** 7.2. Outline of the CENW model (after Kirschbaum 1999).*

The original CENW model runs on a daily time step. Photosynthetic C gain is calculated based on light absorption, temperature, soil water condition, foliar N concentration and any foliage damage due to frost or scorching temperatures during preceding days. Some C is lost by respiration and the remainder utilized for growth, with allocation to different plant organs determined by plant nutrient status, tree height and species-specific allocation factors (Appendix 1).

Allocation fractions for both C and N to different plant organs were calculated daily, based on the assumption that there are constant allocation fractions between certain biomass components (Appendix 1).

N can be taken up from the mineral N pool. N can be supplied by external sources or from the decomposition of organic matter. The nutrient cycle is closed through litter production by the death of trees, or by shedding of plant parts, such as roots, bark, branches and, most importantly, foliage. This transfers C and N to the soil. Organic matter is eventually decomposed, thereby releasing CO_2 to the atmosphere. Any N in excess of microbial requirements can enter the pool of mineral N from where it can be taken up by plants. The decomposition rate is determined by temperature, soil water status and soil organic matter quality based on the CENTURY model (Parton et al 1987).

All plant components other than foliage were assumed to senesce at a constant fractional rate. No seasonal pattern of senescence was assumed. Senescence in dense canopies was assumed to occur when the amount of light reaching the most shaded needles in the canopy fell below a certain threshold. This means that this type of senescence mainly occurred in winter.

The model can be run with constant, observed or simulated climate. For sites where climate data are available, observed data should normally be used. If observed data are not available, it is possible to generate artificial sequences based on observed mean climatic parameters. It is also possible to run the model with constant climatic data. This is principally useful as an analytical tool to investigate the model response to some other perturbation, which is easier to identify in the absence of climatic variability.

The minimum climatic data set consists of daily minimum and maximum temperature, radiation and rainfall. Humidity may be input as either absolute or relative humidity. If relative humidity is supplied it is taken to be the relative humidity at the time of day when mean daytime temperature is reached. When information about humidity cannot be supplied, it is calculated based on the assumption that air is saturated with water vapour at the overnight minimum temperature.

2.2. Application of CENW model in the AFFOREST project

The original CENW model was successfully able to simulate all aspects of tree growth, and simulations were possible with the model run completely by external driving variables, initial soil and stand conditions, standard climatic observations and silvicultural manipulations, no special adjustments of plant internal pools were required.

The model includes essentially all relevant pools and processes that determine forest growth under a range of natural conditions and in response to management manipulations. This completeness was achieved while maintaining individual relationships as straightforward and clear as possible. This simplicity causes some of the remaining discrepancy between modelled and observed responses, such as differences in stem diameters between treatments. Although, the latter is not used for in the new AFFOREST model, this model should reveal whether these discrepancies will be matters of concern for specific model applications or not.

Interestingly, the model was available for a range of applications such as testing the impacts of climate change on forest growth (e.g. Booth et al. 1999; Kirschbaum 1999), assessing the effect of silvicultural manipulations, assessing long-term sustainability, for simulating a range of experimental findings and for explanatory runs that help to understand the reasons for any observed plant responses. In conclusion the CENW model is a good source for the basic modelling processes of the AFFOREST tree model.

3. FOREST/TREE MODEL AS PART OF A METAMODEL

3.1. Introduction

For the purpose of the AFFOREST –project, it was needed to change the model to a fortnightly time step. The newly developed forest/tree model has been developed as a generic model for coniferous trees. In addition to this, the model has been extended generically for deciduous trees by shedding of leaves in autumn and including growth buds in which C can be stored during winter time, and from which re-growth is initiated in spring. In spring, the initial C gives the leaf activity an augmentation for photosynthesis. Also, in the AFFOREST tree model, photosynthetic C gain is calculated based on light absorption, temperature, soil water status and foliar N concentration. Furthermore, C is lost in respiration and the remainder utilized for growth, with allocation to different plant organs determined by plant nutrient status, tree height and species-specific allocation factors. Available N can be taken up from the mineral N soil pool as calculated by the soil model (Chapter 8). The calculated output is used as input for the soil model (Chapter 8).

3.2. Process description of the forest/tree model as part of the Metamodel

A fair part of the process descriptions are taken from Kirschbaum (1999), however, in relation to the forest/tree model a good number of process descriptions are modified or completely new.

Carbon gain. Net photosynthesis is calculated based on the equations given by Sands (1995). Sands used a widely used simple leaf-level photosynthesis model by which assimilation rate can be calculated. Sands was able to develop a set of equations with which it is possible to calculate the photosynthesis based on incident radiation and single-leaf photosynthetic parameters. Daily canopy photosynthesis g according to Sands (1995) is (for model abbreviations see Appendix 1):

$$g = g_R * f_1 / (1 + (g_R / g_B - 1) * f_2) \quad (1)$$

$$g_R = \begin{cases} 1 - (4/(\pi\sqrt{(1 - q^2)})(\arctan(\sqrt{((1-q)/(1+q))}))) & \text{for } q < 1 \\ 1 - ((2/(\pi\sqrt{(q^2-1)}))\ (\ln((1+\sqrt{((q-1)/(q+1))})/(1-\sqrt{((q-1)/(q+1))})))) & \text{for } q > 1 \\ 1 - (2/\pi)) & \text{for } q = 1 \end{cases} \quad (2)$$

$$g_B = \begin{cases} (2/\pi)*q & \text{for } q <= 1 \\ 1 + (2/\pi)*(q - (\sqrt{(q^2-1)}) - (\arcsin(1/q))) & \text{for } q > 1 \end{cases} \quad (3)$$

$$f_1 = 1 + a_1\theta(1 - \theta) + b_1\, \theta^2(1 - \theta)^2 \quad (4)$$

$$f_2 = a_2\, \theta + b_2\, \theta^2 + (1 - a_2 - b_2)\, \theta^3 \quad (5)$$

$$q = (\pi k_1 \alpha Q_a k_2)/(2h(1\text{-}m)A_{max}) \quad (6)$$

Where:

A_{max} = maximum assimilation rate

α = the quantum yield

a_1 = parameter in f_1, which is function of depth in canopy
a_2 = parameter in f_1, which is function of depth in canopy
b_1 = parameter in f_2, which is function of depth in canopy
b_2 = parameter in f_2, which is function of depth in canopy
f_1 = empirical function of θ
f_2 = empirical function of θ
g = total photosynthesis
g_R = light photosynthesis
g_B = dark photosynthesis
h = day length in seconds,
k_1 = the light extinction coefficient
k_2 = a conversion term that converts from total solar radiation to photo synthetically active photon flux density
$1\text{-}m$ = leaf transmissivity,
θ = shape of light response
q = normalized radiation
Q_a = absorbed radiation

The conversion term, k_2 was taken as 2.0 μmol quanta J^{-1} after Sands (1995). Daily photosynthetic C gain, A_d can then be calculated as:

$$A_d = A_{max}\, h\, g\, (1 - \exp(-k_1 L(1-m)))/k_1 \qquad (7)$$

Where:
L = leaf area index.

The function $g\,(1 - \exp(-k_1 L(1-m)))/k_1$ is self shading given by Sands (1995). Total absorbed radiation, Q_a, is calculated as:

$$Q_a = Q_i\,(1-r)\,(1 - \exp(-k_1 L(1 - m))) \qquad (8)$$

Where:
Q_i = incident radiation and
r = the fraction of radiation that is reflected (albedo)

Leaf area index, L, is calculated as:

$$L = S_l\, W_f \qquad (9)$$

Where:
S_l = specific leaf area and
W_f = foliar biomass ($g.m^{-2}$)

The terms α and A_{max} are affected by temperature. It has a strong temperature dependence that can be calculated in a normalized Arrhenius form as (McMurtrie et al. 1992):

$$\Gamma_* = 0.042 exp\,(9.46(T_{day} - 25)/(T_{day} + 273.2)) \tag{10}$$

Where:
0.042 = the value for Γ_* at 25°C (42 μmol.mol^{-1})
T_{day} = mean daytime temperature in Celsius degrees

In the case of the quantum yield, the temperature dependence is only due to the changing ratio of carboxylations to oxygenations, whereas in the case of A_{max} temperature affects the changing ratio of carboxylations to oxygenation as well as the maximum rate at which reactions can be carried out. A_{max} is therefore calculated, following Kirschbaum (1999), as:

$$A_{max} = \alpha_0(c_i - \Gamma_*)/(c_i + 2\Gamma_*) \tag{11}$$

Where:
α_0 = the potential *RuBP* regeneration rate at a given temperature

This assumes that assimilation rate is limited by RuBP regeneration rather than Rubisco activity, which seems reasonable for most conditions (see Kirschbaum, 1999, for further discussion of this point). The temperature dependency of α_0 was calculated with a simple hump function (Kirschbaum 1999), but because of instability caused by the larger time step, i.e. daily vs. fortnight time step, modified into a smooth hump function:

$$T_{Lim} = 1/(1+68exp(-11(T_{temp}-T_{MinLim})/(T_{opt1}-T_{MinLim})))1/(1+68exp(-11(T_{temp}-T_{MaxLim})/(T_{opt2}-T_{MaxLim}))) \tag{12}$$

Where:
T_{MinLim} = the minimum temperatures that allow any photosynthesis
T_{MaxLim} = the maximum temperatures that allow any photosynthesis,
T_{opt1} = lower upper temperature bound that allow optimum photosynthetic rates
T_{opt2} = higher upper temperature bound that allow optimum photosynthetic rates
T_{temp} = mean temperature

The maximum assimilation *MaxAssim* is calculated as a function of foliar N concentration, temperature, and water as:

$$MaxAssim = A_{max}N_{Lim}W_{lim}T_{lim} \tag{13}$$

Where:
A_{max} = the highest photosynthetic rate for that species under non-limiting factors
N_{lim} = a N limitation parameter, defined with a similar function as for temperature, where N_{Lim} is foliar N concentration in the canopy

W_{lim} = a water limitation parameter, also defined with a similar function as for temperature, where *WLim* is water availability in the soil

Soil water balance. The soil is considered as one layer with a specific water holding capacity. Additional water is added to the soil. If the water content exceeds its water-holding capacity, water is lost as deep drainage. Horizontal or any upward movement of water is not modeled. Soil evaporation is restricted to occur from the soil layer (see Chapter 8).

Evapotranspiration. In contrast with the CENW model, evapotranspiration rate (mm/month) as potential evapotranspiration is calculated with a Thorntwait equation instead of the original Penman-Monteith equation (Monteith 1965). (Chapter 8).

$$E_{pot} = 16(10T_{day}/T_w)exp(6.75exp\text{-}7(T_wexp3)\text{-}77.1exp\ 6\ (T_wexp2) + 0.0179T_w + 0.49) \tag{14}$$

Where:

T_w = 12((T_{temp}/5)exp1.514)

C loss. C can be lost through plant respiration, through mortality of individual trees or through senescence of plant organs. C may also be lost through selective removal of trees (harvesting/thinning).

Plant respiration. Respiration is calculated as growth respiration, *Rg,* plus maintenance respiration, *Rm* (e.g. Amthor 1994). Growth respiration is calculated as:

$$Rg = r_g \sum G \tag{15}$$

Where:

r_g = an empirical term that quantifies the amount of C lost in growth respiration per unit of new growth

$\sum G$ = the sum of new C growth of all plant organs.

Maintenance respiration *Rm,* is calculated as:

$$Rm = r_m \int T_{resp} \sum N \tag{16}$$

Where:

r_m = an empirical term that gives the daily respiration rate per unit of N at 25°C

$\sum N$ = the sum of N contained in all plant pools except foliage. Foliage respiration is calculated only for the night-time period, and respiration during the day is included as part of net photosynthetic C gain calculations.

The temperature response function, $\int T_{resp}$ is:

$$\int T_{resp} = exp[-3.166 + 0.001696T_{day}(100 - T_{day})] \quad (17)$$

Mortality. Tree death D_b is modelled as a simple fractional mortality rate. Loss of tree biomass is calculated as:

$$D_b = D_n \int m \quad (18)$$

Where:

D_n = the daily fraction of stems lost

$\int m$ = the ratio of the biomass of dying relative to average sized trees in the stand. Loss of above-ground biomass is assumed to lead to the same relative loss of root biomass.

Senescence. Senescence of plant organs other than foliage and fine roots is calculated as a simple fractional loss, with different empirical loss fractions for different organs. Hence, relative daily foliage senescence rate, S_f, is calculated as:

$$S_f = S_b \quad (19)$$

Where:

S_b = a constant minimum foliage senescence rate. Root and stem senescence is modelled in a similar way.

Allocation. Newly fixed C and N from the soil are initially taken up into plant soluble pools. C for respiration is subtracted from the soluble C pool. The remaining C in the soluble pool can then be utilized for growth. Equivalent calculations are done for N, but for N there is the further restriction that at most as much N can be turned into new growth as corresponds to the new foliage growth rate at maximum foliar N concentration. This limits the extent and rapidity with which plant pools can take up large amounts of N if it suddenly becomes available through fertilization. New growth is then allocated to the different biomass pools based on a number of different considerations. Allocation of C is dealt with first. Allocation to biomass components is based on the assumption that allocation ratios between certain biomass components, such as stem wood, roots and leaves, are constant.

Tree height. Tree height $Tree_h$ is calculated as a function of growth performance:

$$Tree_h = Tree_{h0} + d_h \tag{20}$$

Where:
d_h is the increase in height (m) per time step, which is calculated as follows:

$$d_h = G_{Lim} H_{inc} \tag{21}$$

$$G_{Lim} = N_{Lim} W_{lim} T_{lim} C_{lim} \tag{22}$$

$$H_{inc} = A + B^{(R_t)} \tag{23}$$

Where:
A = species specific height at the start,
B = a species specific increase ratio, and
R_t = the maximum height at time t

N allocation. N allocation is calculated on the basis of the same considerations that govern the allocation of C, but in addition, the N concentration in all plant organs is expressed relative to the C concentration in a C/N ratio.

Soil N dynamics. Available N may come from atmospheric deposition or mineralization of organic N during the decomposition of soil organic matter so that

$$N_{min} = N_{dep} + N_{act} \tag{24}$$

Where N_{min} is the total amount of N becoming available in mineral form, N_{dep} is the amount deposited from the atmosphere, and N_{act} is the amount mineralised from the active (decomposer) pool of organic matter calculated by the soil model in the metamodel - METAFORE (Chapter 8). The rates N_{dep} is user input, with atmospheric input taken to be the same for each day of the simulation and fertilizer being added at specified dates (Chapter 8)

N uptake dynamics. For moderate amounts of N being mineralized, it is assumed that all N is taken up by plants at each time step (minus fractions leached or sequestered in slow organic matter). However, it is assumed that only a maximum amount of N can be taken up by plants during each day. When N is taken up, it is initially added to a soluble plant pool, which can be utilized for subsequent growth. The maximum amount that can be taken up into the soluble plant pool, U_{max}, is given by:

$$U_{max} = X_n \sum N_{max} - \sum N \tag{25}$$

Where:

X_n = an empirical excess N storage ratio,
$\sum N_{max}$= the sum over all plant organs of the maximum amount of N that could be taken up in growth by each pool if that pool had the maximum permissible N concentration depending on C/N ratio.
$\sum N$ = the sum of the amounts actually contained in each pool.

Climatic information. The model can be run with local climate. This is principally useful as a tool to investigate the model response to climatic variability. The minimum climatic data set consists of minimum and maximum temperature, radiation and rainfall. Information about humidity is calculated based on the assumption that air is saturated with water vapor at the overnight minimum temperature. Mean temperature is calculated as the mean of minimum and maximum temperatures. Temperature, T_{temp}, is calculated as follows:

$$T_{temp} = t_{min} + (t_{max}\text{-}t_{min})/2 \qquad (26)$$

Where:
t_{max} = maximum temperature
t_{min} = minimum temperature

3.3. Model parameterization

Model parameterisation was performed using data from the original CENW model,

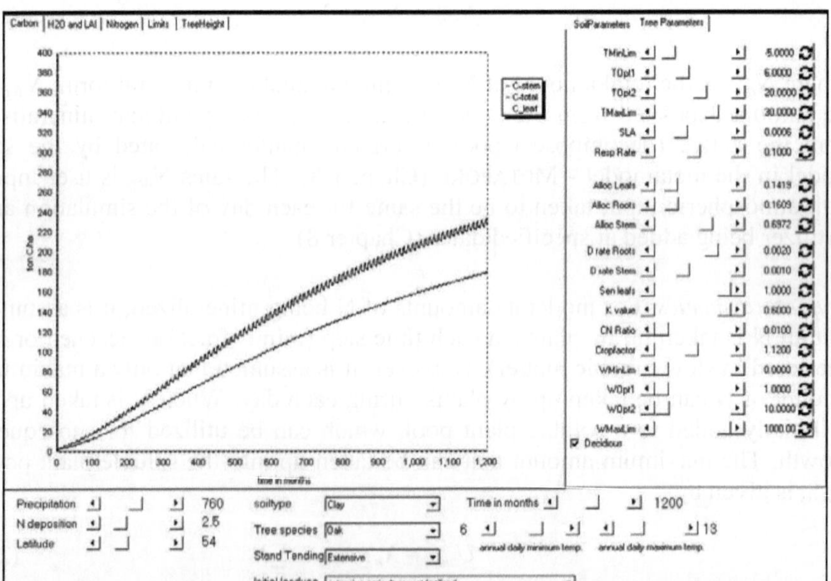

Figure 7.3. *GUI of the forest/tree model including sliders for calibration of the process rates.*

literature data and the use of a GUI in order to see direct responses of the model (Figure 7.3). With this GUI the sensitivity of the variables of the different tree species were tested as well.

Temperature responses. Growth and photosynthetic responses to temperature vary considerably in literature, depending on the growth stage of trees, acclimation stage and whether responses are expressed as the response of photosynthesis or growth. Even for photosynthesis alone, different experiments have yielded very different responses. The choice of parameter values therefore had to be guided as much by the overall plant response to growing temperature as to the specific observations of photosynthetic response to temperature. It was assumed here that the lower threshold mean temperature for C gain was $5^{\circ}C$ that the optimum temperature ranged from 15 to $20^{\circ}C$ and the upper threshold for growth was $30^{\circ}C$. It was also assumed that a certain amount of N was mineralized daily by decomposition of organic matter (unless the addition of C-rich litter was immobilizing N) in addition to atmospheric and fertilizer inputs. The remaining mineral N could either be leached or taken up by plants. Leaching was assumed to only occur when relative water content of the soil layer was in excess of 99% of water holding capacity (Chapter 8)

Allocation. Allocation fractions for both C and N to different plant organs were calculated fortnightly, and were based on the assumption that there are constant allocation fractions between certain biomass compartments, such as stem, branches, roots and leaves (see Appendix 1).

Senescence. All plant components other than foliage were assumed to senesce at a constant fractional rate. Besides the shedding of the leaves of the deciduous trees in autumn, no seasonal pattern of senescence was assumed (see Appendix 1).

4. RESULTS

4.1. Validation

Simulated patterns of water use, N and C dynamics were compared against the observed data from literature in order to test the model. For the simulations shown below, standard climate data were used to give greater transparency to model simulations. Some key input and output data are given in Figure 7.4.

4.2. Validation runs

A comparison between observations from a chronosequence and the results of the AFFOREST model was performed for the oak chronosequence at Vestskoven (DK) (Figure 7.5 a, b, c) (Chapter 1).

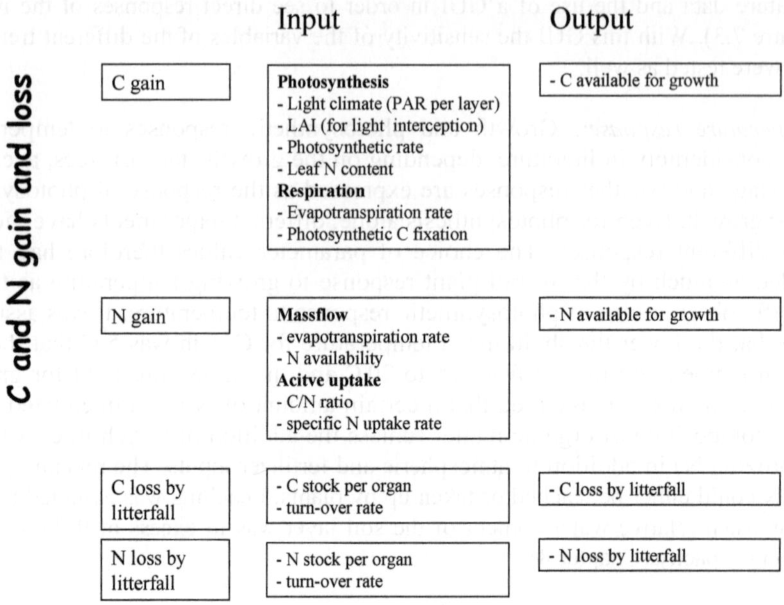

Figure 7.4. *Key input and output data for simulations.*

The simulation of C sequestration in the tree compartments and of tree height is suitable. The data collected in the chronosequences are focused on a limited set of tree

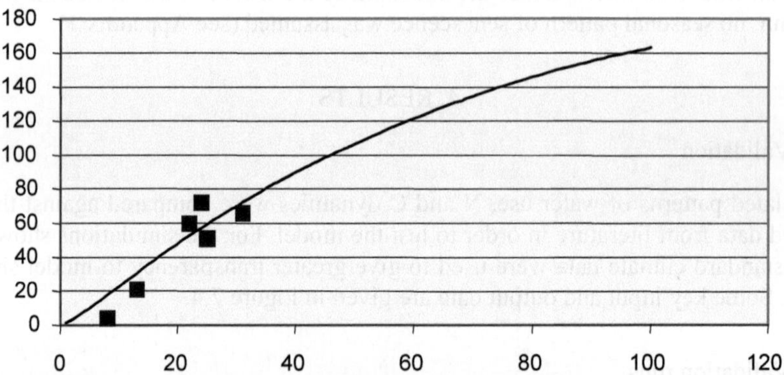

Figure 7.5. a. *Measured (squares) and modelled (line) data for biomass production in an oak chronosequence at Vestskoven in Denmark.*

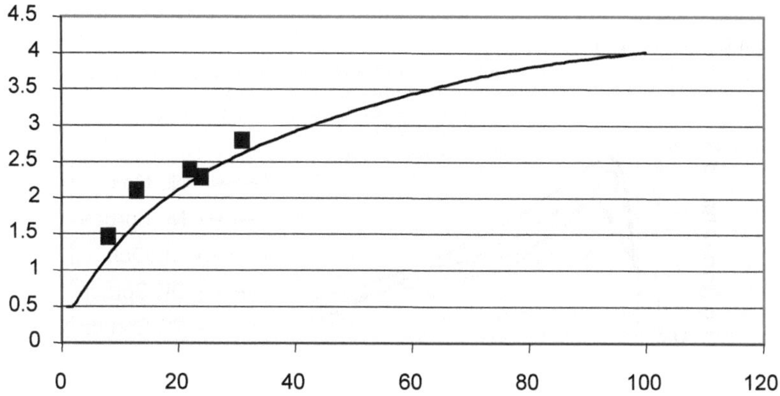

Figure 7.5. b. *Measured (square) and modeled (line) data for litterfall production in an oak chronosequence at Vestskoven in Denmark.*

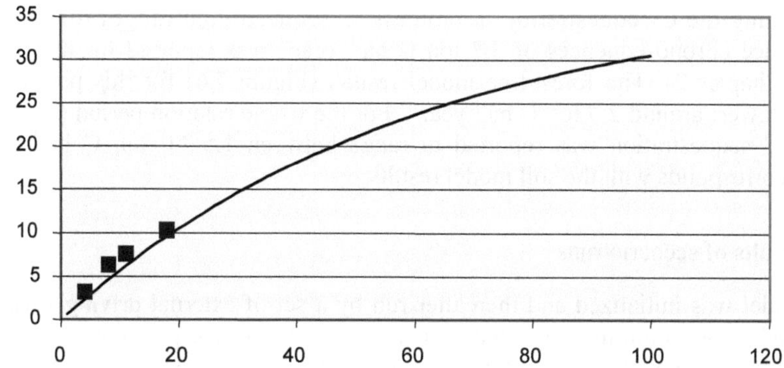

Figure 7.5. c. *Measured (square) and modeled (line) data for tree height in an oak chronosequence at Vestskoven in Denmark.*

species and processes, leading to a limited number of data for validation of the model, i.e. for oak trees. As a consequence, other tree species in the model have not been validated thoroughly. This is an important limitation for the validation of the model as a whole. For the other tree species we instead used the GUI for parameter estimation. Although, the model is not firmly validated for all tree species, the processes are mechanistically implemented in the model so that future adjustment on basis of additional information can be easily done. However, the forest/tree model was also tested in the complete metamodel - METAFORE - by a model to model comparison, literature data and with observations based on chronosequences, and showed good results (Chapter 9).

Modeled C sequestration by the trees for all chronosequences is summarized in Figure 7.6. The highest C sequestration occurred during the first 20 years of forest development. The modeled C sequestration during this period (1.5 to 3 ton C ha^{-1}

year^{-1}) was twice as high as the C sequestration during the period 50 to 100 years (1 to 1.5 ton C ha^{-1} year^{-1}).

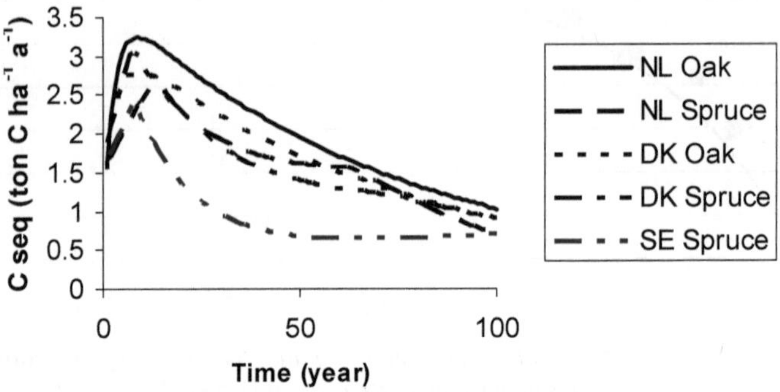

***Figure 7.6**. Modelled C sequestration in the trees for all chronosequences.*

Concerning the C sequestration in biomass, C sequestration ranges for both oak and spruce chronosequences of 3.7 ton C ha^{-1} year^{-1} was reported for the first 40 years (Chapter 2). The forest/tree model results (Figure 7.6) for this period were clearly lower: around 2.5 ton C ha^{-1} year^{-1}. For the whole rotation period (about 100 years) C sequestration was reported to range between 1.5-2.0 ton C ha^{-1} year^{-1}, which corresponds with the soil model results.

4.3. Results of scenario runs

The model was initialized and thereafter run by a set of external driving variables. The model used rainfall, net radiation depending on latitude position and average minimum and maximum temperatures recorded (Table 7.1). Since the Metamodel works with classes, all variables are set to class ranges.

***Table 7.1.** Values for climatic variables for simulation runs. The values are taken from the European climate atlas (ECSN 2005).*

	Precipitation (mm/yr)	Minimum temperature (C^0)	Maximum temperature (C^0)	N deposition (kg ha^{-1} yr^{-1})	Latitude (degrees)
Sweden	700 - 750	-3 - 0	16 -18	10 -15	55 - 60
Denmark	700 - 750	0 – 2	14 -16	10 -15	55 – 60
Netherlands	700 – 750	0 – 2	14 -16	25 -30	50 – 55
Belgium	700 – 750	0 – 2	16 -18	10 -15	50 – 55

From the results in Figure 7.7a to 7.7d, it is obvious that the different trees have different growth rates and as a result the total biomass (above- and below ground) is different after 100 years (Figure 7.7 a-d). In Sweden Oak had the highest total biomass after 100 years (ca. 250 tons per ha), then beech (ca. 180 tons per ha), then spruce (170 tons per ha) and finally pine (ca. 150 tons per ha). In the Netherlands the growth rates were higher and here spruce had the highest biomass (290 tons per ha) (Figure 7.8. a-d). The strongest difference in driving variable is the amount of atmospheric N deposition. The responses of the individual tree species to N deposition are substantial, i.e. the total biomass of all species increased significantly. However, the increase of beech, spruce and of pine is bigger than of oak. In comparison to the simulation results of Sweden, the total biomass of beech increased from approximately 175 to 250 tons per hectare as a result of the increased amount of N deposition. The total biomass of oak stands increased less significantly.

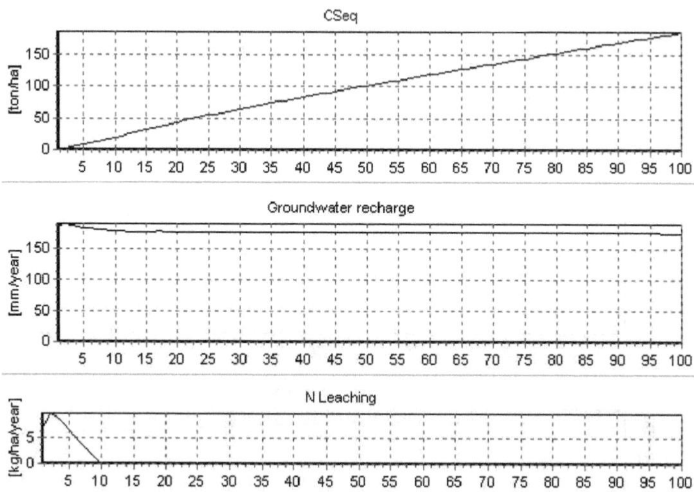

a

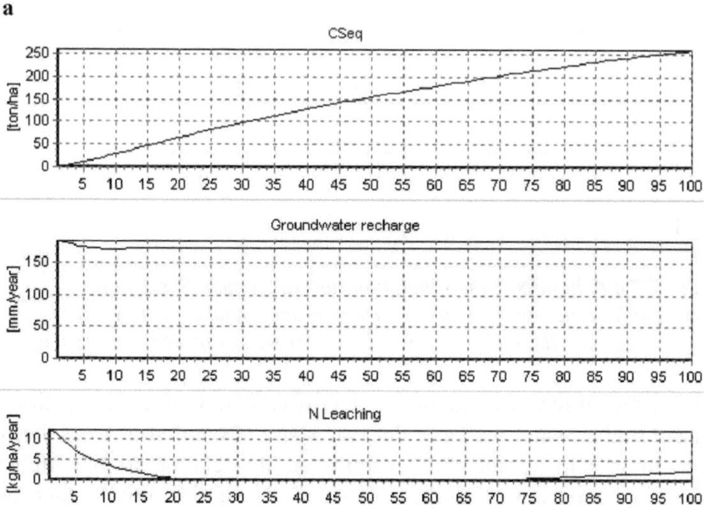

b

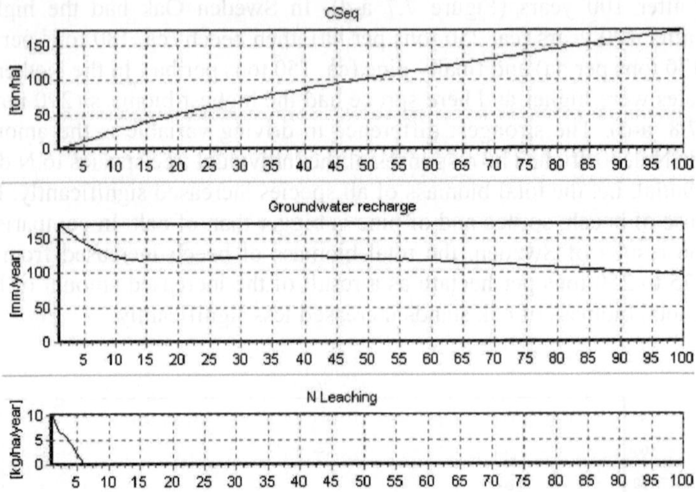

c

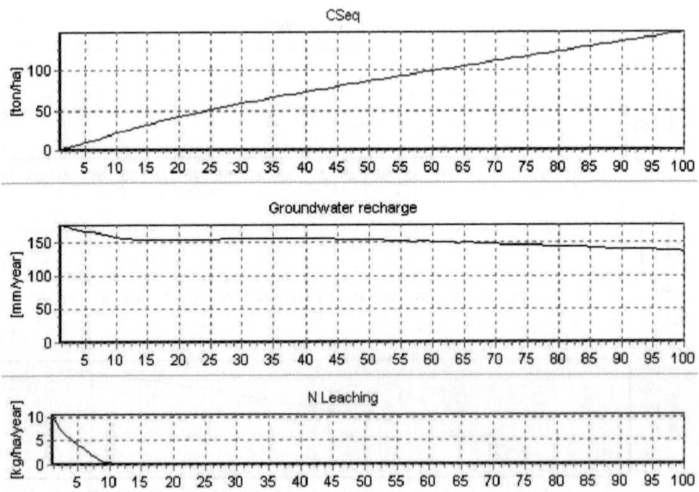

d

***Figure** 7.7 **a-d.** Results of the simulation runs in Sweden. Results are: Top – total C-sequestration (tons ha^{-1} yr^{-1}); middle – groundwater recharge (mm yr^{-1}); and bottom – N-leaching (kg ha^{-1} yr^{-1}) for a) beech, b) oak, c) spruce and d) pine.*

When water recharge was simulated under different tree species, deciduous trees had a significant higher water recharge than the stand of coniferous trees, because of the difference in growth season and evaporation. In relation to the higher biomass production in coniferous tree species, which is significantly higher than in the deciduous tree species, the amount of water recharge under both types of coniferous tree stands (spruce and pine) decreased significantly (>50 cm). Probably, a too fast increase in evapotranspiration was modelled (Chapter 8) and the connection between the forest/tree model, which calculates the potential evapotranspiration during the tree development, and the soil model, which calculates the actual transpiration could be improved.

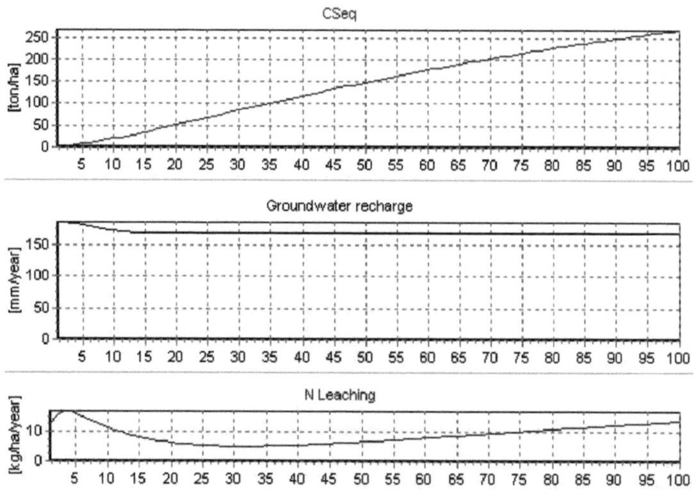

a

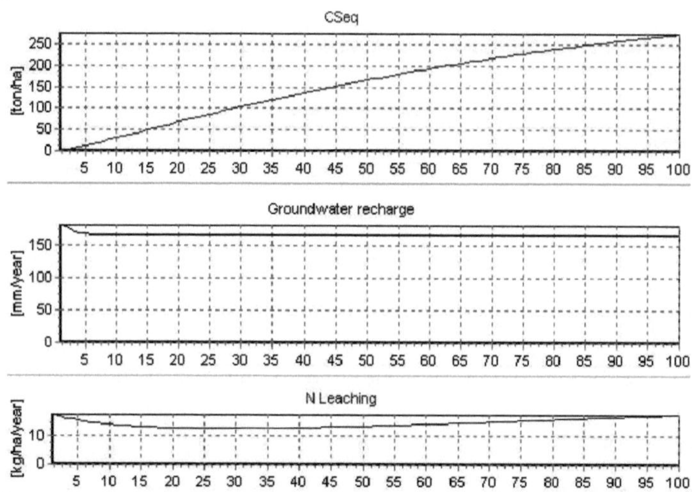

b

Figure 7.8 a-d. *Results of the simulation runs in the Netherlands. Results are: Top – total C-sequestration (tons ha^{-1} yr^{-1}); middle – groundwater recharge (mm yr^{-1}); and bottom – N-leaching (kg ha^{-1} yr^{-1}) for a) beech, b) oak, c) spruce and d) pine.*

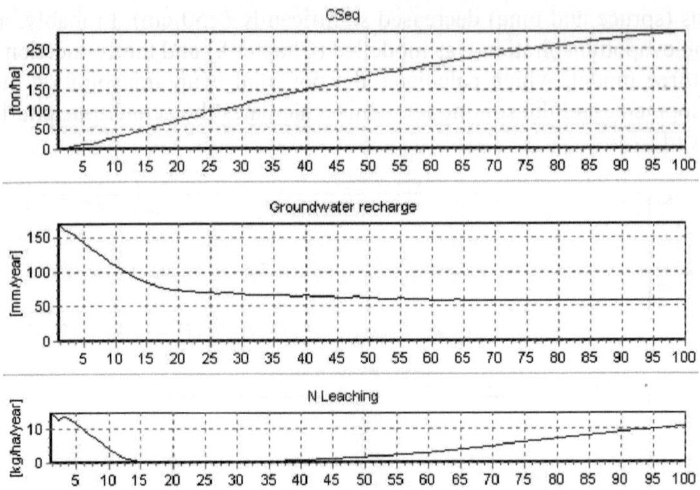

c

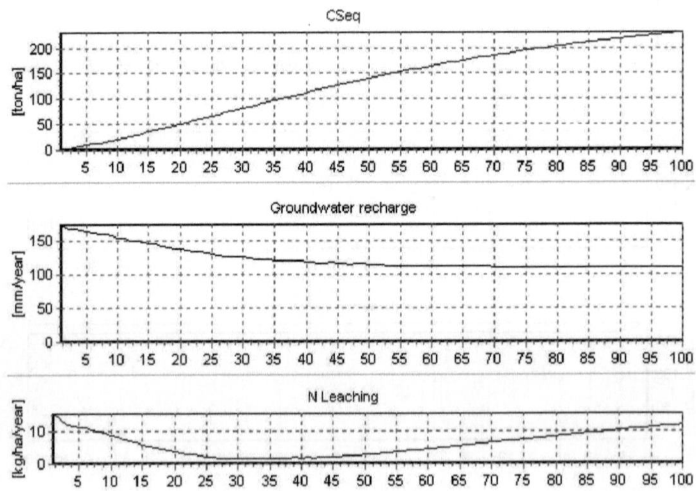

d

Figure 7.8 a-d. *Continued*

The modelled period of enhanced nitrate leaching fluxes was significantly different for the Netherlands and Sweden. In Sweden, this enhanced nitrate leaching was only 5-10 years long as opposed to in the Netherlands where the period of large nitrate leaching after afforestation was more extended (10-25 years). For both deciduous species the modelled period of enhanced nitrate leaching fluxes was significantly higher than for both coniferous tree species.

The amount of nitrate leaching during the period of 100 years was rather similar among the different tree species. In the Netherlands, nitrate leaching was modeled to be low or zero in some years after 15-40 years after afforestation. This is supported by the data from the chronosequences (Chapter 4), however, here the nitrate leaching raised after approximately 20 years after afforestation. The modeled results in nitrate leaching can be expected as also observed in a study by Hansen et al (2006). At the beginning after afforestation, nitrate leaching is rather high due to release of N from former agricultural practice, followed by a gradual decrease due to an increase in uptake and a decrease in mobilization. The lowest nitrate leaching was found half way through the simulated period. When the forests grow older, the nitrate leaching increases again due to decreased growth rate of the trees, and due to an increased N deposition with increased height and roughness length. The results of the simulations of forest stands in Denmark and Belgium show very similar patterns and are only slightly different from those of Sweden.

5. DISCUSSION

There has been a long-standing interest in developing models with which tree stand growth can be simulated (e.g. Goulding 1994), but to date the task has been only partially accomplished. Forest growth is a highly complex process which can be controlled by a variety of environmental and physiological factors, and models that do not comprehensively incorporate all the factors that may affect forest growth can only describe forest growth under a limited set of conditions

Based on the criteria for the aims of the AFFOREST-project, the CENW model was preferred to model linked C, nutrient and water cycles (Kirschbaum 1999). The CENW model has been used as basic model for further development and prototyping in the AFFOREST-project. Generally, it can be concluded that the performance of the tree model is adequate as part of the Metamodel for total biomass and litter production, water transpiration and nitrate uptake and release in beech, oak, spruce and pine tree stands. For water recharge the performance is probably less sufficient (Chapter 8).

The validation and application of the AFFOREST tree model to the AFFOREST oak chronosequence in Sweden showed that C behaviour, i.e. total biomass, litterfall, and tree height can be modelled very well. In comparison to the chronosequence data, a slight decrease in C sequestration was modelled for the first 20 years after afforestation, followed by a slight continuous increase. The slight decrease during the first decades corresponds with data and previous research on changes in C pools in the soil after afforestation of arable land (Vesterdal et al. 2002).

The modelled course in N dynamics reflected in the nitrate leaching seems likely. After afforestation nitrate leaching is rather high due to release of N from former agricultural practice, followed by a gradual decrease due to an increase in uptake and a decrease in mobilisation. After canopy closure, nitrate leaching seems to increase again. This is also supported by the data from the chronosequences (Chapter 4) and by Hansen et al (2006).

The AFFOREST tree part of the Metamodel was tested by comparisons with observations based on chronosequences. The chronosequences appeared to be an excellent means for validating the predicted afforestation impact. However, the problem remains whether these observed afforestation impacts are representative for afforestation impacts in the entire study area. As the AFFOREST-sDSS will be used to identify areas where afforestation will deliver the best environmental performance, the model's capacity for predicting the spatial pattern of afforestation impacts should be checked. As is true for all models, model validation is not an activity leading to an absolute and definite judgement on models adequacy. Therefore, it is absolutely necessary to collect additional data, not only in space but also in time.

6. CONCLUSIONS

In general, it can be concluded that the forest/tree model is an acceptable alternative for more complex process oriented models and that it fulfils the necessarily requirements for use within the Metamodel METAFORE (Chapter 9). The model was able to successfully simulate dynamics in C, N and water across the stands over 5-35 years in the Swedish chronosequence. To model growth with these differing stands is a significant confirmation that the model has successfully captured the essence of growth limitations under natural conditions. Despite the remaining discrepancies, model performance was good for the description of these data. This provides a degree of confidence to applying the model to a variety of hypothetical questions, such as those related to climate change, and investigations of afforestation options. For such simulations, it is essential to incorporate all the relevant feed-backs that interact with the initial response to the environmental change. It is expected that the present model can be used in coming years to be applied to a range of questions of sustainability, climate change and afforestation management.

Acknowledgement. We wish to thank Dr. Miko Kirschbaum from CSIRO, Canberra, Australia for providing the source code of CENW and the possibility for one of the authors (GH) to work in his lab for two months to take apart and reconstruct the CENW model in all its details.

7. REFERENCES

Aber, J. D., & Federer, C. A (1992). A generalized, lumped-parameter model of photosynthesis, evapotranspiration and net primary production in temperate and boreal forest ecosystems. *Oecologia, 92*, 463-474.

Agren, G. I., & Axelsson, B. (1980). PT - A tree growth model. *Ecological Bulletins, 32*, 525-536.

Amthor, J. S. (1994). Plant respiratory responses to the environment and their effects on the carbon balance. In R. E. Wilkinson (Ed.), Plant-environment Interactions, pp. 501-554. New York: Marcel Dekker.

Arp, P. A., & Oja, T. (1992). Acid sulfate/nitrate loading of forest soils: forest biomass and nutrient cycling modelling. In P. Grennfelt & E. Thoerneloef (Eds.), Critical loads for nitrogen - A workshop report, p. 307-357. Copenhagen, Denmark: NORD.

Berdowski, J. J. M., Van Heerden, K., Van Grinsven, H. J. M., Van Minnen, J. G., & De Vries, W. (1991). Model input. In: SOILVEG: A model to evaluate effects of acid atmospheric deposition on soil and forest,Anonymous National Institute of Public Health and Environmental Protection, Research Institue for Nature Management and Winan Staring Centre for Integrated Land, Soil and Water Research, pp. 46-68.

Booth, T. H., Nguyen H. N., Kirschbaum, M. U. F., Hackett, C., & Jovanovic, T. (1999). Assessing possible impacts of climate change on species important for forestry in Vietnam. *Climate and Chemistry, 41,* 109-126.

Botkin, D. B., Janak, J. F., & Wallis, J. R. (1972). Some ecological consequences of a computer model of forest growth. *Journal of Ecology, 60*(3), 849-872.

Busing, R. T. (1991). Busing. A spatial model of forest dynamics. *Vegetatio, 92*, 167-179.

Dewar, R. C., & McMurtrie, R. E. (1996). Sustainable stemwood yield in relation to the nitrogen balance of forest plantations: a model analysis. *Tree Physiology, 16,* 173-182.

Eckersten, H. (1994). Modelling daily growth and nitrogen turnover for a short-rotation forest over several years. *Forest Ecology and Management, 69*, 57-72.

ECSN. (2005). European Climate Support Network. http://eca.knmi.nl/

Friend, D., Shugart, H. H., and Running, S. W. (1993). A physiology-based gap model of forest dynamics. *Ecology, 74*(3), 792-797.

Goulding, C. J. (1994). Development of growth models for *Pinus radiata* in New Zealand-experience with management and process models. *Forest and Ecology Management, 69,* 331-334.

Hansen, K., Rosenqvist, L., Vesterdal, L. & Gundersen, P. (2006). Nitrogen leaching from three afforestation chronosequences on former arable in Denmark. *Global Change Biology (submitted).*

Hauhs, M., A. Kastner-Maresch, & Rost-Siebert, K. (1995). A model relating forest growth to ecosystem-scale budgets of energy and nutrients. *Ecological Modelling, 83*(1-2), 229-243.

Hunt, E. R., Lavigne, M. B., & Franklin, S. E.(1999). Factors controlling the decline of net primary production with stand age for balsam fir in Newfoundland assessed using an ecosystem simulation model. *Ecological Modelling, 122*(3), 151-164.

Jorritsma, I. T. M., Van Hees, A. F. M., & Mohren, G. M. J. (1999). Forest development in relation to ungulate grazing: a modeling approach. *Forest Ecology and Management, 120*(1-3), 23-34.

Kirschbaum, M. U. F. (1999). CenW, a forest growth model with linked carbon, energy, nutrient and water cycles. *Ecological Modelling, 118*, 17–59.

Korzukhin, D., & Ter-Mikaelian, M. T. (1995). An individual tree-based model of competition for light. *Ecological Modelling, 79*(1-3), 221-229.

Krauchi, N. (1995). Application of the model FORSUM to the Solling spruce site. *Ecological Modelling, 83*(1-2), 219-228.

Landsberg, J. J., & Waring, R. H. (1997). A generalised model of forest productivity using simplified concepts of radiation-use efficiency, carbon balance and partitioning. *Forest Ecology and Management, 95*(3), 209-228.

McMurtrie, R. E., Landsberg, J. J., & Linder, S. (1989). Research priorities in field experiments on fast-growing tree plantations: implications of a mathermatical production model. In J. S. Pereira and J. J. Landsberg (Eds.), Biomass Production by Fast-Growing Trees, pp. 181-207. Dordrecht: Kluwer Academic Publishers.

McMurtrie, R. E., Leuning, R., Thompson, W. A., & Wheeler, A. M. (1992). A model of canopy photosynthesis and water use incorporating a mechanistic formulation of leaf C02 exchange. *Forest Ecology and Management, 52,* 261-278.

Mohren, G. M. J. (1987). Simulation of forest growth, applied to Douglas fir stands in the Netherlands, Wageningen. 184 pp.

Monteith, J. L. (1965). Evaporation and environment. Symp. *Society for Experimental Biology. 19,* 205-234.

Parton, W. J., Schimmel, D. S., Cole, C. V., & Ojima, D. S. (1987). Analysis of factors controlling soil organic matter level in Great Plains grasslands. *Journal of American Soil Science Society, 51,* 1173-1179.

Sands, P. J. (1995). Modelling canopy production. II. From single-leaf photosynthesis parameters to daily canopy photosynthesis. *Australian Journal of Plant Physiology, 22,* 603-614.

Schall, P. (1995). Application of the model FIWALD to the Solling spruce site. *Ecological Modelling, 83*, 151-162.

Van Dam, D., & Van Breemen, N. (1995). NICCCE: a model for cycling of nitrogen and carbon isotopes in coniferous forest ecosystems. *Ecological Modelling, 79*(1-3), 255-275.

Van Minnen, J. G., Meijers, R., & Braat, L. C. (1995). Application of the FORSOL model to the spruce site at Solling, Germany. *Ecological Modelling, 83*(1), 197-205.

Van Oene, H., & Agren, G. I. (1995). The application of the model NAP to the Solling spruce site. *Ecological Modelling, 83*(1), 139-149.

Van Oene, H., Berendse, F., & De Kovel, C. G. F. (1999). Model analysis of the effects of historic CO_2 levels and nitrogen inputs on vegetation succession. *Ecological Applications, 9*(3), 920-935.

Vesterdal, L., Ritter, E., & Gundersen, P. (2002). Change in soil organic carbon following afforestation of former arable land. *Forest and Ecology Management, 169*(1-2), 137-147.

Weinstein, A., & Beloin, R. (1990). Evaluating effects of pollutants on integrated tree processes: a model of carbon, water, and nutrient balances. In R. K. Dixon, R. S. Mehldahl, & G. A. Ruark (Eds.), Process modeling of forest growth responses to environmental stress, pp. 313-323. Portland:Timber Press.

Appendix 1. Abbreviations and list with different factors used

A	= species specific height at the start	- m
A_{max}	= maximum assimilation rate under non-limiting factors	- μmol m^{-2} s^{-1}
A_{Pres}	= atmospheric pressure in Pascal	- 1000
α	= the quantum yield	- mol mol^{-1}
α_0	= the potential RuBP regeneration rate at a given temperature	- 0.09 at 25 ^{0}C
a_1	= parameter in f_1, which is function of depth in canopy	- 0.22
a_2	= parameter in f_1, which is function of depth in canopy	- 0.18
b_1	= parameter in f_2, which is function of depth in canopy	- 0.74
b_2	= parameter in f_2, which is function of depth in canopy	- 0.5
B	= a species specific increase ratio	-
D_n	= the daily fraction of stems lost	-
h	= day length in seconds	- s
k_1	= the light extinction coefficient	- 0.5 – 0.6
k_2	= a conversion to active photon flux density	- 2.000.000
L	= leaf area index	- m^{-2} m^{-2}
N_{act}	= the amount mineralised from the active pool of organic matter	- kgN $month^{-1}$
N_{dep}	= the amount deposited from the atmosphere	- kgN $month^{-1}$
N_{lim}	= a N limitation parameter, defined with a similar function as for temperature. where N_{Lim} is foliar N concentration in the canopy	- equation 7.13
N_{min}	= the total amount of N becoming available in mineral form	- kgN $month^{-1}$
$1\text{-}m$	= leaf transmissivity	-
θ	= shape of light response	- 0.7
Q_a	= absorbed total radiation	- MJ m^{-2} d^{-1}
Q_i	= daily incident radiation	- MJ m^{-2} d
q	= normalized radiation	-
P	= Pascal	- 1000
r	= the fraction of radiation that is reflected (albedo)	- 0.125
r_g	= an empirical term that quantifies the amount of C lost in growth respiration per unit of new growth	- 0.0045
r_m	= an empirical term that gives the daily respiration rate per unit of N at 25°C	
R_t	= the maximum height at time t	- m
S_b	= a constant minimum foliage senescence rate. Root and stem senescence is modelled in a similar way	- kg kg^{-1}
S_l	= specific leaf area	- m^2 kg^{-1}
T_{day}	= mean daytime temperature in Celsius degrees	- °C
t_{max}	= mean maximum temperature	- °C
t_{min}	= mean minimum temperature	- °C
T_w	= annual heat index	- 12((T_t/5)exp1.514)
T_{lim}	= a simple hump function	- equation 7.12
T_{MinLim}	= the minimum temperatures that allow any photosynthesis	- °C
T_{MaxLim}	= the maximum temperatures that allow any photosynthesis	- °C
T_{opt1}	= lower upper temperature bound that allow optimum photosynthetic rates	- °C
T_{opt2}	= higher upper temperature bound that allow optimum photosynthetic rates	- °C
T_{temp}	= mean temperature	- °C
W_f	= foliar biomass	- g.m^{-2}
W_{lim}	= water availability in the soil defined	- equation 7.12
X_n	= an empirical excess N storage ratio	-

List with allocation and death rate factors

	Oak	Beech	Spruce	Pine
Allocation Leafs	0.1419	0.1827	0.1541	0.1138
Allocation Roots	0.1609	0.1533	0.0939	0.1369
Allocation Stem	0.6972	0.6641	0.752	0.7493
Death rate Roots	0.002	0.002	0.002	0.002
Death rate Stem	0.001	0.0008	0.002	0.0025
Senescence leafs	1	1	0.015	0.015

CHAPTER 8

MODELLING THE AFFORESTED SYSTEM: THE SOIL AND WATER COMPARTMENT

H. KROS[1], J. MOL[1], B.J. GROENENBERG[1], G.J. REINDS[1], G.W. HEIL[2], W. VAN DEURSEN[3] and W. DE VRIES[1]

[1]Alterra Green World Research, P.O. Box 47, 6700 AA Wageningen, the Netherlands, E-mail: Hans.Kros@wur.nl, [2] Faculty of Science, Institute of Environmental Biology, Sobonnelaan 16, 3584 CA Utrecht, the Netherlands, [3]Carthago Consultancy, Oostzeedijk Beneden 23a, 3062VK Rotterdam, the Netherlands

Abstract. Existing complex mechanistic models require too many input data, which are generally unavailable for application at a regional scale. For the development of a simplified soil model, however, we used existing complex models to start with. This soil model includes all relevant processes in order to simulate the temporal trajectory of carbon (C) sequestrations, nitrate leaching and water recharge. The soil model was derived from the existing models NUCSAM, SMART2 and SMB. For the water balance the existing model WATBAL was used. The principal question was whether the derived simplified soil model is acceptable for use within the AFFOREST project. To test this, the soil model was applied to two afforested oak chronosequences and three spruce chronosequences in Sweden, Denmark and the Netherlands. Validation of the soil and water model was performed by:

- applying the model to the AFFOREST chronosequences and comparing the output on: (i) C sequestration in the soil, (ii) water recharge and (iii) nitrate concentration and nitrate leaching with (a) results from the detailed soil model NUCSAM and (b) measured values in chronosequences
- testing the plausibility of the model behaviour through the evaluation of scenarios on nitrogen (N) deposition levels and temperature for various combinations of soil type and tree species.

Carbon sequestration in the soil is modelled satisfactory but has a tendency to be underestimated. Modelled water recharge fluxes and nitrate leaching fluxes are also in accordance with experimental results reported in literature. In general, we conclude that the soil and water part of the metamodel is an acceptable and even a necessarily alternative for a more complex process oriented model and that it fulfils the necessary requirements for use within the AFFOREST project.

G.W. Heil et al. (eds.), Environmental Effects of Afforestation in North-Western Europe, 175–202.

1. INTRODUCTION

Several studies on the separate effects of afforestation on e.g. water recharge, nitrogen (N) leaching and carbon (C) storage have been performed, but less emphasis has been put on the combined effect of afforestation on changes in C, N and water pools and fluxes. Several studies have been performed on the effect on water recharge (Chapter 3). These studies indicate that water recharge declines with an increase in forest cover. Where the decline in water recharge is generally larger in coniferous forest than in deciduous forest. Effects of afforestation on nitrate leaching have been studied less extensively (Chapter 4). There is however, evidence that leaching from afforested arable land is higher in the younger stage of the succession compared to mature forests (Callesen et al. 1999). Studies on C sequestration following afforestation or development of natural succession forest on former arable land in temperate regions indicate significant increases in soil C stores (Chapter 2). The process of C sequestration in biomass is comparable for afforested stands and any other stand regenerating following disturbance. However, rates of growth and hence rates of C sequestration may be higher for afforested stands because of high nutrient status of former arable soils and less competition from woody shrubs compared to stands planted on clear cuts. The studies on afforestation of arable land (Chapter 2) indicate that C is sequestered faster in biomass than in soils. On the other hand, soil C sequestration may at some sites be expected to go on for much longer time before a steady state between input and decomposition of organic matter is reached after land use change.

The challenge is to build knowledge and capacity regarding afforestation in respect to the combined changes in C, N and water pools and fluxes. Therefore, a spatial Decision Support System (AFFOREST-sDSS) for scenario analysis and environmental impact assessment has been developed (Chapter 10). To evaluate and optimize afforestation in terms of C-sequestration and environmental impacts, a model is needed that integrates present knowledge concerning the water, C and N cycling in forest ecosystems. This chapter describes the development and validation of a simple soil model that is used within the metamodel METAFORE (Chapter 9) for application at both a national and a European scale.

Existing relatively complex mechanistic models require many input data that are generally unavailable for their application on a regional scale. Even if the model structure is correct (or at least adequately representing current knowledge), the uncertainty in the output of complex models may still be large because of the uncertainty in input data. One of the common characteristics of environmental problems such as climate change and air pollution is that they play a role on a local, regional, national, continental and even global scale. It is imperative that the spatial and temporal aspects considered in a model must fit its objectives. Since data availability on a plot scale is relatively large, this scale is chosen as most appropriate to study biogeochemical processes in situ and is thus the most logical level to start with model development. Because of sparser data at larger scale, the scale of the model must however be adapted to the scale of data availability. One possibility is to simplify the model description in such a way that the temporal and spatial resolution is comparable to the resolution of the data. During such a simplification of processes, model results must remain reliable. The reliability can usually be determined by comparing results from the simplified model and the local scale model (e.g. Kros 2002). Presumably, there is an optimal level of model complexity, i.e. a point where the degree of model complexity, e.g. in terms of state variables, match the data resolution and quality, leading to maximal knowledge gain about the modelled system (Jørgensen 1992). Furthermore, simplification is also needed in order to comply with the performance requirements of the AFFOREST-sDSS (Chapter 10).

Therefore, we derived a new simplified soil model for the metamodel METAFORE (Chapter 9). The soil part was derived from the models SMART2 (Kros 2002). and SMB (De Vries et al. 1994). For the water balance the existing model WATBAL (Starr 1999) was used. Validation of the soil and water model was performed by:

- Testing the C and N behaviour of forest soils using literature data on mineralization
- Applying the model to the AFFOREST chronosequences and comparing the output on: (i) C sequestration in the soil, (ii) water recharge and (iii) nitrate concentration and nitrate leaching with (a) modelled results obtained by applying the detailed soil model NUCSAM and (b) estimates derived form observations in various chronosequences
- Testing the plausibility of the model by running the model while using alternating inputs and inspecting the effect on C sequestration in biomass and in the soil, water recharge and nitrate leaching. Scenarios were run for various soil types, tree types, temperatures and N deposition levels.

2. MODELS

In this Chapter, mainly the soil and water model of the metamodel is described. Furthermore, a brief explanation of the complex soil model NUCSAM (Kros 2002) that was used as a reference model is given. The vegetation model, including processes such as potential evapotranspiration, growth, N uptake and litterfall are described in Chapter 7.

2.1. The structure of the soil model in the metamodel environment

The place of the soil model and the interactions with the other models in the metamodel is shown in Figure 8.1. The set up of the approach is the inclusion of simple quantitative relationships focusing on C sequestration in the soil, water recharge and nitrate leaching by use of transfer functions.

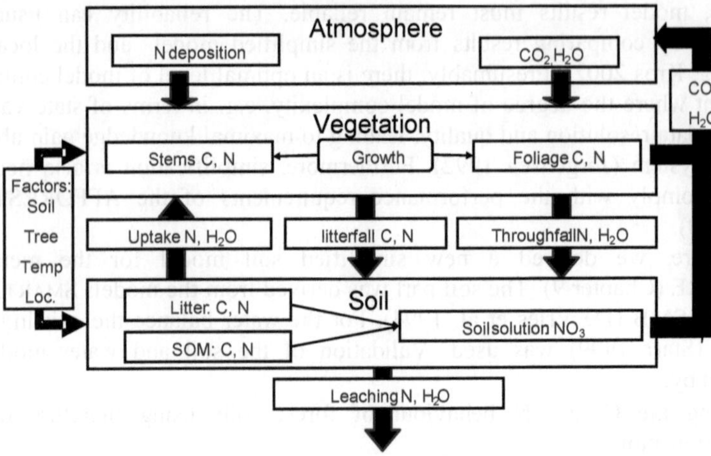

Figure 8.1. *Schematic representation of the soil model (lower part) within the metamodel METAFORE for the AFFOREST-sDSS.*

In most cases these are class transfer functions, e.g. $CN_{soil\ ara}$(sand) = 17, $CN_{soil\ ara}$ (clay) = 11, $CN_{soil\ ara}$ (peat) = 35. The principle equations are given in the following sections.

2.1.1. Water recharge

For the water recharge calculations we used an existing simple water balance model called WATBAL (Starr 1999). WATBAL is a monthly water balance model for forest soils based on the following water balance equation for the rooting zone:

$$P = ET + R \pm \Delta SM \qquad (1)$$

Where:

P = Precipitation in mm month^{-1}

ET = Evapotranspiration in mm month^{-1}

R = Soil water flux in mm month^{-1}

ΔSM = Change in soil water in the rooting zone mm month^{-1}

One of the main advantages of WATBAL is that the model uses input data which are either easily obtained or can be derived from other basic data using transfer functions. It uses relatively simple and readily available climate variables (e.g. precipitation, air temperature and cloudiness) and the available water capacity (AWC) of the soil, which can be derived using transfer functions based on soil texture, bulk density and organic matter content or from the soil moisture curve. It handles snowmelt and sloping sites if the appropriate slope factors are given. Besides estimating values for soil water flux, the components of the water balance are determined: potential (PET) and actual evapotranspiration (AET), soil moisture, snow pack store and snowmelt, as well as global (direct and diffuse) radiation. Evapotranspiration is calculated using an estimation of global radiation using a reference crop equation adjusted by a crop (forest stand) coefficient to take into account the greater evapotranspiration from forest soils. The evapotranspirative withdrawal of soil water may take place at the PET rate or at a reduced rate, the AET rate, depending on a relationship determined by the soil water storage (SM) and the AWC in the rooting zone. Evapotranspirative withdrawal of soil moisture can only take place from the rooting zone where losses from the soil beyond the rooting zone only take place through drainage. If P (+ snowmelt) is in excess of PET, then the excess goes to fill the storage capacity of the soil. If the AWC is filled then any further excess of P (+ snowmelt) goes to form drainage, i.e. the soil water flux from the soil layer in question. WATBAL has been validated for several sites (Starr 1999) with measured soil water fluxes from in-situ (zero-tension) gravity lysimeters and soil moisture content measured with TDR probes.

In WATBAL, the monthly PET and monthly P were used as input. The monthly PET was calculated dynamically by the vegetation model (Chapter 7), whereas the monthly precipitation was derived from the yearly precipitation as stored in within the AFFOREST-sDSS (Chapter 10). The model rescaled the yearly precipitation values to monthly values (see section 3.2).

The monthly PET, as calculated by the vegetation model, was reduced by the interception evaporation (E_i):

$$E_i = f_{i \cdot t} \cdot P_t \tag{2}$$

where:

$f_{i\,t}$ = Tree type dependent interception fraction for month = t (-)
P_t = Precipitation in mm month^{-1}

The monthly interception fraction ($f_{i\,t}$) is directly related with the leaf area index (LAI) as calculated with the vegetation model:

$$f_{i \cdot t} = \begin{cases} \dfrac{LAI_{it}}{LAI_{i \cdot mx}} \cdot f_{i \cdot mx} & \text{for } LAI \leq LAI_{mx} \\ f_{i \cdot mx} & \text{for } LAI > LAI_{mx} \end{cases} \tag{3}$$

where:

$LAI_{i\,t}$ = Leaf area index as calculated by the vegetation model for tree type i (-)
$LAI_{i\,mx}$ = The Leaf area index for tree type i beyond which the interception fraction is at its maximum (-)
$f_{i\,t}$ = Tree type and time dependent interception fraction (-)
$f_{i\,mx}$ = Interception fraction for a mature canopy of tree type i (-)

The monthly throughfall (TF_t) is then calculated as:

$$TF_t = P_t - E_i \quad (4)$$

where:
TF_t =Throughfall in mm month^{-1}

Finally, the remaining energy for the transpiration is then transferred to WATBAL:

$$PT_t = PET_t - E_i \quad (5)$$

where:
PT_t = Potential transpiration in mm month^{-1}
PET_t = Potential evapotranspiration in mm month^{-1}

Within water model the monthly soil cover (SC_t), which was used for snow evaporation, was linked to the AFFOREST-sDSS tree types with:

$$SC_t = \begin{cases} LAI_{it} & \text{for } LAI_i \leq 1 \\ 1 & \text{for } LAI_i > 1 \end{cases} \quad (6)$$

where:
$LAI_{i\,t}$ = Leaf area index as calculated by the vegetation model for tree type i (-)

2.1.2. Mineralization of C and N

The mineralization processes were mainly derived from the model SMART2 (Kros 2002). Contrary to the hydrology, the biogeochemical part of the soil model is calculated with a yearly time step. The accumulation of C in the litter layer is described by two fractions, fresh litter and old litter. Fresh litter can be regarded as one year old litter, whereas old litter is litter older than one year. Mineralization of fresh litter is determined by a fraction of litterfall (as calculated by the vegetation model), and old litter will be mineralized by first order kinetics:

$$\frac{dC_{soil}}{dt} = C_{lf}(t) \cdot (1 - fr_{mifl}) - kr_{miltnat}(T) \cdot C_{soil}(t) - kr_{miagr}(C_{ini} - C_{mn}) \quad (7)$$

where:
C_{soil} = the amount of organic C in the soil (kg ha^{-1})
C_{mn} = the amount of recalcitrant organic C (i.e. the minimum amount of C that always remain in the mineral soil) in the former agricultural soil (kg ha^{-1})
C_{ini} = the initial amount of organic C in the former agricultural soil (kg ha^{-1})
$C_{lf}(t)$ = the litterfall flux (kg ha^{-1} year^{-1})

$fr_{mifl}(T)$ = the mineralization fraction of freshly fallen litter as a function of temperature (-)

$kr_{miltnat}(T)$ = mineralization rate constant for newly formed organic matter (year^{-1}) as a function of temperature (T)

$kr_{miagr}(T)$ = mineralization rate constant for organic matter formed during the agricultural history (year^{-1}) as a function of temperature (T)

Assuming that C_{lf} is constant for the period [0,t], the equation can be solved as:

$$C_{soil}(t) = \frac{1-fr_{mifl}}{kr_{miltnat}} \cdot C_{lf} \cdot (1-\exp(-kr_{miltnat}(T)\cdot t)) + (C_{ini}-C_{mn})\cdot \exp(-kr_{miagr}\cdot t) + C_{mn} \quad (8)$$

In order to account for accumulated N in the mineral soil during agricultural practice, the amount of both C (C_{ini}) and nitrogen accumulated (N_{ini}) in the soil organic matter is transferred to the old litter pool. This ensures that the accumulated N will be mobilised during the first years after afforestation. The mineralised N can either be used for forest growth or leaching/denitrification.

The annual change in C storage in both litter and mineral soil is calculated as:

$$dC_{seqsoil}(t) = C_{soil}(t) - C_{soil}(t-1) \quad (9)$$

Where:

$dC_{seqsoil}(t)$ = the net annual C increment in the whole soil for the considered time period (=time step) (kg ha^{-1} year^{-1})

$C_{soil}(t)$ = the amount of C_{soil} at the end of current year (kg ha^{-1}).

$C_{soil}(t-1)$ = the amount of C_{soil} at the end of previous year (kg ha^{-1}).

Besides C mineralization, N mineralization must be included, because of N-mobilisation and immobilisation. N storage in both litter and mineral soil can be written in the same way:

$$N_{soil}(t) = \frac{1-fr_{mifl}\cdot rf_{CN}}{kr_{milt}\cdot rf_{CN}} \cdot N_{lf}(t)\cdot(1-\exp(-kr_{miltnat}(T)\cdot rf_{CN}\cdot t)) + \\ (N_{ini}-N_{mn})\cdot\exp(-kr_{miltagr}\cdot rf_{CN}\cdot t) + N_{mn} \quad (10)$$

where:

$N_{soil}(t)$ = the amount of organic N in the soil at the end of previous year (kg ha^{-1})

N_{ini} = the initial amount of organic N in the former agricultural soil (kg ha^{-1})

N_{mn} = the amount of recalcitrant organic N (i.e. the minimum amount of N that always remains in the mineral soil) in the former agricultural soil (kg ha^{-1})

rf_{CN} =a C/N ratio dependent reduction fraction (-)

In equation (10) the N mineralization values are reduced at low N contents (high C/N ratios) to account for immobilisation by microbes according to (Janssen 1984):

$$rf_{CN} = \begin{cases} 1 & \text{for } CN_s \leq CN_{mo} \\ 1 - \dfrac{CN_s - CN_{mo}}{DA_{mo} \cdot CN_{mo}} & \text{for } CN_{mo} < CN_s < (1 + DA_{mo}) \cdot CN_{mo} \\ 0 & \text{for } CN_s \geq (1 + DA_{mo}) \cdot CN_{mo} \end{cases} \qquad (11)$$

where:

CN_{mo} = C/N ratio of the micro-organisms decomposing the substrate (-)

CN_s = the C/N ratio of the substrate (fresh litter (s=fl), old litter (s=lt))

DA_{mo} = the dissimilation to assimilation ratio of the decomposing microbes (-).

Values for DA_{mo} and CN_{mo} are related to fungi because they are mainly responsible for mineralization of forest litter.

The effect of temperature on the mineralization rate of litter formed by afforestation was included by using a response function described by Kirschbaum (2000) scaled to a reference temperature of 10°C:

$$rf_T = \cdot e^{3.36\left(\frac{T-40}{T+31.79} - \frac{10-40}{10+41.79}\right)} \qquad (12)$$

This response function was applied to the reference mineralization rate constants of litter formed by afforestation ($kr_{miltnat}$) and the organic mater leftover from former agricultural practise ($kr_{miltagr}$):

$$kr_{milt[nat/agr]}(T) = rf_T \cdot kr_{milt[nat/agr]\ ref} \qquad (13)$$

where:

$kr_{mit[nat/agr]\ ref}$ = the mineralization rate constant for newly formed organic matter (*nat*) and for the organic mater leftover from former agricultural practise (*agr*) at a reference temperature of 10°C.

The annual change in N storage in both litter and mineral soil is calculated as:

$$dN_{seqsoil}(t) = N_{soil}(t) - N_{soil}(t-1) \tag{14}$$

where:

$dN_{seqsoil}(t)$ = the net annual C increment in the whole soil for the considered time period (=time step) (kg ha^{-1} year^{-1})

$N_{soil}(t)$ = the amount of N_{soil} at the end of current year (kg ha^{-1}).

$N_{soil}(t-1)$ = the amount of N_{soil} at the end of previous year (kg ha^{-1}).

The annual mineralization flux ($N_{mi}(t)$) can be derived from the mass balance of litterfall (N in litterfall will either be a part of the organic soil pool or mineralized):

$$N_{mi}(t) = N_{lf}(t) - dN_{seqsoil}(t) \tag{15}$$

where:

$N_{lf}(t)$ = the annual N flux in litterfall (kg ha^{-1} year^{-1})

For a standard soil model application, the C and N stocks in organic matter refer to the total amounts in the litter layer and the top 30 cm of the mineral soil.

2.1.3. Nitrate leaching and denitrification

The nitrate leaching and denitrification processes were mainly based on the SMB model (De Vries et al. 1994).

The nitrate leaching (kg ha^{-1} year^{-1}) is calculated from the mass balance:

$$N_{le}(t) = N_{in}(t) + N_{mi}(t) - N_{gu}(t) - N_{de}(t) \tag{16}$$

where:

$N_{in}(t)$ = the annual atmospheric total N input to the forested ecosystem (kg ha^{-1} year^{-1})

$N_{mi}(t)$ = the annual mineralization flux for the total soil layer (kg ha^{-1} year^{-1})

$N_{gu}(t)$ = the annual total N uptake flux by the forest (kg ha^{-1} year^{-1})

N_{de} = the annual denitrification flux for the total soil layer (kg ha^{-1} year^{-1})

Together with equation (14), equation (15) can also be written as:

$$N_{le}(t) = N_{in}(t) + N_{lf}(t) - N_{seqsoil}(t) - N_{gu}(t) - N_{de}(t) \tag{17}$$

Denitrification in the soil model is described as a linear relation with N leaching:

$$N_{de}(t) = fr_{de} \cdot N_{le}(t) \tag{18}$$

where:
fr_{de} = soil dependent denitrification fraction (Table 8.5)

Substitution of equation (17) into equation (16) gives:

$$N_{le}(t) = \frac{N_{in}(t) + N_{lf}(t) - N_{seqsoil}(t) - N_{gu}(t)}{1 + fr_{de}} \quad (19)$$

Given the nitrate leaching flux and the water recharge flux (GR), the nitrate concentration in leachate (cNO_3 in mg NO_3 l^{-1}) can be calculated as:

$$cNO_3 = \frac{N_{le}}{GR} \cdot 100 \cdot \frac{62}{14} \quad (20)$$

where:
N_{le} = the nitrate leaching flux (kg ha^{-1} $year^{-1}$)
GR = Ground water recharge flux (mm $year^{-1}$)
The factor 100 (62/14) is used to convert from kg N ha^{-1} mm^{-1} to mg NO_3 l^{-1}.

2.1.4. The complex model: NUCSAM

NUCSAM is a process-oriented model that simulates the major hydrological and biogeochemical processes in the forest canopy, organic surface layer and mineral soil (Figure 8.2). It considers evapotranspiration, heat transport, canopy processes, litterfall, mineralization, below and aboveground nutrient uptake, soil processes and solute transport. The change in soil solution and solid phase chemistry is calculated from a set of mass balance equations, describing the input, output and interactions in each compartment. Vertical heterogeneity is taken into account by differentiating between soil layers. Each soil layer is a completely mixed homogeneous compartment of constant density.

Processes in the model are generally described by zero-order and first-order rate equations and equilibrium equations. Most process parameters are described as a function of temperature, soil moisture content and pH. Organic matter is divided over three pools with different decay rates (Groenenberg et al. 1998). The model includes all major ions playing an important role in nutrient cycling and soil acidification: H, Al, Na, K, Mg, Ca, NH_4, NO_3, PO_4, SO_4, Cl and organic anions (RCOO). The model is specially developed for application at forest stands that are intensively monitored for atmospheric deposition, precipitation (meteorological conditions), litterfall and soil solution chemistry.

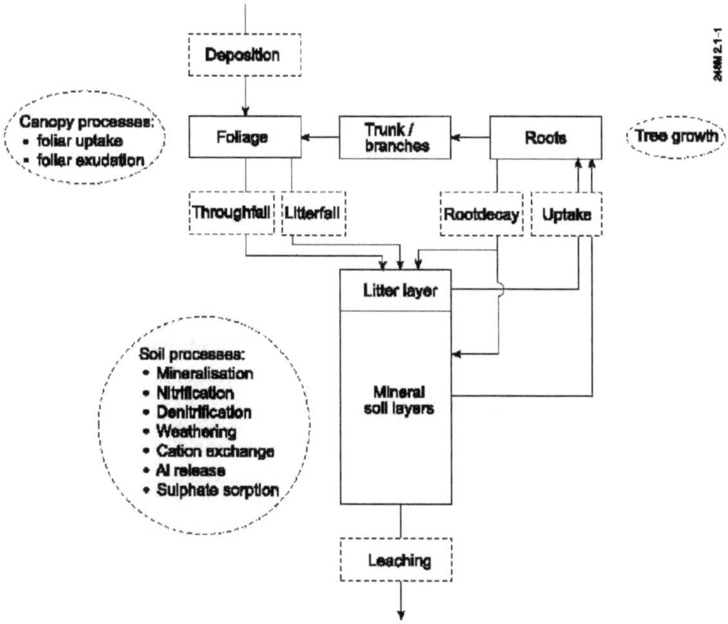

Figure 8.2. *The basic structure of NUCSAM, showing the key pools and fluxes.*

The model inputs include atmospheric deposition, global radiation, precipitation and air temperature. Ideally, the model requires these inputs on a daily basis. However, less detailed input is also conceivable. This is especially true for deposition, which is generally available at a larger time scale. The model computes fluxes and concentrations in the vegetation compartments and the soil layers on a daily basis.

3. DATA DERIVATION

3.1. Atmospheric deposition

The total atmospheric N input consists of the open field deposition related to the land use before afforestation. In the metamodel the open field deposition is transferred to deposition input during the forest development using a tree height relation (Chapter 5):

$$N_{in}(h) = N_{depof} \cdot A \cdot \ln(h) + B \qquad (21)$$

where:

$N_{in}(h)$ = Atmospheric deposition input to a forest with an average tree height of h (m) after conversion of a pasture (kg N ha^{-1} $year^{-1}$)

N_{depof} = Original open field deposition, i.e. grassland or arable land (kg N ha^{-1} year^{-1})
h = Average tree height (m)
A = Regression parameter (kg N ha^{-1} year^{-1})
B = Regression parameter (-)

The values for parameters A and B are given in Table 8.1.

Table 8.1. *Deposition filtering parameters for afforestation of agricultural land.*

Forest type	Original landuse	A (-)	B (kg ha^{-1} year^{-1})
Coniferous	Grassland	0.38	0.88
Deciduous	Grassland	0.37	0.88
Coniferous	Arable	0.30	0.78
Deciduous	Arable	0.30	0.79

The metamodel does not include foliar exudation or foliar uptake (Chapter 5 and 9).

3.2. Hydrology

Within the water model, it is assumed that all water that is leaving the soil compartment fully contributes to the water recharge. This means that the soil water flux R is assumed to be equal with groundwater recharge. This may cause an overestimation of the groundwater recharge, especially in poorly drained soils. The hydrological model uses the following input from the AFFOREST database (Chapter 9):

- Tree species, which were used for the allocation of the interception fractions (fr_i) and maximum LAI (LAI_{mx}) (Table 8.2). Interception fractions for the three distinguished tree types were taken from Kros (2002), whereas the LAI_{mx} was derived from output of the vegetation model (Chapter 7) while assuming that the interception capacity is at its maximum when the forest is ten years old.
- Soil type, which was used for the allocation of the available water capacity (AWC) (Table 8.3).
- Long-term maximum (T_{mx}) and minimum temperature (T_{mn}), which was used for calculating the monthly temperature using a sinus function with T_{mx} as maximum and T_{mn} as minimum.
- Long term average precipitation, which was used for the derivation of the monthly precipitation. A monthly precipitation distribution was derived from a IIASA database (http://www.iiasa.ac.at). This database contains long-term average precipitation pattern on a 0.5×0.5 ° grid, which was aggregated to latitude band with a width of 5°. This yielded a lookup table that delivers monthly precipitation as a function of latitude and the precipitation sum (P_{sum}).
- The monthly PET, which was calculated by the vegetation model (Chapter 7).

Available water capacity (AWC) was implemented as a function of texture and soil wetness class (Table 8.3). This was based on transfer functions of Batjes (1996), who uses the FAO soil map of the world using the average values for the texture classes and soil water regimes in Belgium, the Netherlands, Denmark and Sweden. For the water balance calculations a thickness of the soil compartment of 30 cm was used for shallow soils and 50 cm for others. Simulations were started at the 1st of January of the year of afforestation. The initial amount of snow was estimated using the calculated amount of snow based on the meteorological data in the period October-December of the same year.

Table 8.2. *Interception fractions (f_i) and leaf area index beyond which the interception fraction is at its maximum (LAI_{mx}).*

Tree type	fr_i	LAI_{mx}
	(-)	(-)
Deciduous	0.2	3
Pine	0.3	3
Spruce	0.4	4

Table 8.3. *Available Water Capacity (AWC) use for the combination of soil type and soil wetness class (dry and wet).*

Soil classes [1)]	Soil type	AWC (mm)	
		Dry [2)]	Wet [2)]
1-12	Sand	50	100
13-24	Loam	100	120
25-36	Clay	110	120
37-48	Heavy Clay	130	130
49-60	Peat	500	500

[1)] Numbers refer to the AFFOREST-sDSS (see Chapter 9)

[2)] Dry = Mean highest water table > 200 cm; Wet = Mean highest water table ≤ 200 cm (Chapter 9)

3.3. Soil processes

For the C and N balance calculations, a generic soil depth of 30 cm was used. Mineralization constants were linked to land use classes (Table 8.4). Values were partly derived from Kros (2002) and partly by calibration on measured C pools in the soil. Results of the validation (section 4.1) yielded slightly adapted values for the mineralization rate parameters (fr_{mi} and $kr_{miltnat}$) compared to Kros (2002). Based on the calibration, the dissimilation (DA_{mo}) was set to 5, whereas Kros (2002) used a value of 1.5 (section 4.1).

Table 8.4. *Values for the mineralization parameters for the various land use classes.*

Parameter	Pine	Spruce	Deciduous	Extensive agriculture	Other agricultural classes
fr_{mi}	0.60	0.60	0.60	-	-
$kr_{miltnat}$	0.02	0.02	0.02	0.3	-
$kr_{miltagr}$	-	-	-	-	0.10
DA_{mo}	5	5	5	5	5

Values for all soil parameters (Table 8.5) were mainly derived from the Dutch Soil Information System (BIS) (Bregt et al 1986). For the minimum value for organic matter ($OM_{soil\ mn}$) we used the 5 percentile, as we did for the C/N ratio in the soil during agricultural land use ($CN_{soil\ grs/ara\ exe}$). All other values were median values.

Table 8.5. *Values for the soil parameters for the seven soil types at a depth of 30 cm.*

Parameter [4]	Unit	Sand		(heavy) Clay		Loam	Peat
			Calc [3]		Calc [3]		
$frde$	(-)	0.3	0.3	0.8	0.8	0.8	0.9
CN_{mn}	($kg \cdot kg^{-1}$)	15.0	10.0	15.0	10.0	15.0	15.0
$CN_{soil,\ grs,\ nat}$	($kg \cdot kg^{-1}$)	23.0	15.0	9.0	9.0	11.0[2]	13. [1]
$CN_{soil,\ ara}$	($kg \cdot kg^{-1}$)	17.0	10.0	11.0	9.0	11.0	13. [1]
$CN_{soil,\ grs}$	($kg \cdot kg^{-1}$)	13.0	10.0	11.0	9.0	11.0	13.0
$CN_{soil,\ ara,\ ex}$	($kg \cdot kg^{-1}$)	14.0	10.0	11.0	9.0	11.0	12.0
$CN_{soil,\ grs,\ ex}$	($kg \cdot kg^{-1}$)	12.0	10.0	9.0	9.0	11.[2]	12. [1]
$OM_{soil,\ ara}$	(%)	4.8	1.5	2.3	2.2	2.5	35. [1]
$OM_{soil,\ grs}$	(%)	5.5	2.0	8.0	6.0	3.0	35.0
$OM_{soil,\ grs\ ex}$	(%)	4.1	2.2	7.0	7.0	3.0 [2]	35.0
$OM_{soil,\ ara,\ mn}$	(%)	2.0	1.5	1.6	1.7	1.7	23.0
$OM_{soil,\ grs.\ mn}$	(%)	2.0	2.0	3.0	4.0	3.0	23.0
Rho [1]	($kg\ m^{-3}$)	1370	1480	1180	1340	1470	490

[1] *Overall average from the Dutch Soil Information System, BIS, (Bregt et al. 1986)*
[2] *As grs rest Nat = restored natural grassland*
[3] *Calcareous*
[4] *grs nat = natural grassland; ara = arable land; ara/grs exe = ara or grs that was excessively manured; OM soil ara/grs mn = minimum value for organic matter*

3.4. Data on the chronosequences

An overview of some characteristics of the chronosequences is given in Table 8.6. More details can be found in Chapter 2, 3 and 4. For the application at the chronosequences, location specific values were used for the parameters (Table 8.7).

For other parameters needed, generic values from the AFFOREST-sDSS data were selected based on the applicable soil and vegetation classes. This concerned former land use, soil type, afforestation strategy and minimum and maximum temperature. For the initial C:N ratio in the soil, the value for the corresponding soil and former agricultural land use was assigned (Table 8.5).

Table 8.6. *Overview of the chronosequences used for the model validation and some characteristics.*

Chronosequence	Number of sites	Soil	Deposition
Norway spruce, Halmstad, SE	5	Sand	Low
Norway Spruce, Vestskoven, DK	5	Clay	High
Oak, Vestskoven, DK	6	Clay	High
Oak, Sellingen, NL	4	Sand	High
Norway Spruce, Drenthe, NL	3	Sand	High

Table 8.7. *Overview of the chronosequences used for the model validation and some characteristics.*

Chronosequence	OM_{ar} (%)	C/N (g g^{-1})	$kr_{mi,tree}$ ($year^{-1}$)	fr_{mi} (-)	$kr_{mi\ agr}$ ($year^{-1}$)
Norway spruce, Halmstad, SE	3.3	14	0.02	0.6	0.1
Oak, Vestskoven, DK	5.5	12	0.02	0.7	0.05
Norway Spruce, Vestskoven, DK	5.5	12	0.04	0.8	0.05
Oak, Sellingen, NL	6.0	30	0.02	0.6	0.3
Norway Spruce, Drenthe, NL	5.0	28	0.02	0.6	0.1

For NUCSAM, it was necessary to divide N from former agricultural land use over the L, F and H pools. Therefore, a steady state distribution of the organic mater pool over the L, F and H fractions in the first 30 cm of the mineral soil was calculated. According to this distribution, the initial C and N was distributed over the L, F and H fractions. Below 30 cm, all organic matter was placed in the H pool.

4. MODEL VALIDATION

4.1. Validation using literature data

The rather simple process description on C and N mineralization was validated on a dataset on N dynamics as summarized in Kirschbaum & Paul (2002). Those authors compiled the dataset from literature data on litter types that differed greatly in their quality, and for which C and N dynamics were observed during decomposition. Using these data, the ability of the model to simulate C and N dynamics in these litter types with a wide range in C/N ratios was tested. The model was tested on the relationship between: (i) C loss during decomposition and litter N concentration and (ii) the critical C/N ratio, i.e. the lowest achievable C/N ratio, for different litter types and their decomposability.

For the decomposition rate, we selected two examples that comprise forest litter from Kirschbaum & Paul (2002), i.e. the Cedar, Hemlock and fir litter dataset as collected by (Keenan et al 1996). For the critical C/N ratio we use the data from Berg & Staaf (1981). According to Kirschbaum & Paul (2002), this comparison provided the most critical test of a decomposition model because it gave a quantitative relationship between initial litter quality and litter C and N dynamics. The model was run with temperatures reported for each experiment. For kr_{mi} a generic value of 0.02 year^{-1} was used, whereas fr_{mi} was set to the reported first year litter loss (Kirschbaum & Paul 2002). The value of the dissimilation/assimilation ratio DA_{mo} (Eq. 10) was set to 5, based on calibration and for CN_{mo} a value of 15 was used. For the modelled critical N concentration we used the N concentration in litter at which soil litter could not incorporate additional N without incorporating additional C, i.e. in fact the minimum C/N ratio of the total litter pool that could be achieved by the model. This was calculated by running the model until a steady state for the C and N pools was achieved while remaining the input constant. At this situation, the N mineralization equals the amount of N that enters the system through litterfall. For the dataset of Berg & Staaf (1981) the model was used to simulate the C and N decomposition. Therefore, the modelled critical N concentrations were compared with observed values (Figure 8.3).

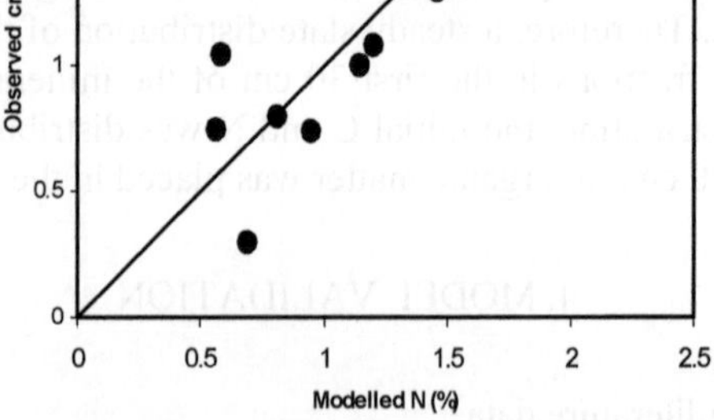

Figure 8.3. Modelled critical N concentration versus predicted critical N concentrations in the data set of Berg & Staaf (1981).

The modelled critical N concentration explained 60% of the observed range in critical N concentration, which is at least as good as was found with Century/CenW: 54% (Kirschbaum & Paul 2002). It was necessary, however, to increase the default assimilation/dissimilation ratio (DA_{mo}) from 1.5 to 5. This means more fungal decomposition (having a ratio of 10) compared to bacterial driven decomposition (having a ratio of 2) (cf. Janssen 1984). The results were quite comparable with CENTURY/CENW (Figure 8.4).

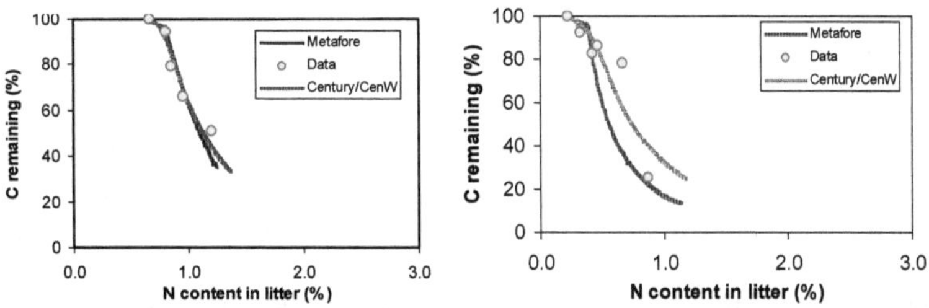

Figure 8.4. *Modelled percentage of remaining C versus N concentration in Cedar litter (left) and Western Helmlock litter (right) with the soil model together with observations and model results of Century/CenW from Kirschbaum & Paul (2002).*

In both CenW and the soil model it was assumed that the decomposition was controlled by litter quality and temperature. For temperature the same relation was used, whereas the formulation related with litter quality was different for both models. Despite this difference both models performed quite comparable. This means that the use of the relatively simple soil model is an acceptable alternative for the relatively complex and data demanding model Century/CenW model for application within the AFFOREST-sDSS. Compared to Century/CenW, the soil model used only one lumped pool of litter production instead of six pools and only two litter pools instead of three.

4.2. Validation towards the detailed soil model NUCSAM

A comparison between observations from a chronosequence and the results of the models NUCSAM and the soil model was performed for the Oak chronosequence of Sellingen (Figure 8.5) and the Norway spruce chronosequence of Drenthe (NL, Sp). For each plot of the chronosequence the model NUCSAM was applied. Results show that modelled Cl concentrations were clearly overestimated in the period up to March 2002. This indicates that the water balance calculations were not adequate (Chapter 3). This overestimation also occurred for the NO_3 concentration. In the period after March 2002, the correspondence with the measurements for both Cl and NO_3 is reasonably good.

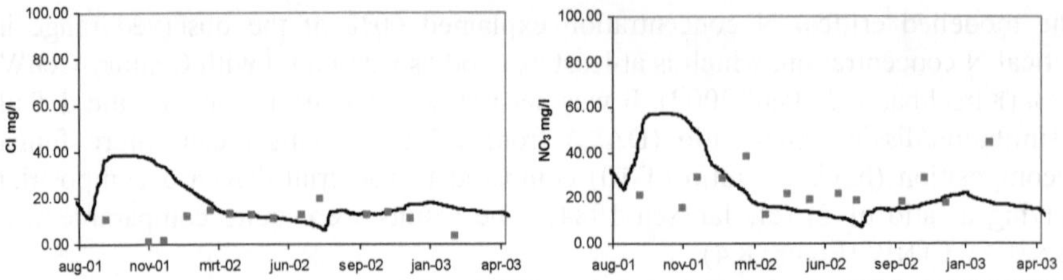

Figure 8.5. Measured and modelled (NUCSAM) Cl (left) and NO_3 concentration (right) at 90 cm depth in plot 3 (11 years old) of the Dutch Oak chronosequence.

The NUCSAM results for each individual plot from the two Dutch chronosequences were averaged for the two monitoring years (2001 and 2002) (Figure 8.6 and 8.7). These results were plotted as individual dots in the figure, since it was not possible to simulate a whole chronosequence at once. Contrary to the soil model, NUCSAM was dependent of the location specific data of each individual plot of the chronosequence. The soil model results, however, are plotted as a line, because this model was especially developed for modelling the whole chronosequence at once. From the comparison for these two chronosequences, it can be concluded that the soil model performs at least as good as NUCSAM.

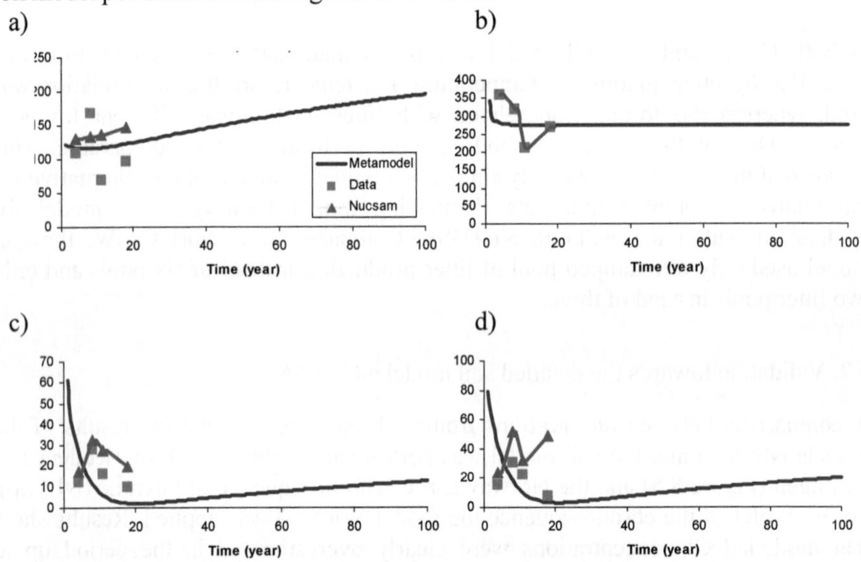

Figure 8.6. a) Observed C sequestration in soil (ton ha^{-1}), b) groundwater recharge (mm $year^{-1}$), c) nitrate leaching (kg ha^{-1} $year^{-1}$) and d) nitrate concentration (mg l^{-1}) in the Dutch oak chronosequence together with the NUCSAM and the soil model simulations.

A notable exception is the NO_3 concentration for oak, where a relatively large deviation between NUCSAM, the soil model and the observations is apparent.

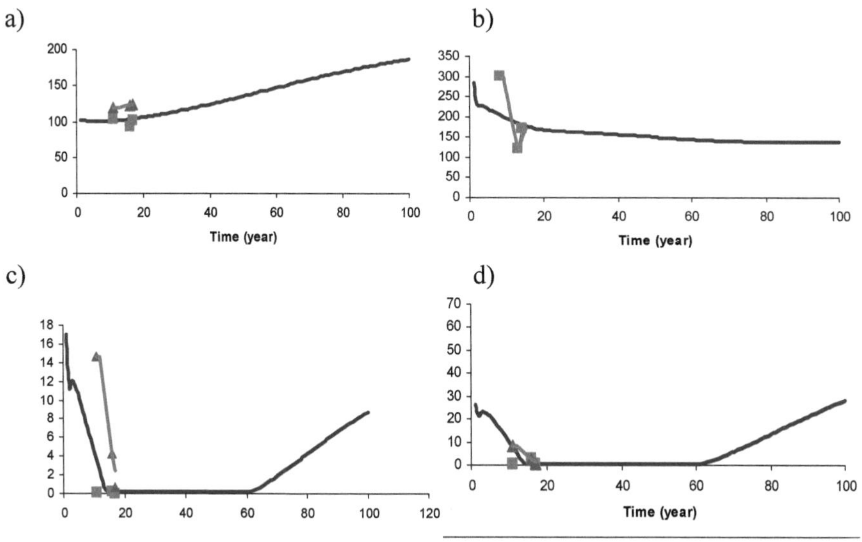

Figure 8.7. *a) Observed C sequestration in soil (ton ha^{-1}), b) groundwater recharge (mm year^{-1}), c) nitrate leaching (kg ha^{-1} year^{-1}) and d) nitrate concentration (mg l^{-1}) in the Dutch Norway spruce chronosequence together with the NUCSAM and the soil model simulations.*

4.3. Validation using chronosequence data in Sweden, Denmark and the Netherlands

4.3.1. Carbon sequestration

The soil model modelled total amount of C in the top 30 cm of the soil, including both litter and organic matter in the mineral soil (Figure 8.8.). Modelled C sequestration by the soil for all chronosequences is summarized in Figure 8.9. Results showed a release of soil C (Cseq soil < 0) during the first 10-20 years after afforestation. The average C sequestration in the soil (period 50-100 years) lies between 50 to 500 kg C ha^{-1} year^{-1}.

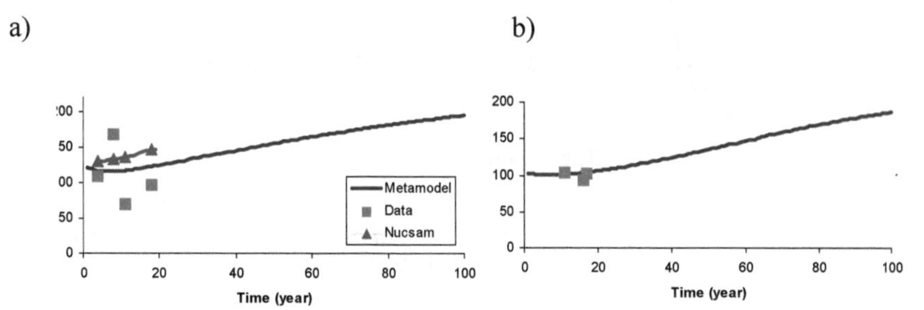

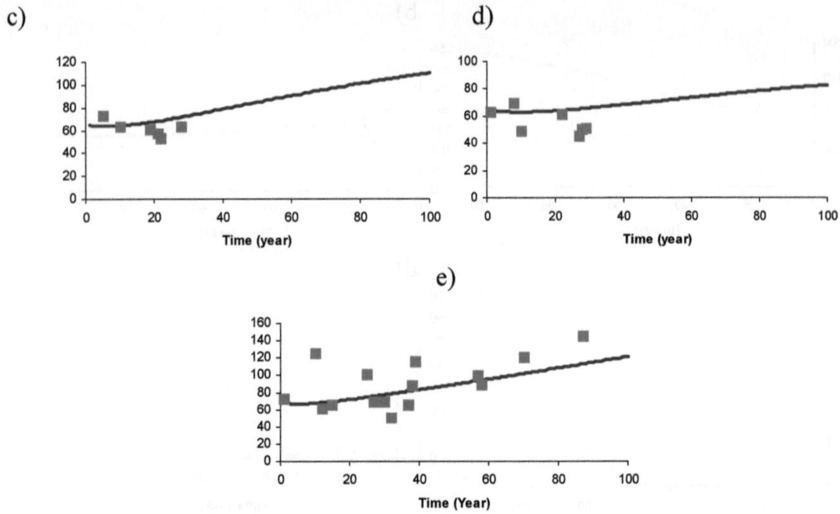

Figure 8.8. *Measured and modelled C stock in the soil for oak and spruce chronosequences in Sweden, Denmark and the Netherlands, where a) oak in the Netherlands, b) Norway spruce in the Netherlands, c) oak at Vestskoven in Denmark, d) Norway spruce at Vestskoven in Denmark, and e) Norway spruce in Sweden.*

Data from the AFFOREST chronosequences (Chapter 2) show that as a minimum, afforestation leads to a constant or an increase in soil C stocks. Except for the first 10 to 20 years, this is confirmed by the soil model simulations (Figure 8.9). For the AFFOREST chronosequences the average soil C sequestration ranges from 0 to 1.3 ton C ha^{-1} $year^{-1}$.

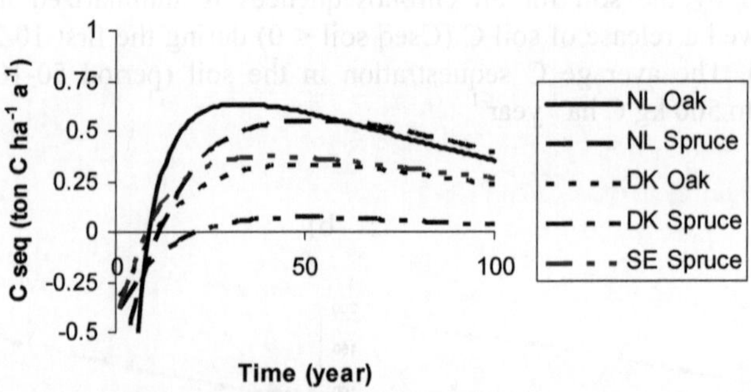

Figure 8.9. *Modelled C sequestration (ton ha^{-1} $year^{-1}$) in the soil for all chronosequences in Sweden, Denmark and the Netherlands.*

The modelled maximum soil C sequestration was around 0.6 ton C ha^{-1} $year^{-1}$. On the other hand (de Vries et al. 2003) reported even lower C sequestration values in European forest soils ranging from 0.1 to 0.4 ton C ha^{-1} $year^{-1}$.

4.3.2. Water recharge

The simulated water recharge data and the by SWAP calculated groundwater recharge fluxes for the inspected chronosequences (Figure 8.10.) refer to the amount of water leaving the soil profile at 90 cm depth. This water could potentially contribute to water recharge, however, it could also be drained laterally (Chapter 3). This means that the reported water leaching fluxes are potential values for groundwater recharge. Because no water fluxes were measured, the reported water fluxes were modelled using the SWAP model (Kroes et al. 2000). The calculation of the water leaching fluxes are described in Chapter 3 and by Van der Salm et al. (2006). Van de Salm et al. (2006) calibrated the SWAP model on all AFFOREST chronosequences. They first calibrated the interception losses based on measured interception data by adjusting the parameters for soil cover and the interception capacity of the canopy. Secondly, they calibrated the soil water fluxes on measured groundwater levels and soil water contents.

Results show that the water recharge for the Dutch oak chronosequence modelled using the soil model compared well with the SWAP results for the individual plots of this sequence. For the two Danish chronosequences, the recharge was overestimated with about 100 mm $year^{-1}$, whereas for the Swedish site the recharge was underestimated with 100 mm $year^{-1}$. As observed in all chronosequences the water recharge decreased during the first 20 years after afforestation. This phenomenon, due to an increase of LAI and height, was also reflected by the soil model.

Water recharge under oak was clearly larger (50 to 150 mm) than under spruce (Figure 8.11.). The period of decreasing water recharge fluxes for oak was relatively small (< 10 years) compared to spruce (about 20 years). Furthermore, the modelled water recharge under oak reached a constant value after 20 years, whereas under spruce it continued to decrease. This is caused by the relatively high actual growth for the spruce sites, which was still linear at this stage (80-100 years; Chapter 2), whereas for oak the actual growth rate decreased after 10 to 20 years.

a)

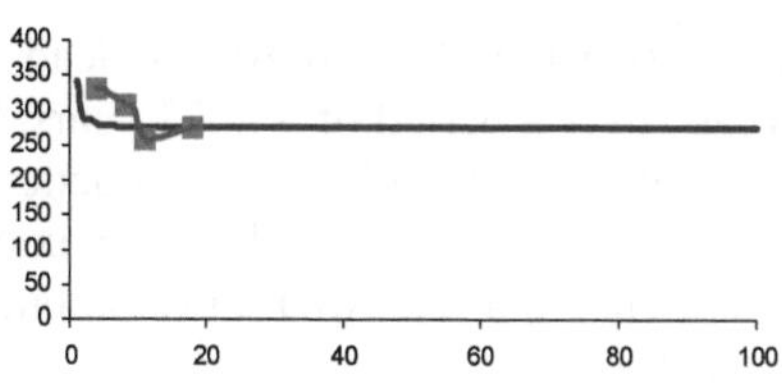

b)

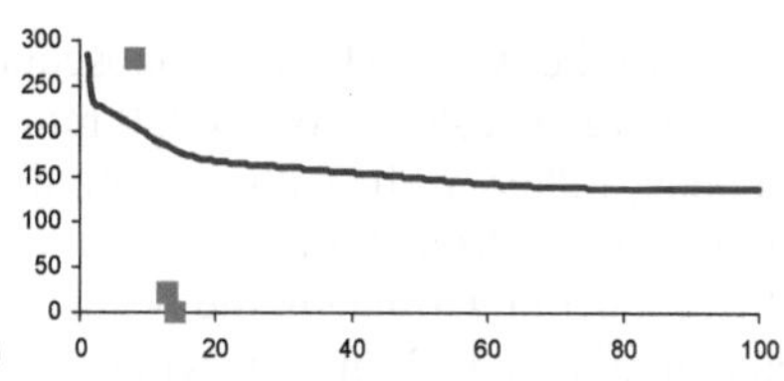

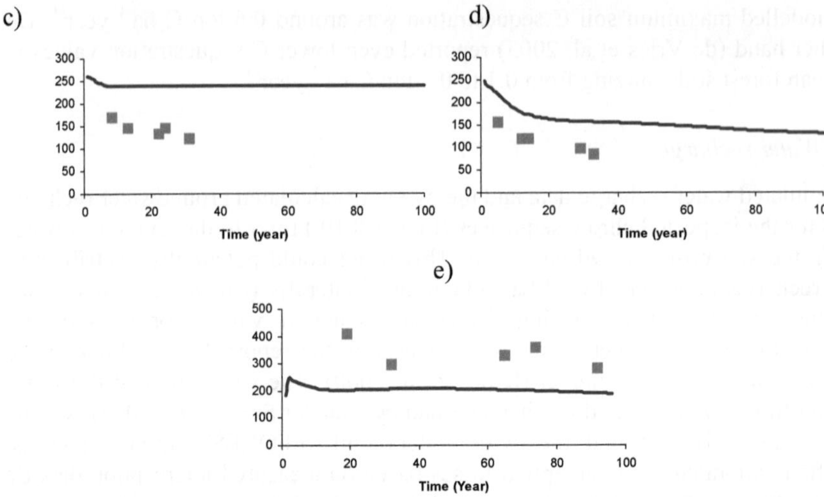

Figure 8.10. Measured and modelled water recharge for oak and spruce chronosequences in Sweden, Denmark and the Netherlands, where a) oak in the Netherlands, b) Norway spruce in the Netherlands, c) oak at Vestskoven in Denmark, d) Norway spruce at Vestskoven in Denmark, and e) Norway spruce in Sweden.

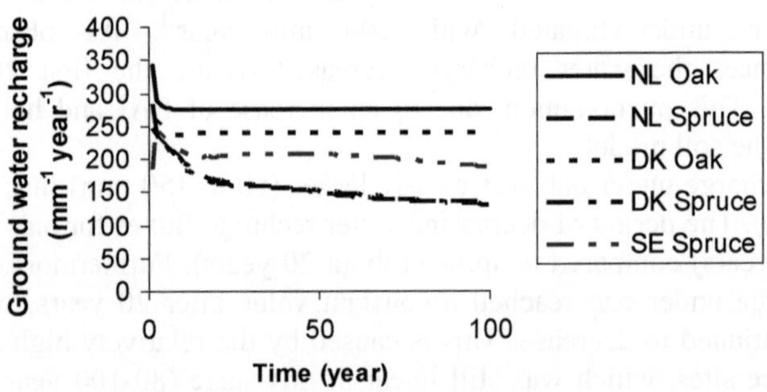

Figure 8.11. *Modelled water recharge (mm year^{-1}) for all chronosequences in Sweden, Denmark and the Netherlands.*

According to the AFFOREST chronosequences (Chapter 3), afforestation leads to a decrease of water recharge of 20-230 mm year^{-1} over a period of 5 to 72 years. This corresponds rather well with the modelled effects for all chronosequences, a reduction ranging from 50 to 150 mm yr^{-1} (Figure 8.11). Although, the maximum modelled reduction (spruce in the Netherlands) remained somewhat lower about 150 mm compared to the 230 mm as reported in Chapter 3.

4.3.3. Nitrate concentration and nitrate leaching

The simulated and measured NO_3 concentrations for the inspected chronosequences are given in Figure 8.12.

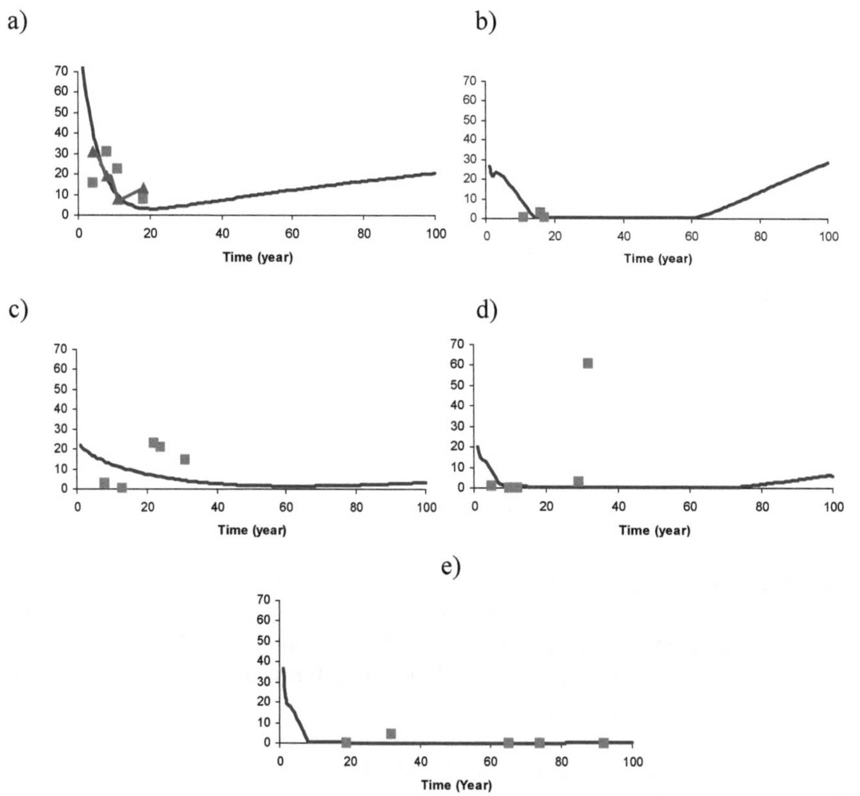

Figure 8.12. *Measured and modelled nitrate concentrations (mg l^{-1}) for oak and spruce chronosequences in Sweden, Denmark and the Netherlands, where a) oak in the Netherlands, b) Norway spruce in the Netherlands, c) oak at Vestskoven in Denmark, d) Norway spruce at Vestskoven in Denmark, and e) Norway spruce in Sweden.*

The water model was able to model the observed NO_3 concentrations in the chronosequences quite reasonably (Figure 8.12). The general trend started with relatively high concentrations just after afforestation followed by a gradual decrease during the forest development. After a period of several years, the model simulated a gradual increase in the NO_3 concentration, due to a decrease in net growth and an increase in N deposition to the stand. Apparently this effect started earlier for the oak sequences. For the spruce chronosequence in Sweden, this effect was even absent, which means that the forest is still accumulating N. For the Spruce sequence in the Netherlands, the modelled increase in NO_3 was remarkably high. This was caused by a combination of a gradually decreasing growth and gradually increasing mineralization. For the nitrate fluxes this effect is clearly smaller (see Figure 8.12).

In Chapter 4, it was concluded that afforestation will cause a rather fast reduction in NO_3 concentrations within the first five years after planting. At sites with high N status and high N deposition, however, the stands start leaching NO_3 again after canopy closure when the N demand is decreasing. The results of the water model application as presented in Figure 8.13 illustrate this.

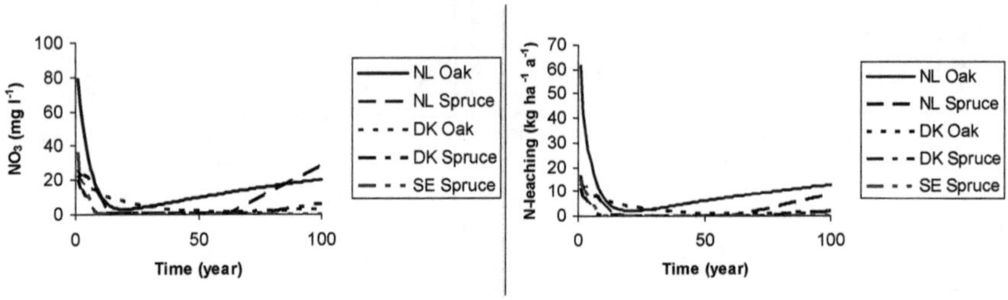

Figure 8.13. *Modelled nitrate concentration (mg l^{-1}) (left) and nitrate leaching flux (kg ha^{-1} $year^{-1}$) (right) for all chronosequences.*

5. DISCUSSION AND CONCLUSION

In general, the performance of the soil model was adequate for soil organic matter and nitrate leaching. For water recharge the performance was less adequate. Table 8.8 gives an overview of the performance of the soil model while applied to the chronosequences. When comparing the soil model performance to results in Chapter 2, 3 and 4, it can be concluded that:

- Carbon sequestration in the soil was modelled satisfactory but had a tendency to be underestimated.
- Modelled water recharge fluxes correspond with the field results
- Nitrate leaching fluxes were in accordance with the field results

We conclude that the soil model is an acceptable alternative for more complex process oriented model and that it fulfils the necessarily requirements for use within the METAFORE and AFFOREST-sDSS (Chapter 9 and 10).

***Table 8.8.** Summary of the performance[1)] of the soil model applied to the AFFOREST chronosequences.*

Chronosequence	Soil OM	Recharge	Nitrate
NL Oak	+/- ↓	+	+
NL Spruce	+	+/- ↑	+
DK Oak	+	- ↓	+
DK Spruce	+	+/- ↑	+
SE Spruce	+	+/- ↓	+

[1)] Performance; +: good; - : bad; +/-: moderate
↓: underestimation; ↑: overestimation

5.1. Carbon sequestration

The validation and application of the soil model to the AFFOREST chronosequences showed that C behaviour can be modelled adequately. For the first 20 years after afforestation, a slight decrease in C sequestration was modelled followed by a slight continuous increase. The slight decrease during the first decades corresponded with the results in Chapter 12 and previous research on changes in C pools in the soil after afforestation of arable land (Vesterdal et al 2002).

An important assumption was that we only took C in the first 30 cm of the mineral soil and the litter layer into account. Thus, a steady state for C and N was assumed for the subsoil. However, since the pool in the subsoil could be rather high relatively low, decomposition rates may result in large C losses. For example, 0.1% organic matter over 1 m and a bulk density of 1500 kg m^{-3}, yields 15 ton organic material ha^{-1} or 7.5 ton C ha^{-1}. A decomposition rate of 1% yields a loss of 75 kg C ha^{-1} $year^{-1}$. It is obvious that this high amount could not be delivered in one year. Combining this with a widely observed constant organic matter content in the subsoil over time, means that the decomposition rate in the subsoil should be very low.

5.2. Water recharge

The chronosequences studied within the AFFOREST project showed a decrease in annual water recharge of 20 to 230 mm yr^{-1} resulting from an increase in forest age over a period of 5 to 72 years. This corresponds rather well with the modelled effects for all chronosequences (50 to 150 mm yr^{-1}). This reduction due to the change in land use was modelled previously by Rijtema & De Vries (1994). They found comparable reductions for the transition from arable/grass land to forest (both deciduous and coniferous), that ranged from 40 to 150 mm yr^{1}.

Although, the maximum modelled reduction (spruce in the Netherlands) remained somewhat lower (about 150 mm) compared to the 230 mm as reported in Chapter 3. Furthermore, the initial and final levels for oak were modelled correctly. However, a too fast increase in evapotranspiration was modelled (oak in the Netherlands). For spruce in the Netherlands and Denmark, the water recharge was overestimated compared to the results from the SWAP model, whereas it was underestimated for spruce in Sweden. This means that the connection between the vegetation model, which calculates the potential evapotranspiration trajectory during the forest succession, and the soil model, which calculates the actual transpiration, could be improved.

5.3. Nitrate leaching

The modelled trajectory in nitrate leaching seemed to be plausible. At the beginning after afforestation, the nitrate leaching was rather high due to release of N from former agricultural land use, followed by a gradual decrease due to an increase in uptake and a decrease in mobilisation. The lowest nitrate leaching was observed half way through the rotation. When the forest is getting older (> 60 years), the nitrate concentration and leaching started to increase due to a decrease in uptake and an incerase in N deposition.

The modelled period of enhanced nitrate leaching fluxes lies between 10 (spruce) and 20 (oak) years. This is supported by the data from the oak chronosequences in Chapter 4. The spruce chronosequences do not show an increased initial leaching at all. The modelled period of enhanced nitrate leaching fluxes for spruce is clearly shorter than for oak but not absent.

5.4. Comparison of NUCSAM and CENW with the soil model

The soil model results for decomposition rate and critical C/N were quite comparable with CenW (Chapter 7) and literature data. In both CenW and the soil model it was assumed that the decomposition was controlled by litter quality and temperature. For temperature, the same relation was used, whereas the formulation related with litter quality was different for both models. Despite this difference both models performed quite comparable.

Experience and comparison with a model such as NUCSAM showed that the model helps to summarise and integrate results from individual disciplines and provides a multidisciplinary perspective on complex systems. The detailed nutrient cycling and soil acidification model NUCSAM was built to simulate effects of atmospheric deposition on soil solution chemistry on a site scale on a daily basis in different soils layers. At individual plots the agreement between observed and simulated changes in soil solution chemistry was reasonably good. However, when using the results from the individual plots for the reconstruction of the chronosequence, NUCSAM did not perform better than the soil model. Furthermore, a sensible application of NUCSAM is only feasible for individual chronosequence plots that have complete datasets. This implies that, besides several practical aspects such as computation time, data handling etc., the lack of good quality data is a crucial limiting factor for using a model such as NUCSAM at larger spatial scale. Accordingly, for regional applications, model simplification is inevitable.

5.5. Capability of the soil model in the AFFOREST-sDSS

The soil model was tested by model to model comparisons, literature data and with observations based on chronosequences. These chronosequences appeared to be an excellent means for validating the predicted afforestation impact. In general we conclude that the soil model is an acceptable and even a necessarily alternative for more complex process oriented model and that it fulfils the necessary requirements for use within the AFFOREST-sDSS. However, the problem remains whether these observed afforestation impacts are representative for afforestation impacts in the entire study area. As the AFFOREST-sDSS will be used to identify areas where afforestation will deliver the best environmental performance, the model's capacity for predicting the spatial pattern of afforestation impacts should be checked. As is true for all models, model validation is not an activity leading to an absolute and definite judgement on models adequacy. Therefore, it is absolutely necessary to collect additional data, not only in space but also in time.

Acknowledgement. This work was supported by the European Commission (contract) and the Dutch Agricultural Research Programme (DWK programme 384). We wish to thank Dr. Miko Kirschbaum from CSIRO, Australia for providing the validation data on C an N decomposition and help on interpreting the results. Our colleague Jan-Cees Voogd from Alterra is acknowledged for data handling and producing the graphs.

6. REFERENCES

Batjes, N. H. (1996). Development of a world data set of soil water retention properties using pedotransfer rules. *Geoderma, 71*(1-2), 31-52.

Berg, B., & H. Staaf (1981). Leaching, accumulation and release of nitrogen in decomposing forest litter. In F. E. Clark & T. Rosswall (Eds.), Terrestrial Nitrogen Cycles. Ecological Bulletin 33. Stockholm, pp. 163-178.

Bregt, A. K., Denneboom, J., Stolp, J., & Vleeshouwer, J. J. (1986). BIS, an information system for soil survey and research. Transactions of the XIII Congress of International Society of Soil Science, Hamburg August 1986, Volume III.

Callesen, I., Raulund-Rasmussen, K., Gundersen, P., & Stryhn, H. (1999). Nitrate concentrations in soil solutions below Danish forests. *Forest Ecology and Management, 114*(1), 71-82.

De Vries, W., Reinds, G. J., & Posch, M. (1994). Assessment of critical loads and their exceedance on European forests using a one-layer steady-state model. *Water, Air, and Soil Pollution, 72*(1/4), 357-394.

de Vries, W., van der Salm, C., Reinds, G. J., Dise, N. B., Gundersen, P., Erisman, J. W., & Posch, M. (2003). Assessment of the dynamics in nitrogen and carbon sequestration of European forest soils In: Alterra-rapport 818, Alterra Wageningen, 147 pp.

Groenenberg, J. E., de Vries, W., & Kros, J. (1998). Simulation of the long-term carbon and nitrogen dynamics in Dutch forest soils under Scots pine. *Hydrology and Earth System Sciences, 2*(4), 439-449.

Janssen, B. H. (1984). A simple method for calculating decomposition and accumulation of 'young' soil organic matter. *Plant Soil, 76,* 297-304.

Jørgensen, S. E. (1992). Integration of ecosystem theories. Dordrecht: Kluwer Academic Publishers.

Keenan, R. J., Prescott, C. E., Kimmins, J. P., Pastor, J., & Dewey, B. (1996). Litter decomposition in western red cedar and western hemlock forests on northern Vancouver Island, British Columbia. *Canadian Journal of Botany, 29*(74), 1626-1634.

Kirschbaum, M. U. F. (2000). Will changes in soil organic carbon act as a positive or negative feedback on global warming? *Biogeochemistry, 48*(1), 21-52.

Kirschbaum, M. U. F., & Paul, K. I. (2002). Modelling C and N dynamics in forest soils with a modified version of the CENTURY model. *Soil Biology and Biochemistry, 34*(3), 341-354.

Kroes, J. G., Wesseling, J. G., & van Dam, J. C. (2000). Integrated modelling of the soil-water-atmosphere-plant system using the model SWAP 2.0 an overview of theory and an application. *Hydrology Processes, 14*(11/12), 1993-2002.

Kros, J. (2002). Evaluation of biogeochemical models at local and regional scale. PhD thesis, Wageningen University, Wageningen, The Netherlands.

Rijtema, P. E., & de Vries, W. (1994). Differences in precipitation excess and nitrogen leaching from agricultural lands and forest plantations. *Biomass and Bioenergy, 6,* 103-113.

Starr, M. (1999). WATBAL: a model for estimating monthly water balance components, including soil water fluxes. In S. Kleemola & M. Forsius (Eds.), 8th Annual report 1999. UN ECE ICP Intergrated Monitoring. The Finnish Environment 325. Finnish Environment Institute, Helsinki, Finland, pp. 31-35.

Van der Salm, C., Denier van der Gon, H., Wieggers, R., Bleeker, A., & van den Toorn, A. (2006). The effect of afforestation on water recharge and nitrogen leaching in The Netherlands. *Forest Ecology and Management, 221*(1-3), 170-182.

Vesterdal, L., Ritter, E., & Gundersen, P. (2002). Change in soil organic carbon following afforestation of former arable land. *Forest Ecology and Management, 169*(1-2), 137-147.

CHAPTER 9

METAFORE: THE AFFOREST DEPOSITION-SOIL-WATER-VEGETATION METAMODEL

W. VAN DEURSEN[1], J. MOL[2], G.W. HEIL[3] and H. KROS[2]

[1]Carthago Consultancy, Oostzeedijk Beneden 23a, 3062VK Rotterdam, the Netherlands, E-mail: wvandeursen@carthago.nl, [2]Alterra Green World Research, P.O. Box 47, 6700 AA Wageningen, E-mail: Hans.Kros@wur.nl, [3] Faculty of Science, Institute of Environmental Biology, Sobonnelaan 16, 3584 CA Utrecht, the Netherlands

Abstract. This chapter describes the development of the METAFORE metamodel for the AFFOREST project, focusing on aspects that are important in defining the role of the metamodel in the entire system. Two modes are distinguished: one in which the METAFORE metamodel operates in a batch mode for generating the AFFOREST-sDSS tables for decision support, and another mode of operating with an extended user interface and extended possibilities for evaluating detailed results. The various detailed process-based models are from different sources and each of the institutes had experts on the processes that were modelled. From these models, the individual partners developed meta-descriptions of their parts of the system. The task of the METAFORE metamodel was to combine all this knowledge into a single model executable, and assure a correct calculation of the values needed for the AFFOREST database and the AFFOREST-sDSS. The design of the METAFORE distinguishes a metamodel framework and the model components or submodels. The metamodel framework focuses on the interface and communication between the different submodels and it is responsible for the communication between the submodels. In this design, the submodels are merely servers, waiting to be initialized or called to perform one step of the simulation (i.e. one month or one year of the simulation). To do this, each submodel has only a limited set of exposed methods. Although the detailed process models, as part of their scientific development process, have been extensively validated and calibrated, this does not automatically assure a proper simulation of the processes by the metamodel. During the entire process of the development of METAFORE, the simulation behaviour of the METAFORE modules were constantly tested against the detailed process models. In the end, METAFORE has been developed as a simplification of the detailed models with a lesser demand on data. This means that the detailed behaviour in the process models can at best result in similar but aggregated behaviour in the METAFORE metamodel. It is concluded that the results are satisfactory.

G.W. Heil et al. (eds.), Environmental Effects of Afforestation in North-Western Europe, 203–225.

1. INTRODUCTION

This chapter describes the development of the METAFORE metamodel for the AFFOREST project. The relation between the metamodel and the other AFFOREST AFFOREST-sDSS system components are in briefly described in section 2. The process of combining the various supporting detailed models into one metamodel is described in section 4. The supporting detailed process models and their development are not subject of this chapter. The description of these detailed models is found in Chapter 5, 7 and 8. The modes of operation for the METAFORE metamodel are described in section 5. Two modes are distinguished: one in which the METAFORE metamodel operates in a batch mode for generating the AFFOREST-sDSS tables for decision support, and one mode of operating with an extended user interface and extended possibilities for evaluating detailed results. The process of quality and robustness control for the metamodel development is described in section 6, while the appendices contain the model code list, the methodology for developing expert rules, discussion papers for the quality management and afforestation strategies.

2. SYSTEM DESCRIPTION OF THE AFFOREST-SDSS

The entire AFFOREST-sDSS system is described in detail in Chapter 10. This section focuses on some of the aspects that are important in defining the role of the metamodel in the entire system. In the AFFOREST system (Figure 9.1), we distinguish the following components:

- Raw GIS database, for storing the spatial site-specific data (Chapter 10)
- GIS core, the engine for relating the GIS database with the simulation models and the AFFOREST-sDSS (Chapter 10)
- The AFFOREST-sDSS, the system for evaluating the afforestation strategies and optimizing solutions (Chapter 10)
- The simulation models

The raw GIS database consists of a table of site specific characteristics for all 'potentially to be afforested' locations in the AFFOREST region. This table is linked to a set of maps, thus, the system is capable of producing maps of the site characteristics.

The AFFOREST-sDSS is the component that is responsible for the (multi-criteria) evaluation of the user questions. In the AFFOREST-sDSS, the user can define or search for afforestation strategies, described as the action of what is being proposed to be done with the selected region; the afforestation objectives, as what is the aim or the goal the user is trying to reach; the environmental performance as being the actual behaviour of the system in terms of carbon (C) sequestration, nitrate leaching and groundwater recharge; the location site conditions or site specific characteristics as the boundary conditions of the simulations. The location site conditions also describe the initial situation, what is the starting point of the simulations. The AFFOREST-sDSS component is described in more detail in Chapter 10.

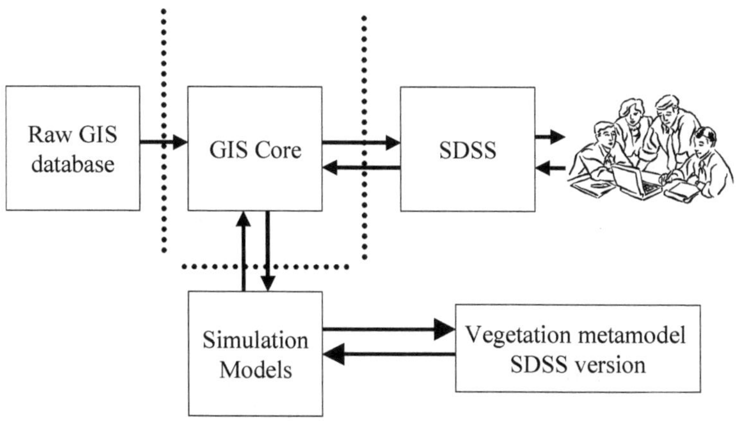

Figure 9.1. Overview of the AFFOREST system.

The GIS core is managing the flow of information between the several components of the AFFOREST-sDSS system.

The task of the simulation models is to provide a full set of model results (environmental performance) for every pixel and for every afforestation strategy. This means that the simulation models will create a database with environmental performance of all locations under the various afforestation strategies. Once all model results for all locations (pixels) are available, the AFFOREST-sDSS can search through these results to find optimal solutions for the afforestation questions of the end user. In almost all cases, the AFFOREST-sDSS does not need to initiate any simulation runs and does not interact directly with the simulation models (the metamodel). Most of the optimization questions the AFFOREST-sDSS needs to provide is an answer that can be solved by analyzing the results of the entire set of simulation model runs (Chapter 10).

The simulation model used in the AFFOREST-sDSS is a metamodel, a simplified model based on existing complex process models (Chapters 5, 7 and 8).

3. DEVELOPING METAFORE: UPSCALING FROM COMPLEX PROCESS MODELS TO A SIMPLIFIED INTEGRATED METAMODEL

One of the common characteristics of environmental problems such as climate change and air pollution, but also for the problems that are addressed in the AFFOREST project, is that they play a role on a local, regional, national, continental and even global scale. It is imperative that the spatial and temporal aspects considered in a model must fit its objectives. In practice, however, an ideal fit is difficult to achieve, because model input data (e.g. initial conditions and parameters) are often limited or even unknown at the relevant scale. Especially, at large spatial scales, many model parameters can not be measured directly. A general problem that arises from applying a plot scale model on a larger scale is the parameterisation. The more parameters a model contains, the less likely it is that they can be derived either directly from available data or indirectly by using pedo-transfer functions. In addition, when particular parameters can only be obtained by calibration, identification problems may thwart the calibration.

Since data availability on a plot scale is relatively large, it is most appropriate to study biogeochemical processes in situ and it is thus the most logical level to start with in model development. Because of sparse data at a larger scale, the scale of the model must however be adapted to the scale of data availability. One possibility is to simplify the model description in such a way that the temporal and spatial resolution is comparable to the resolution of the data. During such a simplification of processes, model results must remain reliable. The reliability can usually be determined by comparing results from the simplified model and the local scale model (e.g. Kros 2002). Presumably, there is an optimal level of model complexity, i.e. a point where the degree of model complexity, e.g. in terms of state variables, match the data resolution and quality, leading to maximal knowledge gain about the modelled system (Jørgensen 1992). Since environmental systems are regarded as complex, 'increased complexity in models is often interpreted as evidence of closer approximation to reality' (Oreskes 2000) but in words of Jansen (1998) 'a model should be made no more complex than can be supported by the available brains, computers and data'.

On a site scale, quantitative process based mechanistic plot-scale model with a high degree of process knowledge, spatial (vertical) and temporal resolution. Generally, those models do run with a relatively short time step, typically on a daily or shorter basis. These plot-scale models are calibrated and validated on detailed high time resolution data of C, N and water fluxes that are gathered at the monitoring sites. Calibration is performed by minimising the uncertainty and difference between observations and model results by calibrating poorly defined model parameters (calibration) whereas validation includes the comparison of model results with (another) high resolution data-set.

A disadvantage of these relatively complex mechanistic models is, however, that input data for their application on a regional or continental scale is generally incomplete. So, even if the model structure is correct (or at least adequately representing current knowledge), the uncertainty in the output of complex models may still be large because of the uncertainty of input data. There is thus a trade off between detail and reliability of information obtained and regional applicability. Consequently, the desired degree of spatial resolution in model output is a factor of crucial importance when selecting the level of detail in both the model formulation and its input data. A larger application scale justifies the development of a simpler model.

Therefore, in the AFFOREST project we adapted the approach using of simpler or simplified models with relatively small data requirements at continental scale, with a relatively high degree of certainty, above complex models with large data requirements that are difficult to fulfil. Because no model was available that met these criteria, a new simplified model, called METAFORE was derived. This model, which was derived from existing models, includes all relevant processes in order to simulate C sequestration, nitrate leaching, and groundwater recharge. The vegetation part was derived from the complex CenW model (Kirschbaum 1999), whereas the soil part was derived from the complex model Nucsam (Kros 2002).

4. COMBINING THE VARIOUS MODELS IN ONE INTEGRATED MODEL METAFORE

4.1. Overview of the METAFORE integrated model

The various complex process-based models are described in Chapters 5, 7 and 8. These complex process models were developed separately by different experts at different institutions, and their development was not part of the AFFOREST project. For each complex process model, the respective expert developed a set of meta-descriptions for the processes. From the meta-descriptions, metamodel components or submodels were developed, implementing the meta-descriptions for the processes in that compartment. These metamodel components were combined into one metamodel framework, responsible for the communication between the model components. The result of these steps is the METAFORE integrated metamodel (Figure 9.2).

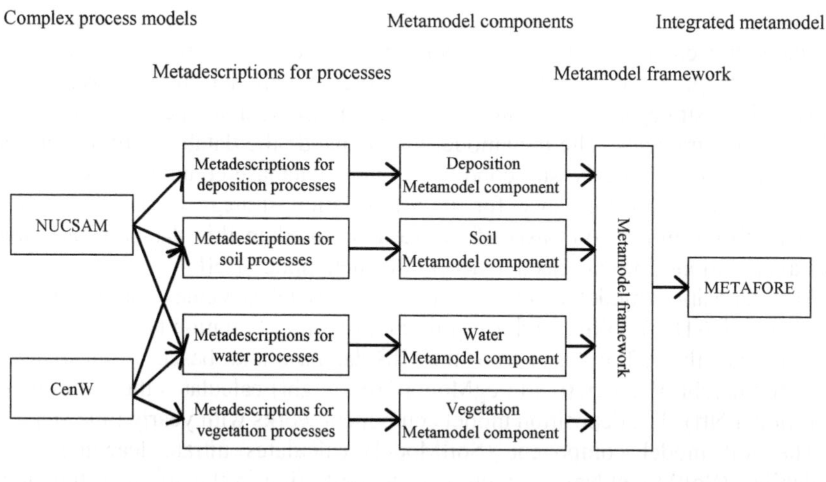

Figure 9.2. *The development process of the METAFORE metamodel.*

Starting from the complex process models as the most detailed description of the processes, the experts responsible for these complex models developed a set of meta-descriptions. These meta-descriptions capture the mechanics of the detailed models for a limited set of environmental parameters. They can be seen as a simplified model approach, with larger timesteps, and only a limited set of input parameters. The meta-description captures only the major processes and controls of the complex models. The meta-descriptions neglect the details that were considered not important for the modeling purpose in the AFFOREST project. The task of the METAFORE metamodel was to combine these meta-descriptions into a single model executable, and to assure a correct calculation of the values needed in the AFFOREST database and the AFFOREST-sDSS.

The design of the METAFORE overall metamodel distinguishes a metamodel framework and the model components or submodels. The metamodel framework focuses on the interface and communication between the different submodels. This requires a detailed definition of the sub-models, a definition of the tasks assigned to each submodel and a complete description of the interaction between the submodels.

In this design, interaction between the model components is the responsibility of the metamodel framework. Model components never interact directly with each other, but ask their input data from the metamodel framework and present their results to this framework. Individual model components are thus not dependent on specific implementations of the other model components. They solely rely on interface definitions. For the METAFORE, the following submodels were identified (Table 9.1):

- The deposition model component (DepModel)
- The soil model component (SoilModel)
- The water model component (WaterModel)
- The vegetation model component (VegModel)

The metamodel gets the location specific site conditions from the GIS database and calculates the environmental performance for a specific afforestation strategy. In fact, for all pixels in the GIS database, the metamodel iterates over all the afforestation strategies and calculates the environmental performance of all afforestation strategies. The metamodel thus expands the database from a database with only location specific site conditions into a database containing for all pixels the environmental performance for all afforestation strategies. The AFFOREST-sDSS performs mostly on basis of the results stored in this database, and under normal operations does not need to access the metamodel itself.

The deposition model component (DepModel) takes values for N deposition (AtmosphericNDep), initial land use (IniLanduse) and vegetation type (derived from iterating over the AffStrategy) from the GIS database and, based on these and the simulated height of vegetation (VegModel.TreeHeight) calculates the N throughfall (DepModel.Ntf). The deposition model component works with yearly time steps.

The soil model component (SoilModel) calculates nitrate leaching and N availability (SoilModel.Nav) as well as C sequestration in the soil as a function of litterfall (VegModel.Clf for C and VegModel.Nlf for N) and N throughfall and calculates C sequestration in the soil based on soil characteristics, vegetation type, temperature and initial land use. The soil model component works with yearly time steps.

The water model component (WaterModel) calculates the status of the groundwater recharge (WaterModel.GWR) based on precipitation, evapotranspiration demand of the system VegModel.Pet), and vegetation and soil characteristics (AffStrategy and SoilClass). The water model component works with monthly time steps.

The vegetation model component (VegModel) simulates vegetation development based on site characteristics (SoilClass, Latitude, Temperature) and N availability and calculates C sequestration in the vegetation. The vegetation model component works with monthly time steps.

There are three supporting metamodel components (Table 9.1):

- The input data component, which encapsulates the access to the GIS database
- The output data component, which controls the writing of the results to the AFFOREST-sDSS database tables
- The user interface for direct analysis of user defined model cases (section 5)

Table 9.1. *The relation between the metamodel components.*

Model component	Model Time step	Input from other components	Input from GIS database	Output to sDSS tables
Deposition model	Year	VegModel. TreeHeight	Atmospheric NDep IniLanduse AffStrategy	
Soil model	Year	DepModel.Ntf WaterModel. GWR VegModel.Clf VegModel.Nlf	MinTemp MaxTemp SoilClass IniLandUse AffStrategy	N leaching C seque-stration soil
Water model	Month	VegModel.Pet VegModel.LAI VegModel.Month Temp	Precipitation Latitude SoilClass AffStrategy	Groundwater recharge
Vegetation model	Month	SoilModel.Nav WaterModel. Wupt	MinTemp MaxTemp AffStrategy Latitude	C seque-stration vegetation

4.2. The metamodel framework: Controlling the metamodel components

The metamodel framework is responsible for the communication between the submodels. In this design, the submodels are merely servers, waiting to be initialized or called to perform one step of the simulation (i.e. one month or one year of the simulation). To do this, each submodel has only a limited set of exposed methods. Each model component has a method called Init which is used by the model component to initialize all variables for the simulation run (using the parameters provided by the Input data component), and each model component has one or more Step methods, which allow the model component to make one simulation step. Two

Step methods were necessary in the soil and the vegetation model component. These two methods, called Step1 and Step2, solve problems in calculation sequences, which are related to the interaction between calculations in the two components.

Since the model components were developed individually by various partners and derived from existing models, the models do not work with the same time step. Two of the model components are designed to work with a yearly time step, while two others work with a monthly time step. It is the task of the metamodel framework to bridge this gap in time steps. The metamodel framework implements two loops, one over the years in which the Deposition model and the Soil model are calculated. Within this yearly loop the framework implements a monthly loop in which the Water model component and the Vegetation model component are calculated. A set of additional variables is used to calculate yearly values from the monthly results.

At the end of the simulation of each year, the *environmental performance* values C sequestration, nitrate leaching and groundwater recharge are written to the AFFOREST-sDSS tables, using the Output data component.

The actual model components are implemented as dynamic link libraries (dll's) which are initiated and linked by the metamodel frame. Thus, in principle it is possible to create an updated version of one of the model components without affecting the implementation in the other components. The pseudo code of the METAMODEL framework is listed in Appendix 1.

5. INTERFACING THE METAMODEL WITH THE OTHER AFFOREST-SDSS COMPONENTS

METAFORE is used in two modes within the AFFOREST-sDSS system. The first mode is batch mode where METAFORE is used to create the AFFOREST-sDSS tables as a basis for finding optimal solutions to user questions (section 5.1.). The second mode of operation of METAFORE is a direct simulation of a specific afforestation strategy for a specified location (Appendix 4). This means that METAFORE reads site specific data for that location and simulates the soil-water-plant processes for the location. This mode of operation is described in section 5.2.

5.1. METAFORE batch mode: Interfacing the metamodel with the AFFOREST-sDSS

The batch mode of operation for the METAFORE is used to create all AFFOREST-sDSS tables necessary for the multicriteria decision making. The mode is illustrated in Figure 9.3. METAFORE processes all records of the GIS database in batch mode and produces environmental performance tables for each location for each of the 36 defined afforestation strategies. These tables are used by the AFFOREST-sDSS for comparing the effectivity of strategies to meet certain user-defined criteria. METAFORE needs to create these tables only once and they are stored in the AFFOREST-sDSS system. This approach assures that all results of all model runs are readily available for the AFFOREST-sDSS, and the task of the AFFOREST-sDSS is now simply searching through these tables for evaluating user questions. In this design, only changes in the input data (meaning changes in site specific characteristics) require a new model run.

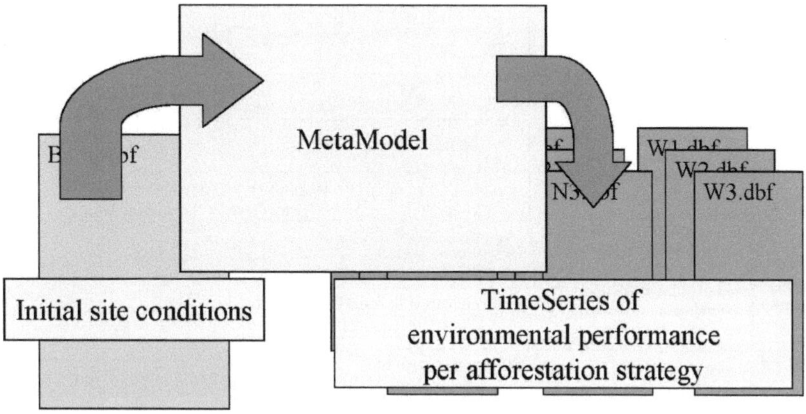

Figure 9.3. *The metamodel METAFORE interfacing with the GIS database to create AFFOREST-sDSS tables.*

5.2. METAFORE user interface mode: Direct user interaction with the metamodel

For detailed inspection of the simulation runs for specific locations, METAFORE can also be invoked from the maps in the AFFOREST-sDSS. The "invoke metamodel tool" in the AFFOREST-sDSS allows for selecting locations for which METAFORE will simulate the relevant processes (Figure 9.4).

Figure 9.4. *Access to METAFORE from the AFFOREST-sDSS shell.*

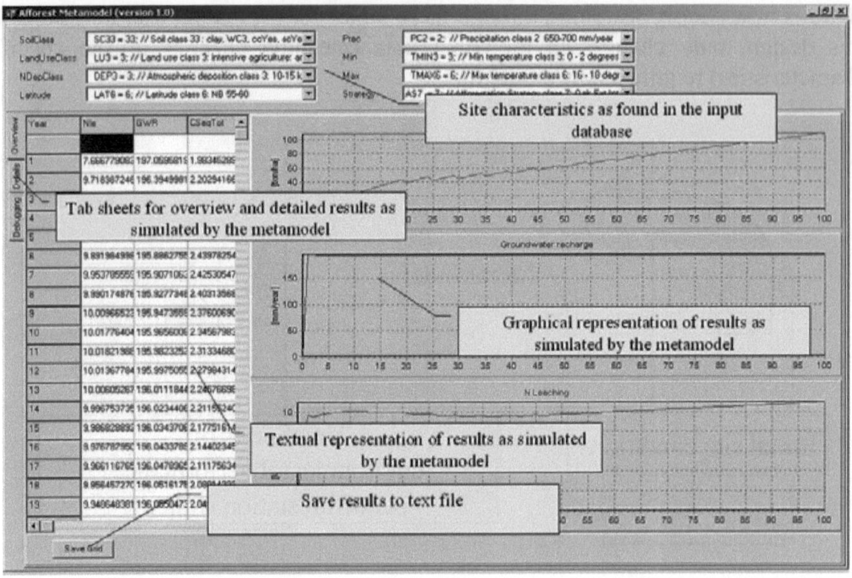

Figure 9.5. *The user interface to METAFORE.*

METAFORE is now in an active, user interface enabled mode, the screen as in Figure 9.5. In this mode, the user can find sensitivity to the site specific characteristics (questions such as how would this system perform on clay instead of sand), analyzing the differences between afforestation strategies or finding optimal strategies for the selected location. For further analysis, METAFORE allows for detailed inspection of the results and storing results for analysis in other programs. Detailed reports and graphs of this simulation are produced, and the user can change any of the (location specific) site characteristics or afforestation strategy and perform sensitivity analysis in this manner. Whenever a map is visible in the system this mode of operation is available, so any pixel may be selected and analyzed in detail.

6. MANAGING THE QUALITY AND ROBUSTNESS OF METAFORE

Because of the area extent of the application of METAFORE, it is entirely infeasible to validate such a metamodel on this scale with field experiments. To do this, we should have had field data covering 100 years of vegetation growth for each of the combinations of major forest systems in the AFFOREST region, major soil types and climate types. Although we do have access to some datasets, their coverage is completely insufficient for validating the entire range of METAFORE. We adopted a more pragmatic approach for validating the development of METAFORE. Three important components may be distinguished in this approach:

- Testing METAFORE against existing field data
- Testing METAFORE against simulations with the detailed models
- Testing METAFORE against existing knowledge of experts

The validating of the individual modules is described in the chapters 5, 7 and 8. Here, the general approach is briefly discussed.

6.1. Testing METAFORE against existing field data

The chronosequences that have been developed for this project were analyzed with METAFORE. METAFORE was able to describe the vegetation growth and the soil chemistry sufficiently accurate. This testing of METAFORE against the chronosequences has been described in more detail in Chapter 7 and 8.

6.2. Testing METAFORE against simulations with the detailed models

During the entire process of the development of METAFORE, the simulation behavior of the METAFORE modules was constantly tested against the detailed process models. We constructed some predefined cases, in which we defined soil characteristics, climate variables and afforestation strategy. During the entire process of development of METAFORE, the model behavior and the simulation results of the detailed process models had to be similar to the results of METAFORE. This ensured that for those cases in which we did have sufficient data to run the detailed models, we could check whether METAFORE was actually capable of reproducing this behavior. It should be noted that METAFORE is not capable of reproducing the detailed models exactly. After all, METAFORE was developed as a simplification of the detailed models with a lesser demand on data. This means that the detailed behavior in the process models can at best result in similar but aggregated behavior in the METAFORE metamodel.

7. INFLUENCE OF TREE SPECIES, SOIL TYPE, TEMPERATURE AND N DEPOSITION

To gain insight in the plausibility of the model the effect of tree species, soil type, temperature and N deposition on C sequestration in biomass and in the soil, water recharge and nitrate leaching was investigated. Therefore, the model was run for various combinations of soil types, tree species, temperature and N deposition (Table 9.2).

Table 9.2. *Overview of the evaluated combination of soil, vegetation, temperature and deposition.*

Name	Soil	Vegetation	Temperature [1] (T_{mx}, °C)	Deposition [1] (kg N ha^{-1} $year^{-1}$)
Reference	Sand dry	Oak	15	22.5
'Low'	Sand wet	Spruce	7	2.5
'High'	Clay	Pine	21	37.5

[1] Values refer to the centre of the used classes in the AFFOREST-sDSS (Chapter 10)

The modelled effect of tree species is presented in Figure 9.6. These results show that tree species mainly influences groundwater recharge and nitrate leaching, whereas the effect on C sequestration, both in soil and vegetation, was less pronounced. Soil type (Figure 9.7) mainly influence water recharge and to a lower extent nitrate leaching. There was no effect on C sequestration in the vegetation, whereas for the C sequestration in the soil there was an effect only during the first 20 years.

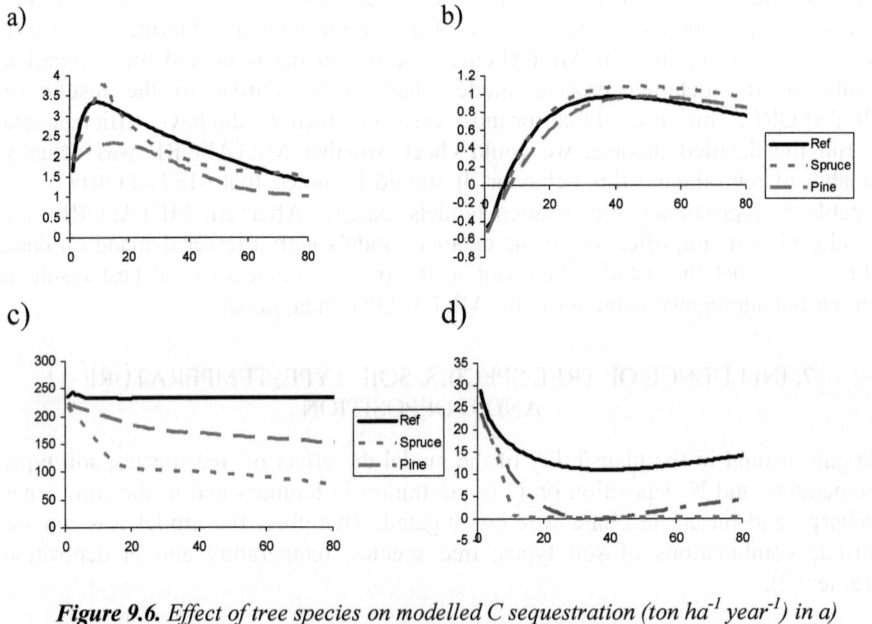

Figure 9.6. *Effect of tree species on modelled C sequestration (ton ha^{-1} $year^{-1}$) in a) vegetation and b) soil, c) groundwater recharge (mm $year^{-1}$) and d) nitrate leaching (kg ha^{-1} $year^{-1}$).*

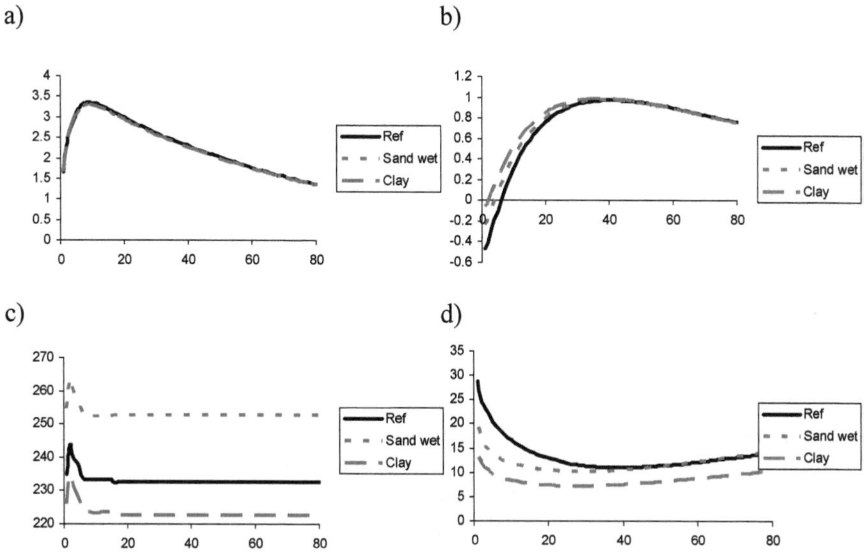

Figure 9.7. *Effect of soil type on modelled C sequestration (ton ha^{-1} year^{-1}) in a) vegetation and b) soil, c) groundwater recharge (mm year^{-1}) and d) nitrate leaching (kg ha^{-1} year^{-1}). charge and nitrate leaching.*

Temperature (Figure 9.8) slightly influences the C sequestration in the biomass, especially during the first 20 years. After 80 years there was hardly any affect of temperature left, whereas the opposite was true for the C sequestration in soil. In the beginning there were no differences, but after 80 years results showed a pronounced difference. Under cold circumstances (T_{mx}= 21°C) the C sequestration was twice as high as under warm circumstances (T_{mx} = 7 °C) 500 vs 100 kg C ha^{-1} year^{-1}.

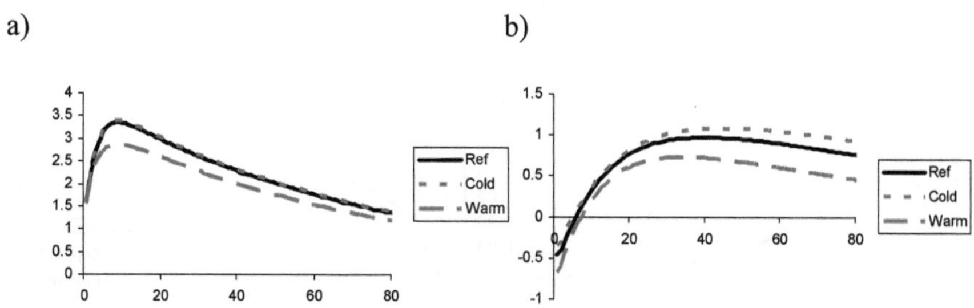

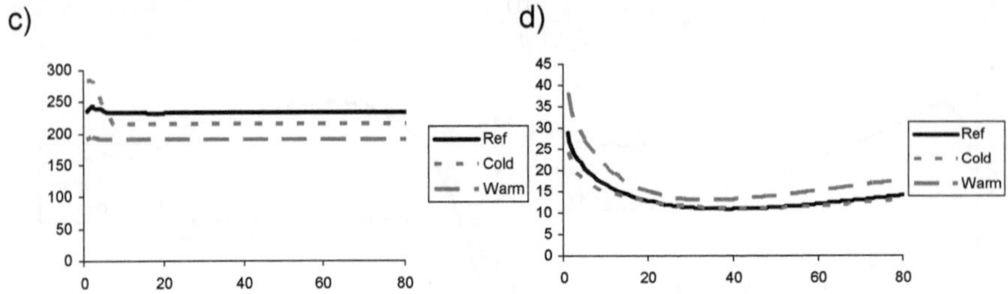

Figure 9.8. Effect of temperature on modelled C sequestration (ton ha^{-1} year^{-1}) in a) vegetation and b) soil, c) groundwater recharge (mm year^{-1}) and d) nitrate leaching (kg ha^{-1} year^{-1}).

Figure 9.9 shows that there was only an effect on C sequestration, both in soil and vegetation, at a low (2.5 kg N ha^{-1} year^{-1}) deposition level. The model showed no visible effect between 22.5 kg N (Ref.) and 37.7 kg N (high). This is an underestimation compared to other researchers, e.g. Nadelhoffer et al. (1999) assumed that always 5% of the additional N deposition will be used for additional growth at deposition levels higher than 10 kg N ha^{-1}. This is in the same range as reported by De Vries et al. (2003), who found values ranging between 3.5 and 7%. Because the C sequestration in the soil is driven by litterfall, the model also showed no effect of enhanced N deposition on soil C sequestration at deposition rates above 22.5 kg N kg ha^{-1}

At low N deposition, the N availability was too low to maintain growth resulting in a collapse of the forest growth. This was also reflected in the water recharge with lower growth resulting in lower transpiration and eventually in higher water recharge. As expected, higher deposition resulted in higher nitrate leaching. All additional deposited N that is between the Reference and the High Deposition, i.e. 15 kg N (37.5-22.5) was not retained by the system and leached. Whether this is realistic is questionable.

The effect of former land use (not shown) only seriously affected the nitrate leaching. The effects on C sequestration and water recharge were negligible.

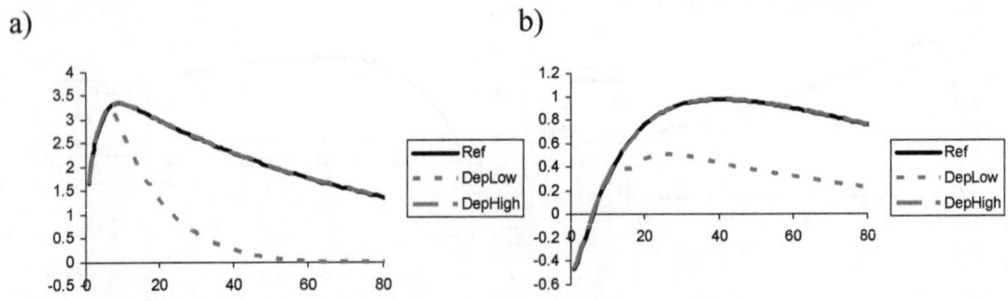

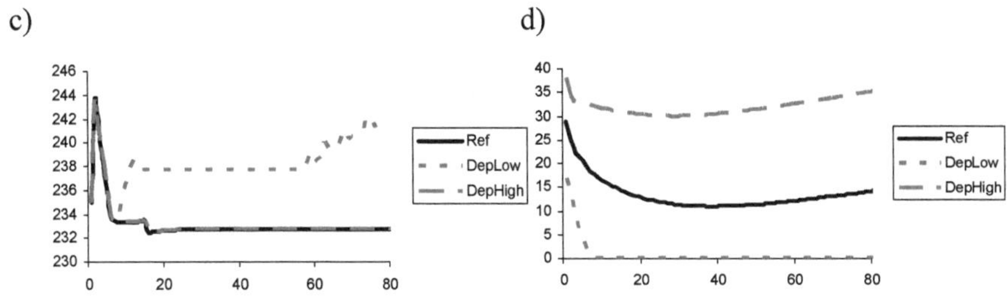

Figure 9.9. *Effect of N deposition on modelled C sequestration (ton ha^{-1} $year^{-1}$) in a) vegetation and b) soil, c) groundwater recharge (mm $year^{-1}$) and d) nitrate leaching (kg ha^{-1} $year^{-1}$).*

8. TESTING METAFORE AGAINST EXISTING KNOWLEDGE OF EXPERTS

Another procedure we followed to assess the quality of METAFORE was the derivation of rules of thumbs meaning knowledge that experts have on ecosystems. This knowledge is inside the heads of the experts, and not very explicitly available. In order to assemble a list of common expert knowledge on this type of rules, we organized an interactive session in which all experts were asked to answer a large number of simple questions about ecosystems. Each question is actually one statement about a fact of ecosystem development, and the answers to the questions were used to score the agreement amongst experts on the validation of the statement. This procedure resulted in a long list of facts on ecosystem modeling. The next step was to evaluate the METAFORE results against this list. The results were satisfactory. The discussion of the scores of the expert questionnaire revealed the issue of 'power of expression' of the simulation results. How significant are differences in numerical results of various simulation runs. The aim was to find error bands around the simulation results and use this information in the AFFOREST-sDSS to analyze whether the statements made by the AFFOREST-sDSS are strong statements (afforestation strategy A is preferred over afforestation strategy B, and the results of A and B are significantly different) or weak statements (although the simulation results suggest that afforestation strategy A is better then afforestation strategy B, the results are so similar that no strong differentiation can be made between these results). However, the discussion revealed that far too little numerical data (actually measured data) is available to support this analysis or to construct the error bands (Appendices 2 and 3).

9. CONCLUSIONS

Although the detailed process models, as part of their scientific development process, have been extensively validated and calibrated, this does not automatically ensure a proper simulation of the processes by the metamodel. During the entire process of development of METAFORE, the model behavior and the simulation results of the detailed process models had to be similar to the results of

METAFORE. This means that the detailed behavior in the process models can at best result in comparable but blended behavior in the METAFORE metamodel. The assumptions made that the generalization of the processes and the application to a larger region can be supported. Obviously, METAFORE is not capable of reproducing the detailed models exactly. However, in general we conclude that the metamodel METAFORE is an acceptable and a beneficial alternative for more complex process oriented model and that it fulfils the necessary requirements for use within the AFFOREST-sDSS.

8. REFERENCES

De Vries, W., van der Salm, C., Reinds, G. J., Dise, N. B., Gundersen, P., Erisman, J. W., & Posch, M. (2003). Assessment of the dynamics in nitrogen and carbon sequestration of European forest soils In: Alterra-rapport 818, Alterra Wageningen, 147 pp.

Jansen, M. J. W. (1998). Prediction error through modelling concepts and uncertainty from basic data. *Nutrient Cycling of Agroecosystems, 50,* 247-253.

Jørgensen, S. E. (1992). Integration of ecosystem theories. Dordrecht: Kluwer.

Kirschbaum, M. U. F. (1999). CenW, a forest growth model with linked carbon, energy, nutrient and water cycles. *Ecological Modelling, 118,* 17-59.

Kros, J. (2002). Evaluation of biogeochemical models at local and regional scale. Doctoral thesis Landbouw en Milieuwetenschappen., Wageningen Universiteit (Netherlands).

Nadelhoffer, K. J., Emmett, B. A., Gundersen, P., Kjønaas, O. J., Koopmans, C. J., Schleppi, P., Tietema, A., & Wright, R. F. (1999). Nitrogen deposition makes a minor contribution to carbon sequestration in temperate forests. *Nature, 398,* 145 -148.

Oreskes, N. (2000). Why believe a computer? Models, Measures, and Meaning in the Natural World. In J. S. Schneiderman (Ed.), The earth around us. San Fransisco: W. H. Freeman.

APPENDIX 1. THE PSEUDO CODE FOR THE METAMODEL FRAMEWORK

```
InputData.Init;
 OutputData.Init;

 DepModel.NDep:=InputData.AtmosphericNDep;
 DepModel.IniLandUse:=InputData.IniLandUse;
 DepModel.AffStrategy:= InputData.AffStrategy;
 DepModel.Init;

 VegModel.MinTemp := InputData.MinTemp;
 VegModel.MaxTemp := InputData.MaxTemp;
 VegModel.AffStrategy:= InputData.AffStrategy;
 VegModel.LatitudeClass:=InputData.LatitudeClass;
 VegModel.Init;

 SoilModel.Temp := InputData.Temp;
 SoilModel.SoilClassID := InputData.SoilClass;
 SoilModel.IniLandUseClass := InputData.IniLandUse;
 SoilModel.AffStrategy := InputData.AffStrategy;
 SoilModel.Init;

 WaterModel.Precipitation := InputData.Precipitation;
 WaterModel.Lat := InputData.Latitude;
 WaterModel.SoilClassID := InputData.SoilClass;
 WaterModel.TreeSpecies := InputData.AffStrategy;
 WaterModel.Init;

 CSeqTot:=0;
 for year:=1 to 100 do
  begin
   YearNupt := 0.;
   YearClf := 0.;
   YearCseqVeg := 0.;
   YearGWR := 0.;
   YearNlf:=0;

   DepModel.TreeHeight := VegModel.TreeHeight;
   DepModel.Step;
   SoilModel.Ntf:=DepModel.Ntf ;
   SoilModel.Nlf := YearNlf;
   SoilModel.ayears := year;
   SoilModel.Step1;

   Nav_tot := SoilModel.Nav;
   VegModel.NavYear:=Nav_tot;

   for month:= January to December do
    begin
     VegModel.Step1;

     WaterModel.MonthTemp:= Vegmodel.MonthTemp;
     WaterModel.PEt:= Vegmodel.PEt;
     WaterModel.Month := month;
     WaterModel.LAI := VegModel.LAI;
     WaterModel.Step;
```

```
      VegModel.Wupt:= WaterModel.Wupt;

      VegModel.Step2;

      YearNupt := YearNupt + VegModel.NUpt;
      YearClf := YearClf + VegModel.Clf;
      YearNlf := YearNlf + VegModel.Nlf;
      YearCseqVeg := YearCseqVeg + VegModel.Cseq;
      YearGWR := YearGWR + WaterModel.GWR;
     end;
    SoilModel.Clf := YearClf;
    SoilModel.Nlf := YearNlf;
    SoilModel.GWR  := YearGWR;
    SoilModel.Nupt := YearNupt;
    SoilModel.Step2;
    CSeqTot:=CSeqTot+SoilModel.CSeq + YearCSeqVeg;

    OutputData.NValues:= SoilModel.Nle;
    OutputData.H2OValues:= YearGWR;
    OutputData.CValues:= SoilModel.CSeq + YearCSeqVeg;
    OutputData.Step;
   end;

end;
```

APPENDIX 2. THE METHODOLOGY FOR DEVELOPING EXPERT RULES

During an interactive session with all the members of the AFFOREST team, the following list of statements was distributed. The scores of validity the experts gave to these statements were used to evaluate the validity of the METAFORE metamodel.

STATEMENTS FOR THE DEVELOPMENT OF THE RULES OF THUMB FOR THE AFFOREST PROJECT

Introduction

The following statements are meant to establish and test assumptions about the effects of afforestation on C sequestration, N leaching and groundwater recharge. Ideally, these assumptions should be based on extensive modelling exercises with thoroughly tested and validated, very detailed models. In practise, these models are lacking, and the models that were developed within the course of the AFFOREST project lack sufficient data to become entirely and optimally calibrated and validated. Out of necessity, our trust in the good behaviour of the model is partly based on field experience and expert knowledge of the participants of the AFFOREST project. This questionnaire is meant to make explicit some of this expert knowledge and field experience.

Each of the statements can be considered an individual hypothesis about facts and knowledge of afforested systems. As with all hypotheses, each individual one can either be true or false, or something in-between. The statements may contain explicit mentioning of countries. This should be interpreted as Sweden being further north and Belgium being the South boundary of our systems, rather than an exact determination of location.

The statements were not created to be true; they were created to be clear. A score of 1: The opposite of the statement is generally true, thus reveals just as much information as a score of 5: The statement is generally true.

The task of each of the members of the AFFOREST team was to score each of the statements as being true or false (or half true). This will enable us to explicitly express the common knowledge of the entire team.

Scoring table

Each of the following statements should be scored with one of the following scores.

1. The statement can not be understood (phrasing is incorrect)
2. The opposite of the statement is generally true (Generally this statement is NOT true)
3. The statement is assumed to be not true, although exceptions exist
4. Neutral: the statement is neither true nor the opposite is true
5. The statement is assumed to be true, although exceptions exist
6. Statement is generally true

Statements

Statement 1: Pine trees have a higher Cseq (tons/ha/year) than oak trees under same circumstances

Statement 2: In Sweden, pine trees have a higher Cseq (tons/ha/year) than oak trees under the same circumstances

Statement 3: Spruce trees have a higher Cseq (tons/ha/year) than pine trees under similar circumstances

Statement 4: On sandy soils, spruce trees have a higher Cseq (tons/ha/year) than pine trees under similar circumstances

Statement 4: No matter which tree species, Cseq (tons/ha/year) in Belgium is higher than Cseq (tons/ha/year) in Sweden

Statement 5: In situations with a precipitation surplus, a positive groundwater recharge and sandy soils, Pine trees will have a larger effect on Groundwater Recharge than Beech trees

Statement 6: In situations with a precipitation surplus, a positive groundwater recharge and clayey soils, Pine trees will have a larger effect on Groundwater Recharge than Beech trees.
Statement 7: In situations in which N leaching occurs, Pine trees will have a larger N leaching than oak trees
Statement 8: Oak trees are less susceptible for drought conditions on clayey soils than on sandy soils
Statement 9: Pine trees are less susceptible for drought conditions on clayey soils than on sandy soils
Statement 10: In Sweden, Pine trees on suitable soils with suitable water will have obtained a larger Cseq after 20 years than Oak trees on suitable soils with suitable water.
Statement 11: In the Netherlands, Pine trees on suitable soils with suitable water will have obtained a larger Cseq after 20 years than Oak trees on suitable soils with suitable water.
Statement 12: In Sweden, Pine trees on suitable soils with suitable water will have obtained a larger Cseq after 70 years than Oak trees on suitable soils with suitable water.
Statement 13: In the Netherlands, Pine trees on suitable soils with suitable water will have obtained a larger Cseq after 70 years than Oak trees on suitable soils with suitable water.
Statement 14: In Denmark, Pine trees on sandy soils will be more susceptible to drought than Oak trees.
Statement 15: In Denmark, Pine trees on a sandy soil will have a larger groundwater recharge than Beech trees in similar circumstances.
Statement 16: In Belgium, Cseq on clayey soils will be larger for coniferous trees than for deciduous trees
Statement 17: In Belgium, Cseq on sandy soils will be larger for coniferous trees than on clayey soils
Statement 18: In Belgium, Cseq on sandy soils will be larger for deciduous trees than on clayey soils
Statement 19: In Belgium, Cseq on sandy soils will be larger for coniferous trees than for deciduous trees
Statement 20: In Sweden, Cseq on clayey soils will be larger for coniferous trees than for deciduous trees
Statement 21: In Sweden, Cseq on sandy soils will be larger for coniferous trees than on clayey soils
Statement 22: In Sweden, Cseq on sandy soils will be larger for deciduous trees than on clayey soils
Statement 23: In Sweden, Cseq on sandy soils will be larger for coniferous trees than for deciduous trees
Statement 24: Pine on sand will sequester C more in Sweden than in Belgium
Statement 25: Oak on sand will give a greater groundwater recharge in Belgium compared to Sweden.
Statement 26: Pine on clay will lead to larger N leaching in Denmark than in Belgium
Statement 27: Oak on clay will sequester more C in Belgium than in Denmark.
Statement 28: Spruce trees in Sweden on clay soils result in higher groundwater recharges than Beech.
Statement 29: Spruce trees in the Netherlands on clay soils result in higher groundwater recharges than Beech.
Statement 30: Oak trees in the Netherlands on sandy soils result in higher N leaching than Pine trees
Statement 31: For Pines N-leaching on sandy soils will be larger than on clay soils.
Statement 32: For Oak N-leaching on sandy soils will be larger than on clay soils
Statement 33: There is a larger difference between coniferous and deciduous than within these groups.
Statement 34: Total amount of C sequestration for coniferous species is higher than for deciduous species after 100 year.
Statement 35: Yearly C increase before maturity is higher for coniferous than deciduous trees.

APPENDIX 3. THE DISCUSSION NOTES ON POWER OF EXPRESSION OF THE METAFORE MODEL

ACCURACY AND POWER OF EXPRESSION IN THE AFFOREST PROJECT

1. The AFFOREST-sDSS consists of a database with geographical data, a metamodel able of calculating environmental performance for the habitats defined by this geographical data and an multicriteria analysis procedure which can find solutions that satisfy multiple criteria. This entire system is accessed through a AFFOREST-sDSS. This AFFOREST-sDSS allows you to find optimal solutions for various questions, ranging from questions such as 'where should I afforest in my country to obtain maximum C sequestration' to 'what afforestation strategy should I choose if I want to get a minimum C sequestration of xxx tons/ha and I want to use only the sandy soils'.
2. The internals of the system evaluates all possible afforestation strategies on all relevant pixels, and then searches through these strategies to find optimal solutions. To do this, the system uses Environmental Performance time series, which contain the behaviour for C-sequestration, N-leaching and groundwater recharge through the years. These Environmental Performance time series are created by the metamodel, and an Environmental Performance time series exist for each combination of site specific input data and afforestation strategies. The Environmental Performance time series are the key component of the entire AFFOREST-sDSS since they are the values used to select optimal strategies and locations.
3. A few times during the AFFOREST project we have had discussions on the accuracy of the AFFOREST-sDSS system and the reliability of the results produced by this system. There seems to be a huge need to do something with this subject: how reliable are the results, and how do we want to communicate this reliability with the end user.
4. There have been a number of meetings in which we agreed to proceed by doing sensitivity analysis on the metamodel. Given the numerical value of each of the parameters in the model, we could easily assign values in the range of plus or minus 10 percent to these parameters, and then analyze the effect this would have on the Environmental Performance. This basically means we want to go for a Monte Carlo analysis of the metamodel. This type of uncertainty analysis is what I would like to call input data driven uncertainty analysis.
5. Problems with this Monte Carlo approach:
   - We don't know the actual distributions, so even for the parameter values that could have in principle a distribution curve associated with them, in practice we don't know this distribution.
   - This doesn't allow for analysing the effects of an error in the categorical data, since by definition they don't have a distribution curve associated with them
6. Analysis of the problem:
   - Actual numerical value that comes out of the model
   - Consider the whole system as a decision making system
   - Relative value of the outcome relative to the other outcomes
   - Error added by the multi criteria analysis
7. Instead of the input data driven uncertainty analysis with its associated problem, this paper proposes a result oriented uncertainty analysis. This analysis will not yield uncertainty bands around the absolute values of results, but will yield significance indicators of proposed measures.
8. In the decision making context of AFFOREST, the relative values of the Environmental Performance time series are more important than the actual values: we are looking for the best, the highest or lowest.
9. If we find this 'best, highest or lowest' solution, we need to be able to say something about the significance of this alternative as being the 'best, highest or lowest'. Is it significantly better than the second best, then we can actually make the statement and take responsibility for it. If it is not significantly better than the second best, then the statement we are making is not that this is the best alternative, but the second best should be included as 'equally good'.
10. From this perspective, error analysis, uncertainty analysis and reliability evaluation should start at the back end of the system: how much difference in Environmental Performance is considered to be significantly different.

11. Examples:
    - Consider case 1 after 35 years of afforestation produces a result of 340 ton/ha of sequestered C, and case 2 produces a result of 335 ton/ha. Although case 1 has a higher value than case 2, it is clear that this small difference is not supported by the data input and the models we used. The statement we are making here is: Case 1 is the highest, but case 2 is not significantly different from case 1.
    - Consider case 1, with an afforestation strategy of xx after 35 years of afforestation produces a result of 720 ton/ha of sequestered C, and case 2 produces 280 ton/ha. This large difference surely makes case 1 significantly different from case 2 and the afforestation strategy of case 1 is definitely producing higher results then strategy 2. The statement we are making is: Case 1 is the highest.
    - Consider case 1 after 35 years results in 450 ton/ha and case 2 results in 390 tons/ha. Is this difference significant? Can we say by using our models we should believe a difference of 60 tons/ha is enough to really say that the strategy of case 1 yields better results than the strategy of case 2.
12. The core modelling system of AFFOREST produces three types (or three axis or three dimensions) of output: on water recharge, on nitrate leaching and on c sequestration. These are the Environmental Performance time series. For each of these dimensions of the output, we (= the AFFOREST experts, not the AFFOREST end users) should be able to define a threshold above which differences in results are significantly different from each other. By definition, every comparison of two cases with a difference under the defined threshold would give a non-significant difference.
13. The meaning of this significance threshold is rather straightforward and should be treated as such. Any comparison between two not significant cases can yield a result but should yield an indication saying that these cases do not significantly differ.
14. If we evaluate cases in which we are interested in the C sequestration of certain alternatives, then the result should be indicated as being insignificant (or the same) for all alternatives that are within the defined threshold value for C sequestration of the highest alternative. Since we are interested in C sequestration in this case, we don't have to consider the threshold values for nitrate leaching and groundwater recharge. These dimensions are not part of the solution anyhow.
15. If we consider cases in which we analyze carbon sequestration combined with groundwater recharge, then we should take the significance on each dimension as a guiding principle. If we consider 2 cases which don't perform significantly different for C and N, but do perform significantly different for groundwater, then the result are two significantly different cases. Only those cases that do not differ significantly in all dimensions analyzed should show up as not significantly different.
16. The important issue now is to define significance levels for the Environmental Performance time series. For each dimension C-sequestration, N-leaching and Groundwater recharge a numerical value for this significance level should be defined. The definition of this value should be guided by the 'gut-feeling' of the modellers and experts.
17. Very concrete task should be to discuss this idea, decide on whether this is the uncertainty analysis we will offer in the AFFOREST-sDSS, and if so, we should find some time to discuss this, and establish values for the threshold.

APPENDIX 4. AFFORESTATION STRATEGIES: RECOMMENDATIONS TO METAMODELLERS HOW TO INTEGRATE THE AFFOREST AFFORESTATION STRATEGIES INTO THE AFFOREST METAMODEL

General considerations

Afforestation strategies (AS) in AFFOREST include Tree species choice (TS), Site preparation (SP) and Stand tending (ST) (figure 1). Rotation length is considered minimum 100 years. AFFOREST does not consider what happens after this 100 years. The site quality and the initial land use may have an influence on tree species choice and site preparation but these are neglected (figure 1).

Tree species choice

Tree species choice is fully user defined. The metamodel models biomass growth as a function of the tree species choice and abiotic factors of climate and soil. For extremely unsuitable site factors the metamodel might give a feedback to the end user that the tree species is unsuitable for the considered site (figure 2).

Incorporating stand preparation in the metamodel

As illustrated in figure 1 stand tending is considered not influenced by other factors and is fully user defined. There are three levels: low, moderate and high. But what do they mean in environmental terms?

Step 1. Site preparation concerns three major activities: soil preparation, fertilizing and vegetation control. Their environmental impact is calculated as follows: The end user has the choice: (a) he answers to the site prep questionnaire if he doesn't know. (b) he chooses low, moderate or high for the three major activities separately or (c) he directly chooses low, moderate or high for site preparation as a whole.

Step 2. The metamodel uses this information to model environmental impact. Response curves show the response in terms of C sequestration, nitrate leaching and groundwater recharge of low, moderate and high levels of site preparation or its components (example figure 7).

Incorporating stand tending in the metamodel

Stand tending is restricted to thinning operations. Pruning and other operations are not taken into account. The user interface allows choosing between low, moderate and high stand tending intensity. Stand tending is depending on tree species and site quality (figure 1). It means for example that moderate thinning in spruce on good sites will take more wood than on bad sites or that for the same site, moderate thinning in spruce will take different amounts of wood than in pine. Stand tending intensity is also depending on management objectives or even management styles. Such management styles may be country or region specific. It means that low thinning in NL does not necessarily mean the same as in Sweden. For example, it is known that in central Europe thinning in conifers have a high frequency of once every 3-8 years, while in Scandinavia for the same site quality, this tending frequency may be substantially lower. Therefore it would be recommendable to make country specific user defined tending schemes.

CHAPTER 10

SUPPORT OF DECISIONS ON AFFORESTATION IN NORTH-WESTERN EUROPE WITH THE AFFOREST-sDSS

J. VAN ORSHOVEN[1], S. GILLIAMS[2], B. MUYS[1], J. STENDAHL[3], H. SKOV-PETERSEN[4] and W. VAN DEURSEN[5]

[1]*Katholieke Universiteit Leuven, Division Forest Nature and Landscape, Celestijnenlaan 200E, 3001 Leuven, Belgium, E-mail: jos.vanorshoven@biw.kuleuven.be,* [2] *VITO, Boeretang 200, 2400 Mol, Belgium,* [3]*Swedish Agricultural University, Box 7001, 750 07 Uppsala, Sweden,* [4] *Forest and Landscape, Denmark, Hørsholm Kongevej 11, 2970 Hørsholm, Denmark,* [5]*Carthago Consultancy, Oostzeedijk Beneden 23a, 3062VK Rotterdam, the Netherlands*

Abstract. A spatial decision support system (AFFOREST-sDSS) has been built to address 'Where', 'How', 'How long' and 'What if' questions related to the environmental performance (EP) of afforestation of agricultural land in north-western Europe at two spatial resolutions. EP is defined in terms of carbon (C) sequestration in biomass and soil, nitrate losses through leaching and groundwater recharge, as determined by 36 possible afforestation management practices. Management practices are a combination of tree species choice, site preparation and thinning regime. Time series of EP are pre-computed by means of the metamodel METAFORE and stored in a spatial database in a GIS-environment. The GIS has been upgraded to a AFFOREST-sDSS by incorporation of a Goal Programming tool which allows for optimisation based on the three components of EP. A user-interface was developed to allow non-specialist users making use of the AFFOREST-sDSS. The output of the AFFOREST-sDSS is complementary to and should be used together with other sources of information like empirical evidence, expertise and scientific literature. The AFFOREST-sDSS shows a promising road to the integrated valorisation of knowledge and technical capabilities to better integrate environmental concerns in the planning and management of human interventions in the rural environment.

G.W. Heil et al. (eds.), Environmental Effects of Afforestation in North-Western Europe, 227–247.

1. FROM MODELLING CAPABILITY TO SPATIAL DECISION SUPPORT SYSTEM

The approach taken in the AFFOREST project as described in the previous chapters leads to a meta-modelling capability for assessing the environmental steady state of initial systems on the one hand and the transient environmental state of afforested systems on the other hand. In this chapter, we describe how this modelling capability is integrated in a computerised spatial Decision Support System (AFFOREST-sDSS). First we introduce the concept of Decision Support System (DSS) in general and sDSS in particular. Next we enumerate the requirements set for the AFFOREST-sDSS and describe how these have been addressed in the design and implementation phase of the system. Herewith we put emphasis on the non-model components of the AFFOREST-sDSS since the metamodel is amply described in chapter 9. We also illustrate the functionality of the system.

2. SPATIAL DECISION SUPPORT SYSTEMS

In the late 1970ies, the first computer-based DSSs were developed, mainly to support ill-structured management problems. From then onwards, DSSs have become increasingly important in a wide variety of fields, ranging from industry, the military, agriculture, forestry, health care to education (Eom & Lee 1990; Vacik et al. 2003).

Because tools for decision support have been developed in widely varying domains, an unequivocal definition of DSS is hard to find. According to Janssen (1992), definitions of DSS range from restrictive, e.g. interactive computer-based systems that help decision-makers utilize data and models to solve unstructured problems (Gorry & Morton 1971), to extremely open, e.g. any system that makes some contribution to decision making (Sprague & Watson 1989).

Sprague (1990) defined three technology levels to structure the domain of DSS development:

- The first technology level relates to a Specific DSS which can be described as an IT-application or dedicated information system that allows decision makers to deal with a specific set of interrelated problems;
- The DSS Generator is situated at the second technology level. It consists of a software package that provides a set of capabilities to build a specific DSS quickly and inexpensively (Turban 1995). Electronic spreadsheet packages and integrated office packages can be considered as limited DSS generators. They provide capabilities for model manipulation, (spreadsheet) model building, database management and dialogue management (menus) capabilities (O'Brian 1990);
- The third technology level is termed DSS Tools. It is the most fundamental level of technology applied to develop a DSS. It consists of software utilities and tools which enable and facilitate the development of both specific DSSs or DSS generators. Examples include graphic editors, query systems, random number generators etc. (Sprague & Carlson 1982).

In the case of decisions to be taken with a spatial 'where' dimension, a geographic information system (GIS) software can be used as a DSS Generator. The GIS-software provides tools that rely on spatial databases and a spatially-enabled interface. A DSS integrating GIS capabilities may be termed a spatial Decision Support System and abbreviated to 'sDSS' (Craig & Moyer 1991; Densham 1991; Moon 1992; NCGIA 1992). Techniques such as 'Dynamic Data Exchange', 'Object Linking and Embedding' and 'Open DataBase Connectivity' allow the transfer of data from the GIS-environment to the other sDSS-components, providing facilities not found in the GIS, and then back to the GIS for storage and visualization. The challenge for developers of comprehensive sDSS is to achieve an appropriate synthesis of modelling techniques, interface and database approaches, drawn from both the GIS and specialized domains.

An important common feature of DSS is that they are built to provide end users with interactive support of their decision-making processes. Hence, end user's requirements are essential in DSS development.

3. USER-REQUIREMENTS FOR THE AFFOREST-SDSS

The targeted users of the AFFOREST-sDSS are spatial planners active in national and regional administrations, and afforestation policy planners and managers of afforestation projects. At the start of the AFFOREST project, consultation with potential users from Sweden, Denmark, the Netherlands and Flanders, Belgium was organised.

Both categories of users require support to select sites and choose among alternative options for afforestation management and afforestation duration with a view to optimise the impacts or performance of afforestation. For the AFFOREST-sDSS three environmental impacts are included. These are carbon (C) storage in biomass and soil, nitrate leaching to deeper soil layers and groundwater bodies and water recharge of those bodies. Since these impacts are highly dependent on spatially variable site conditions, the information is stored, processed and visualized in a GIS-environment, which justifies the term sDSS.

The actual impacts those users aim to optimise are the changes of one, two or three components of the environmental performance of the afforested systems as compared to the environmental performance of the corresponding initial system. The user's degrees of freedom for optimisation are the selection of the initial systems, the afforestation strategy and the time length of afforestation. As such, a user may rely on the expected changes under an afforestation scenario for deciding on what initial system to plant the trees ('where' question), which species to plant and how to manage the forest ('how' question) and how long to keep the trees in place before harvesting and possible replanting ('how long' question).

For spatial planners and afforestation policy planners emphasis will, on the other hand, be put on spatially generalised 'where' questions, managers of afforestation projects, will typically require more spatial and temporal detail and will also need advice regarding tree species, stand preparation and stand management.

4. CONCEPTUAL AND TECHNICAL COMPONENTS OF AFFOREST-SDSS

4.1. Assessment of the change of environmental performance

In order to assess the environmental impact of a given afforestation management in a given initial system after a given number of years, the modelled environmental performance of the agricultural land under its continued initial use and management must be compared with the modelled environmental performance of the same land after the considered period of afforestation. This comparison to the agricultural baseline is what we term 'Environmental change assessment due to afforestation'. The environmental impact of the baseline is considered in steady state, which means that it does not change over time. Assessing changes due to afforestation can be done for individual years or cumulatively over several years.

When individual years are studied, the baseline performance of the initial system, expressed in terms of annual C sequestration (kg m^{-2} yr^{-1}, set equal to zero due to the assumption of steady state), rate of nitrate leaching (kg m^{-2} yr^{-1}) and rate of water recharge (mm yr^{-1}) is compared with the C sequestration and nitrate and water recharge rates for the considered year after afforestation. Computation of differences in rates between the initial situation and the situation in the considered year after afforestation is a straightforward operation.

The evaluation of changes in a cumulative temporal perspective is probably more relevant. In this case, C gains or losses are evaluated by comparing the initial agricultural soil C stock (kg m^{-2}) with the C stock in soil and biomass after the number of considered years since afforestation. For assessment of changes in water recharge and nitrate leaching, first the total amount of nitrates and water that would have been lost during, say, 50 years of continuing the initial agricultural situation, has to be computed. Since under agriculture, annual leaching of nitrate and water is assumed to be constant, these constant values are multiplied with the number of years and expressed as kg nitrate m^{-2} and mm water for the total period of 50 years. Next, the total amounts of nitrate and water leached under the growing forest are obtained by summing the modelled 50 annual (non constant) values, again expressed as kg nitrate m^{-2} and mm water. The difference between the two summed values informs about the avoided loss of nitrates (Figure 10.1) and the reduction (or increase) of water recharge respectively.

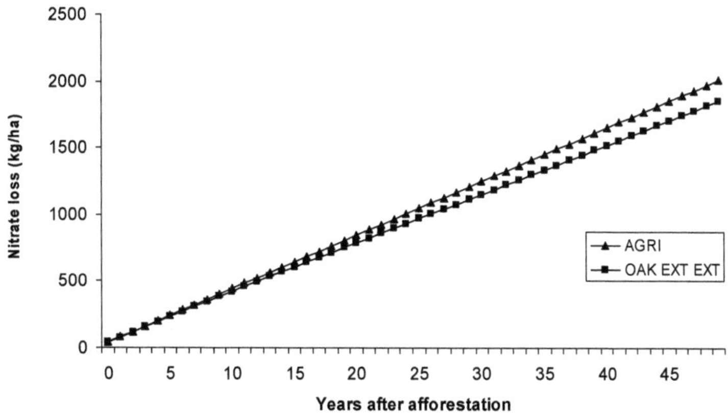

Figure 10.1. *Comparison of cumulative nitrate losses between continued agriculture and afforestation with extensively managed oak for one initial system. The total difference over 50 years is 155 kg.ha*$^{-1}$

4.2. Spatial resolution

To accommodate for the two distinct categories of envisaged users, two corresponding levels of spatial resolution have been identified. At the more detailed project planning level, a spatial resolution of 1 hectare (100 meter by 100 meter) was selected. Since for the Netherlands, the available spatial input datasets do not support such spatial detail, a resolution of 6.25 hectares (250 meter by 250 meter) was retained. For policy planning at the national or regional scale a spatial resolution of 1 km^2 (1.000 meter by 1.000 meter) was selected. As a result of this double resolution separate geographical databases have been developed: a single coarse resolution (1 km^2) database covering all four countries, and four fine resolution databases, one for each country (1 ha resolution, 6.25 for the Netherlands). The database is provided in Arc/Info GRID format.

4.3. The AFFOREST database

The metamodel METAFORE provides the modelling capability to compute the environmental performance with a yearly time step up to 100 years after afforestation for each initial system in the study area at both selected spatial resolutions. METAFORE is derived from calibrated and validated mechanistic models.

For each initial system, input data for METAFORE are retrieved from a spatial input database, complemented by a database with non-spatially explicit crop-, tree- and management-related model parameters. Model output for each initial system is stored into a spatio-temporal output database.

4.3.1. Spatial Input database

The AFFOREST geographical database provides the basic information used to model the development over time of the environmental state variables as a consequence of different afforestation alternatives. It covers eight input variables, including initial land use, soil type, climate, N deposition and latitude (Table 10.1). The original raw data were combined into classes appropriate for the AFFOREST-sDSS:

- Land use is used in the first place to separate different types of agricultural land from areas that are not available for afforestation. For the category of agricultural land, the previous land use is important since it co-determines the initial C and N status of the soil at the time of afforestation.
- Soil type influences many soil processes. The properties contained by the database are soil texture class, soil wetness class, presence of calcareous material and soil depth class.
- Climate includes annual mean values on precipitation, temperature, maximum temperature, and minimum temperature. The annual mean values are based on recent, usually 30-year, standard meteorological period and the maximum/minimum values refer to monthly values for the same 30-year period.
- Nitrogen deposition is modeled from emission data and spatial data on land use using Eutrend (Chapter 5).

The input data used in the AFFOREST database is the best data available taking into account the administrative and commercial restrictions, or data that could be acquired at small cost. Thus, it does not always represent data of the highest resolution available.

For each of the eight spatial input variables the raw data were categorized into classes suitable for the AFFOREST-sDSS in order to minimize data storage and to facilitate swift data operations. The soil and land use data, which were already categorical, were reclassified into relevant classes for modeling the soil processes related to afforestation. The other data, e.g. the climate data, which consisted of continuous values, were classified into a limited number of classes that were chosen to encompass the major range of variability in north-western Europe.

The processing of the geographical input data depended on the type of data delivered: vector data, point data, or raster data.

Table 10.1. *Classification scheme for the eight input variables in the AFFOREST spatial database.*

Land Use Class	Description
1	extensive agriculture; not fertilized
2	intensive agriculture; pasture
3	intensive agriculture; arable land
4	intensive agriculture with excessive manuring; pasture
5	intensive agriculture with excessive manuring; arable land

Soil Class	Texture[1)]	Wetness[2)]	Calcareous[3)]	Shallowness[4)]
1	Sand	<50 cm	Y	Y
2	Sand	<50 cm	Y	N
3	Sand	<50 cm	N	Y
4	Sand	<50 cm	N	N
5	Sand	50-200 cm	Y	Y
6	Sand	50-200 cm	Y	N
.	.	.	.	.
.	.	.	.	.
59	Peat	>200 cm	N	Y
60	Peat	>200 cm	N	N

Mean Precipitation Class	(mm)	Mean Temperature Class	(°C)	Maximum Temp. Class	(°C)
1	<650	1	0-2	1	6-8
2	650-700	2	2-4	2	8-10
3	700-750	3	4-6	3	10-12
4	750-800	4	6-8	4	12-14
5	800-850	5	8-10	5	14-16
6	850-900	6	10-12	6	16-18
7	900-950			7	18-20
8	950-1000			8	20-22
9	1000-1050			9	22-24
10	>1050				

Minimum Temp. Class	(°C)	Nitrogen Deposition Class	(kg ha^{-1})	Latitude Class	(°N)
1	<-3	1	0-5	1	30-35
2	-3-0	2	5-10	2	35-40
3	0-2	3	10-15	3	40-45
4	2-4	4	15-20	4	45-50
5	>4	5	20-25	5	50-55
		6	25-30	6	55-60
		7	30-35	7	60-65
		8	35-40	8	65-70
		9	>40	9	70-75

1) Texture classes: sand, loam, clay, heavy clay and peat
2) Wetness classes: < 50 cm, 50-200 cm, > 200 cm depth to the groundwater level
3) Calcareous classes: yes, no
4) Shallowness (obstacles in the root zone between 0-80): yes, no

Polygon data
The datasets that were provided in vector format contained mainly categorized polygon data, especially on land use and soil type, but also some meteorological data were provided as isoclines maps. The categorized data were converted into rasters of 100 meters cell size (250 for the Netherlands) by applying a straightforward vector to raster transformation, which assigns the spatially most abundant class to each cell in the raster. Subsequently, the rasters were reclassified into the AFFOREST classes. The contour data were converted directly to a raster of appropriate cell size by a contour to raster function. Some preprocessing was necessary where the contoured features were presented as zones of equal value and the zone values did not comply with the class limits in the AFFOREST classification scheme. In this case the data were first converted to raster and subsequently re-contoured into an appropriate contour interval. Finally, a contour to raster operation was made to get a raster of the required cell size, which was reclassified according to the AFFOREST classes.

Point data
The point data sources were mainly data from meteorological stations that were used to estimate climate variables. The conversion of the point data to raster was made by inverse distance weighting interpolation.

Raster data
Most raw data for the AFFOREST database was delivered in raster format, and they were both categorical data, including data on land use and soil type, as well as continuous data, e.g. climate variables. Although the final database was in raster format it was necessary that all input data were converted into rasters with the same origin and cell size, which enabled to combine the rasters into one database covering several variables. The main operation to accomplish this was to resample the input rasters to the right size and origin.

The principle for resampling categorized data from a smaller cell size (e.g. 25 m or 50 m) to a larger cell size (100 m or 1 km) was to calculate the majority class within regular blocks of e.g. 10 by 10 cells and assigning all cells in the block the value of the majority class. Secondly, the raster was resampled to the required cell size, which corresponded to the block size. However, in Arc/Info this operation fails to give any value when no class is in majority within the block. Hence, an intermediate step was added, which was to calculate the majority class using irregular blocks (Figure 10.2) for those cells where no majority class could be defined by using regular blocks.

In case the resampling was made from a larger to a smaller cell size, a similar approach was taken and the majority class was given to each cell in the new raster. For continuous data in raster format the original data were resampled to a new grid cells size by bi-linear interpolation from the nearby raster cells.

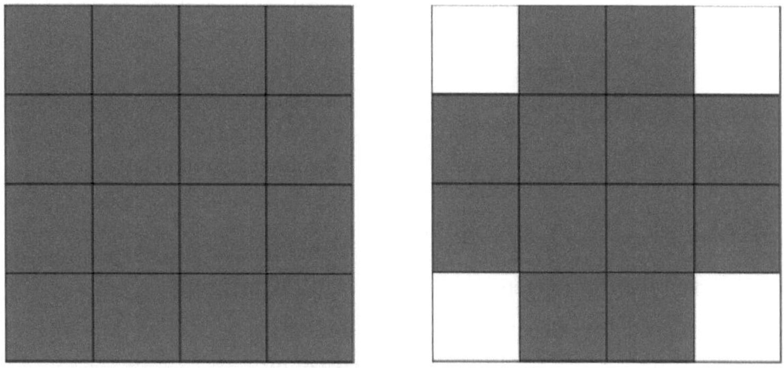

A. B.

Figure 10.2. *Example of a 4 by 4 regular (A) and irregular block (B) used for resampling rasters to a larger cell size.*

Combining data into the final AFFOREST database

The raw input data, which had been processed into common 100-meter rasters (250 meter for the Netherlands) of the same origin, were combined into a multi-layered database including all eight input variables. The attribute table for the raster databases has a many-to-one relation, which means that one row in the attribute file may relate to many raster cells with identical properties. The number of rows in the attribute table corresponds to the number of combinations of site characteristics that are present in the database. This was made for each country, which resulted in four high resolution national databases.

The low resolution European database was constructed based on the four national AFFOREST databases. This involved (1) to project the national databases into a common geographical reference system and (2) to resample the data to a 1 km cell size. This projection step was necessary since the coordinates in each national geographical database was according to the national reference systems. A suitable projection system has to provide an acceptable geographical representation for the whole area. The guidelines of EUREF were used to find a suitable reference system. The selected reference system used in the AFFOREST European database is based on Lambert's azimuthal projection (Table 10.2), which represent the area extent correctly and is an ETRS89 compliant reference system. In the national databases, the data were kept in the original national reference systems.

To project the rasters between geographical reference systems was not straightforward since the original square grid lattice could not be maintained. This introduces some distortion to the projected map file. In order to minimize errors in this process, the 100 m national raster databases were resampled to a finer cell size (10 or 25 meters) before the projection was made to the common reference system. Subsequently, the projected grids were converted to a 1 km raster by first calculating the majority class within 1 km blocks, and subsequently resampling them to 1 km cell size. Finally, they were merged into one common European database.

As an example the combined land use and soil texture maps in the AFFOREST European database are given in Color Plates 4 and 5.

Table 10.2. *Reference system used for the low resolution European database.*

Projection	Lamberts azimuthal equal area
Spheroid	GRS80
Units	Meters
Radius of the sphere of reference	6378137 meters
Longitude of center of projection	9° 0' 0''
Latitude of center of projection	48° 0' 0''
False easting	0 meters
False northing	0 meters

In both the high and low resolution databases, an initial system is a unique combination of class variables expressing initial land use, soil type, climatic condition, initial N deposition and latitude. The attribute 'spatial' connected to this database component means that the position of the initial system in the landscape is known and stored in the database.

To bridge the gap between the available classified information on the initial systems and the parameter requirements of the metamodel, the database is complemented by a set of lookup tables to extend the initial system class identification and location with the parameters required to run the metamodel.

4.3.2. Non-spatial model input data

In addition to spatially explicit model parameters linked to the initial systems, METAFORE (Chapter 9) requires generic parameters related to the chosen tree species (pine, spruce, oak and beech) and management options (low, medium and high intensity of stand preparation; low, medium and high intensity of thinning – Appendix 4 in Chapter 9) to simulate tree growth and associated environmental state variables. Those parameters are hard-coded in METAFORE but can be considered as being retrieved from a non-spatial input database.

4.3.3. Spatio-temporal output database

The output database contains the environmental state variables making up the environmental performance for all initial and all afforested systems. These outputs were pre-calculated using METAFORE from the spatial and non-spatial input databases.

The attribute 'spatio-temporal' for this output database indicates that not only the position of the initial and afforested system in the landscape is known but that also the values expressing the environmental state of all initial systems and of all possible afforested systems are pre-calculated and stored for 100 years after afforestation.

4.4. From database to sDSS

The model output database (section 4.3.3.) has been combined with an extended GIS-engine and GIS-based interface. This combination can be regarded as the AFFOREST-GIS. It presents the capability to query the database in various ways, to perform spatial analysis and present the query and analysis results in maps, tables, charts etc. The AFFOREST-sDSS is the result of the further extension of the interface and engine to cope with multi-objective and optimization questions. A comprehensive schema of the AFFOREST-sDSS and its building blocks is given in Figure 10.3.

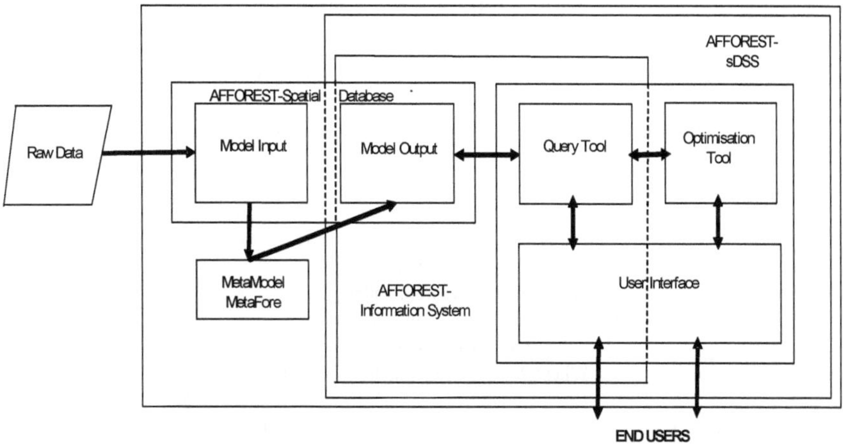

Figure 10.3. *Components of the AFFOREST-sDSS.*

4.4.1. End-user questions and database queries

The type of questions, which can be handled by any DSS is conditioned by the content of the underlying database and or the scope of the underlying model. For the AFFOREST-sDSS, the database content is the result of the options taken when designing the project, the afforestation system analysis (Gilliams et al. 2005a) and the modeling capability. The data content goes back to the modeled relationships between 4 components:

- The initial agricultural system, characterised by its location, agricultural land use, other site characteristics and the baseline environmental performance;

- The afforestation management including the actions taken at the time of planting (site preparation, tree species selection) and during tree growth (management);
- The time after afforestation expressed with a time step of one year up to 100 years after afforestation;
- The afforested system, described by the location and characteristics of the corresponding initial system and by the environmental state variables resulting from the application of a given afforestation management during a given time to the given initial system.

Questions or queries can be constructed by means of one or more of these 4 components while the result of the query should provide information on the components not specified in the query. Table 10.3 displays all possible queries based on the data content of the spatio-temporal output database.

Table 10.3. *Overview of all query types possible in AFFOREST-sDSS.*

Question type	Specified input (I) and obtained output (O)			
	Location, site characteristics and environmental performance of initial system	Location, site characteristics and environmental performance of afforested system	Afforestation Management	Time
	Where ?	How much ?	How ?	How long ?
1	I	O	O	O
2	O	X	O	O
3	O	O	X	O
4	O	O	O	X
5	X	X	O	O
6	X	O	X	O
7	O	X	X	O
8	X	O	O	X
9	O	X	O	X
10	O	O	X	X
11	X	X	X	O
12	X	X	O	X
13	X	O	X	X
14	O	X	X	X

The most straightforward questions are of the 3-1 type: 3 components are specified to parameterise the query (Questions types 11-14 in Table 10.3). The 4th component is the result of the query. Such a query is further termed a single component question as opposed to multi-component questions which are of the 2-2, 1-3 or 0-4 type.

A typical example of a single component question is

> "the search for those initial systems yielding a C stock larger than A kg m^{-2} AND having a cumulative nitrate leaching less than B kg m^{-2} AND characterised by a cumulative groundwater recharge larger than C mm when submitted to a given afforestation management during a given time lag."

A typical multi-component question of the 1-3-type encompasses

> "the search for those afforested systems (i.e. combination of initial system, afforestation management and years after afforestation) characterised by a C stock larger than A kg m^{-2} AND having a cumulative nitrate leaching less than B kg m^{-2} AND characterised by a cumulative groundwater recharge larger than C mm."

Both types of queries use threshold values and operators (equal to, larger than), possibly combined with Boolean operators (AND, OR, NOT, XOR) to specify the targeted afforested systems.

To fully accommodate and exploit the multi-dimensional character of the environmental performance of initial and afforested systems, the query possibilities have been implemented to address questions in both a mono- and a multicriteria analysis approach. Whereas for most of both approaches, regular database query facilities, as implemented in (geo-) information systems, suffice to find answers to the questions, this is not the case for queries in which multi-criteria are interdependent. Such queries require advanced processing commonly not found in information systems but which are typical for 'true' decision support systems.

Single component single criteria questions

A typical example of a single component, single criteria question is:

> "Which afforestation management results in the highest C stock 50 years after afforestation of a given initial system?"

Three components are specified being the initial system, the time after afforestation and a threshold value for one environmental state variable characterising the afforested system. The unspecified component is the afforestation management. The answer to this question is found by comparing the C stock for the given initial system at year 50 between the 36 management practices (each stored in one separate table).

Single component multi-criteria questions

The multifold dimension of the environmental performance of initial and afforested systems leads to a second category of single component questions, i.e. single-component, multi-criteria questions. In the most simple case, this type of questions can be decomposed into several single component, single criteria questions, which have to be solved sequentially.

> "How much C will be stored in and how much N will have been lost from an afforested system after 50 years when installed on a given initial system using a given afforestation management?"

The only unknown component is related to the afforested system. However two environmental state variables of the afforested system are being asked for. The question can be solved by addressing sequentially the two corresponding single component, single criteria questions.

If however, the answer to the question depends upon the combination of the multiple criteria, the sequential decomposition approach is not possible as is illustrated by the following example question.

> "Which afforestation management results in the highest possible C stock *and* the lowest possible cumulated nitrate leaching 50 years after afforestation of a given initial system?"

Since it is possible and even likely that the afforestation practice, which yields the highest C stock, is not the same as the one resulting in the lowest nitrate leaching, a sequential solution of the question would lead to two conflicting answers. This type of questions needs some kind of optimization, i.e. a mechanism to trade off between the required highest C stock and lowest nitrate leaching, resulting most probably in the selection of a management practice which delivers neither the absolutely highest possible C stock neither the absolutely lowest possible nitrate leaching but rather the best possible compromise between both criteria.

In order to allow the user to address single component, multi-criteria questions by means of the AFFOREST-sDSS, the Goal Programming (GP) technique has been implemented (see Gilliams et al. 2005b). GP has been introduced by Charnes & Cooper (1961) as a simple linear technique. In the AFFOREST-sDSS the GP-interval approach is used as introduced by Charnes and Collomb (1972) and Ignizio (1974). If in the above example question, the maximum value for C sequestration and the minimum value for nitrate leaching correspond to the same afforestation management, the optimal solution is straightforward and no further querying is needed. However, when the afforestation management practices do not coincide, a suboptimal solution needs to be pursued. This is performed assuming certain tolerance intervals around the maximum and the minimum of each environmental state variable separately. The size of the interval is depending on the weight users set for the variables in question. If in a first step no sub-optimal solution is found, in subsequent steps, the intervals are progressively enlarged according to the weights. As soon as an afforestation management practice is found which is common to C and N, the sub-optimal solution is identified. This approach using intervals implies that more than one management approach may match the query criteria. For a given state variable, it now becomes also possible to assess the difference between the sub-optimal solution found in the single component, multi-criteria approach and the result of the corresponding single component, single criteria question. This can either be relative (e.g. the suboptimal solution lies within a 5% interval of the optimal solution) or it can be absolute (e.g. the difference between the suboptimal and the optimal solution for C sequestration is 5 t ha^{-1}).

The Goal Programming approach is illustrated in Table 10.4 using the previous example question. The table shows part of the AFFOREST output database. It represents in a combined way extracts for one initial system from six tables from the database: 3 tables, one for each of 3 afforestation management options, with pre-computed time series of C contents and 3 tables, again one for each of 3 considered afforestation management options, with pre-computed time series of nitrate leaching rates, converted into cumulative nitrate leaching. Carbon stocks and amounts of nitrate leached are expressed in relative terms with respect to the initial system.

Considering the basic data there is no best solution to the example question since with regard to C, management option #2 is better than the others, and with regard to nitrate, management option #3 is performing best.

The question can be answered, however, when tolerance intervals are applied to the carbon and the nitrate data. In a first step a 5% interval is used for both carbon and nitrate, this in case that the end user gives equal weight to both decision criteria. For carbon, 'tolerance' means that we are satisfied with a carbon stock which is 5% lower than the maximal one, while for nitrate, a loss of 5% higher than the minimum loss is accepted. As a result of the creation of intervals, the management options #2 and #3 provide similar results for carbon, i.e. the intervals are overlapping. For nitrate, option #3 still returns lower losses than the other options. From these findings the DSS will propose option #3 to the user as the optimal one.

Although a solution is found using the 5% interval, the case of a 10% interval is presented for the sake of illustration. If a 10% tolerance would have been used, all three management options would have been considered as being equal regarding carbon. With respect to Nitrate, options #2 and #3 can no longer be distinguished. The recommendation from this exercise is that both afforestation management options #2 and #3 are applicable to achieve the set objectives.

Table 10.4. *Illustration of the Goal Programming approach used in AFFOREST-sDSS.*

	IniSys	Aff Mngmnt	C_0	C_1	$C_{..}$	C_{25}	C_{50}	N_0	N_1	$N_{..}$	N_{25}	N_{50}
Basic data												
	1	1	100	99	132	155	152	100	95	1200	1980	3600
	1	2	100	102	125	145	166	100	100	1070	2000	3000
	1	3	100	101	144	155	165	100	105	990	1500	2800
5% interval (downward for C, upward for N)												
	1	1	100	NA	NA	NA	152-144	100	NA	NA	NA	3600-3780
	1	2	100	NA	NA	NA	166-156	100	NA	NA	NA	3000-3150
	1	3	100	NA	NA	NA	165-157	100	NA	NA	NA	2800-2940
10% interval (downward for C, upward for N)												
	1	1	100	NA	NA	NA	152-136	100	NA	NA	NA	3600-3960
	1	2	100	NA	NA	NA	166-149	100	NA	NA	NA	3000-3300
	1	3	100	NA	NA	NA	165-148	100	NA	NA	NA	2800-3080

IniSys = Initial system class
Aff Mngmnt = Afforestation management option
C_0 = Carbon stock in the initial system
C_1 = Carbon stock in the afforested system after 1 year of afforestation
$C_{..}$ = Carbon stock in the afforested system after N years of afforestation
C_{25} = Carbon stock in the afforested system after 1 year of afforestation
C_{50} = Carbon stock in the afforested system after 1 year of afforestation
N_0 = Amount of nitrate leached per year from the initial system
N_1 = Cumulative amount of nitrate leached from the afforested system after 1 year of afforestation

N = Cumulative amount of nitrate leached from the afforested system after N years of afforestation
N_{25} = Cumulative amount of nitrate leached from the afforested system after 25 years of afforestation
N_{50} = Cumulative amount of nitrate leached from the afforested system after 50 years of afforestation
NA = Not Applicable

Multiple component single criteria questions

Questions of the multi-component type return more open answers than the single component questions. Only 2, 1 or (theoretically) even no component at all 3 are specified so that the query will come up with the other 2, 3 or 4 components in the output. A typical 2-2 example could be:

> "How will C stocks evolve over time when a given initial system is afforested according to a given management practice?"

The result is one record pertaining to the specified initial system out of one of the 36 C tables (the one for the specified management practice). It can be displayed as a graph with a single C as a function of time curve.

A 1-3 question may look like:

> "How will C stocks evolve over time when a given initial system is afforested using whatever afforestation management practice?"

Here the result is a graph as Color Plate 6 with 36 C as a function of time curves, one for each management practice.

Multiple component multi-criteria questions

With regard to the component 'environmental performance of the afforested system' a further distinction of the multiple component questions can be made between single and multi-criteria questions. When several criteria are formulated independently from one another, this type of questions can be solved by reformulating it as a sequence of single criteria questions:

> "What will be the C stock and groundwater recharge for a given afforestation management practice after a given time?"

Here, the results are given in two maps covering all initial systems displaying C stocks on the one hand and amount of recharged water on the other hand.

When several criteria are inherently linked, an optimization approach is necessary to address a question like:

> "Which combination of afforestation practice and time lag results in the highest possible C stock *and* the lowest possible nitrate leaching for a given initial system?"

4.4.2. User interface

The interface, programmed in the ArcView-GIS scripting language AVENUE, guides the user through the query formulation and presents the results in an easy-to-interpret way. An extended description of the user interface is incorporated in the AFFOREST-sDSS tutorial, which is a separate deliverable of the AFFOREST project and available on the project's web site (www.sl.kvl.dk/afforest).

Query formulation

The AFFOREST-sDSS is designed in the first place to handle single component questions of both the single- and multi-criteria types, including optimization by means of the Goal Programming approach. The standard functionality of the AFFOREST-sDSS comprises indeed the solution of four 3-1 type of questions ('where', 'how' 'how long' and 'how much') where the how much part (environmental performance) can be limited to one state variable or to more than one, simultaneously considered, state variables, i.e. in an optimization exercise. In addition, also multi-component questions can be accommodated, the only limitation being that optimization in the presence of multiple criteria is not possible.

Query formulation through the Graphical User Interface (GUI) of the AFFOREST-sDSS requires the user to go through a set of customised dialog windows that guide through the system and allow to structure the afforestation question into one of the types the system can handle. Figure 10.4 illustrates the work flow.

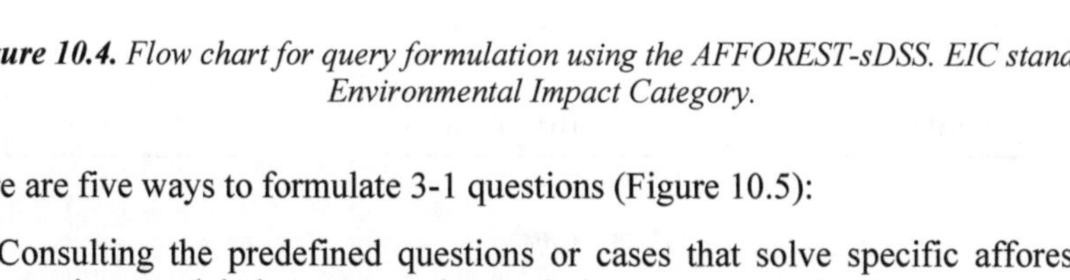

Figure 10.4. *Flow chart for query formulation using the AFFOREST-sDSS. EIC stands for Environmental Impact Category.*

There are five ways to formulate 3-1 questions (Figure 10.5):

- Consulting the predefined questions or cases that solve specific afforestation questions and help users understand the system so that they can use it to formulate and solve their own specific questions;
- When 'Where' questions are formulated all components can be specified except for the initial system;
- When 'How much' questions are composed all components can be specified except for the environmental performance;

- When 'How' questions are asked all components can be specified except for the afforestation management;
- When 'How long' questions are formulated all components can be specified except for the evaluation time.

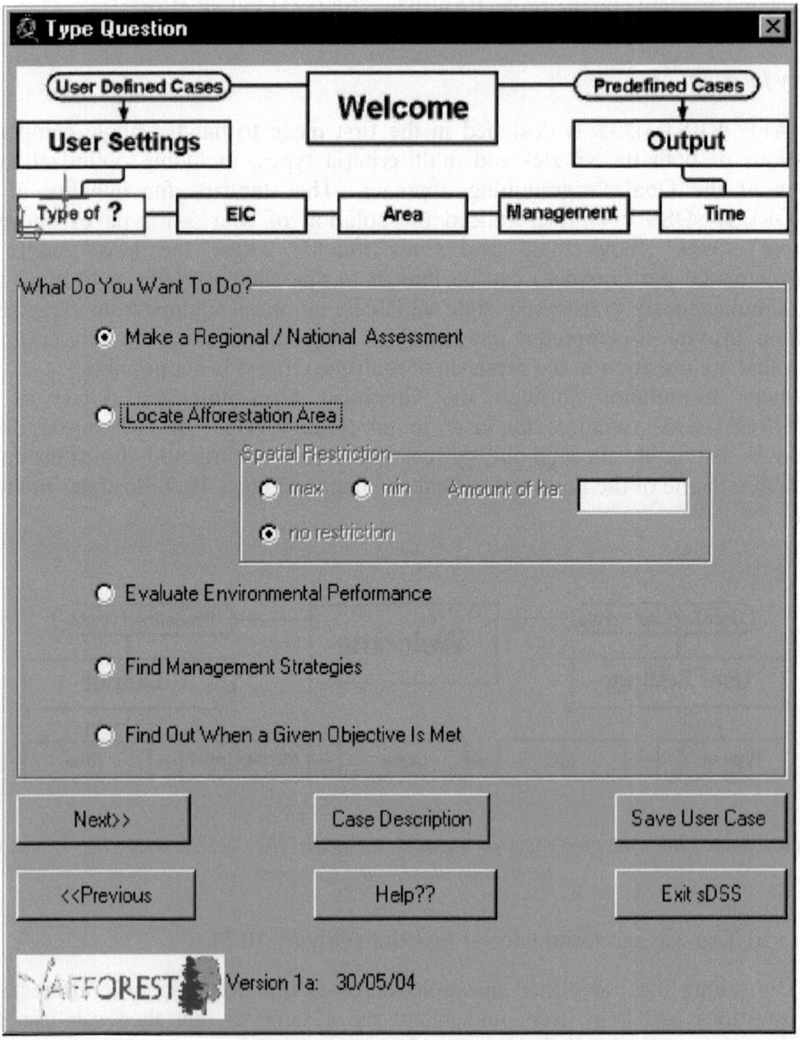

Figure 10.5. *Main menu for query formulation.*

If the question is a 2-2 or a 1-3 question, components that are not specified become part of the answer, like in the example question

> "Which afforestation management needs to be applied in order to reach the best environmental performance 25 years after afforestation on sandy soils?"

This is a 3-1 question and all windows except for the management window need to be addressed by the user. If the time window is not set the question is transformed into a 2-2 question which can be formulated as

> "Which afforestation strategy yields the highest environmental performance and when will this happen, when afforesting on sandy soils ?".

Presentation of results

The query result can be presented in the form of maps, displaying e.g. for each initial system the management leading to the highest possible C stock after a given number of years or as tables, e.g. a list of afforestation management options with the initial systems to which they can be applied to yield the threshold environmental states. A graph option is not included in the programming code of the AFFOREST-sDSS but graphs can be produced using the ArcView-GIS standard functionality to show e.g. evolution of environmental state for one or more variables with time.

The red pixels in Color Plate 7 correspond to sandy initial systems within a municipality in Flanders, Belgium, where, after 25 years, maximal C sequestration is obtained when oak is planted after extensive site preparation and with extensive tending of the growing trees. In the output table the characteristics of the initial system and the results can be found.

5. RELIABILITY OF THE AFFOREST-SDSS

The database underlying the AFFOREST-sDSS has been derived by applying a simplified process-based tree growth model fed by geo-referenced parameters derived from a generalized spatial database on the one hand and by parameters regarding tree behaviour and afforestation practice from chronosequence experiments, literature and expert knowledge on the other hand. It is obvious that the model output presents shortcomings. However, general tendencies and spatial patterns are found to be consistent with expert knowledge and with observations made on chronosequences.

The output of the AFFOREST-sDSS can however not be the single source of information for afforestation policy makers and afforestation planners and managers to take decisions. The output is meant as a formalised support for the decision making process. Therefore, guidelines have been drafted based not only on AFFOREST-sDSS output but also on expert knowledge and results of experiments (Chapter 11).

6. CONCLUSIONS

In the AFFOREST-sDSS, the capability of process-based models to simulate the environmental impacts of afforestation of agricultural land, has been enhanced with the capabilities of GIS for spatially explicit modelling and with the capabilities of multicriteria analysis for decision support. The resulting AFFOREST-sDSS provides a user-friendly framework and engine to address single and multi-component questions regarding the environmental performance (EP) of afforestation, dealing with 'where', 'how', 'how long' and 'what if' questions. Since EP is defined as the combination of C sequestration, nitrate leaching and groundwater recharge, the possibility to address multi-objective problems using the goal programming approach has been incorporated.

The answers provided by the AFFOREST-sDSS complement empirical evidence and scientific expertise with respect to the support of policy and management decisions. Unlike evidence and expertise, the process-based model enables the AFFOREST-sDSS to assess unobserved or potential combinations of site conditions, afforestation management and duration. Other added value compared to the sole use of the metamodel encompasses the fact that 'Where' questions can be explicitly answered and that 'Where', 'How' and 'How long' questions involving the optimisation of multiple environmental criteria and resulting in suboptimal solutions can be managed.

All components of the AFFOREST-sDSS have scope for further improvement: spatial input data, database structure, query system, optimisation algorithm, user's interface. Nevertheless, the AFFOREST-sDSS shows a promising road to the integrated valorisation of knowledge and technical capabilities to better integrate environmental concerns in the planning and management of human interventions in the rural environment.

7. REFERENCES

Charnes, A., & Collomb, B. (1972). Optimal economic stabilization policy: Linear goal-interval programming models. *Socio-Economic Planning Science, 6,* 431-435.

Charnes, A., & Cooper, W. W. (1961). Management Models and Industrial Applications of Linear Programming. New York: Wiley.

Craig, W. J., & Moyer, D. D. (1991). Progress on the Research Agenda: URISA '90. *URISA Journal, 3:1,* 90-96.

Densham, P. J. (1991). Spatial Decision Support Systems. In: Maguire, D.J., M.F. Goodchild, and D.W. Rhind, (Eds.). Geographical Information Systems: Principles and Applications, vol.1, London: Langman, 403-412.

Eom, S. B., & Lee, S. M. (1990). A survey of decision support system applications (1971- April 1988). *Interface, 20:3,* 65-79.

Gilliams, S., J. Van Orshoven, B. Muys, H. Kros, G.W. Heil & Van Deursen, W. (2005a). AFFOREST-sDSS: a metamodel based spatial decision support system for afforestation of agricultural land. *New Forests, 30,* 33-53.

Gilliams, S., D. Raymaekers, B. Muys, & Van Orshoven, J. (2005b). Comparing multiple criteria decision methods to extend a geographical information system on afforestation. *Computers and electronics in agriculture, 49,* 142-158.

Gorry, A., & Morton, M. S. S. (1971). A framework for information systems. *Sloan Management Review, 13,* 55-70.

Ignizio, J. P. (1974). Interval Goal Programming and Applications. Working paper, Pennsylvania State University.

Janssen, R. (1992). Multiobjective decision support for environmental management. Dordrecht: Kluwer Academic Publishers.

Moon, G. (1992). Capabilities Needed in Spatial Decision Support Systems. In: GIS/LIS '92 Proceedings, San Jose, Calf, USA 2: 594-600.

NCGIA. (1992). A Research Agenda for Geographic Information and Analysis. Technical Report 92-7. The National Center for Geographic Information and Analysis.

O'Brian, J. A. (1990). Management information systems: a managerial end user perspective. Boston: Richard, D. Irwin Inc.

Sprague, R. H., & Carlson, E. D. (1982). Building effective decision support systems. Englewood Cliffs, N.J.: Prentice-Hall.

Sprague, R. H., & Watson, H. J. (1989). Decision Support Systems: Putting theory into practice. London: Prentice-Hall International.

Turban, E. (1995). Decision support and expert systems: management support systems.. Englewood Cliffs, N.J.: Prentice-Hall.

Vacik H., Lexer M. J., Rauscher M. H., Reynolds K. M., & Brooks R. T. (Eds.) (2003). Decision support for multiple purpose forestry.. A transdisciplinary conference on the development and application of decision support tools for forest management. April 23-25, 2003, University of Natural Resources and Applied Life Sciences, Vienna, Austria, CD-Rom Proceedings.

CHAPTER 11

GUIDELINES FOR PLANNING AFFORESTATION OF FORMER ARABLE LAND

K. HANSEN[1], L. VESTERDAL[1], B. MUYS[2], S. GILLIAMS[3], L. ROSENQVIST[4], C. VAN DER SALM[5], M. ELEMANS[6], H. DENIER VAN DER GON[7], P. GUNDERSEN[1], M-B. JOHANSSON[4], J. VAN ORSHOVEN[2], G. HEIL[8], H. KROS[5], A. BLEEKER[9], W. VAN DEURSEN[10] and J. STENDAHL[4]

[1]Forest and Landscape Denmark, KVL, Hørsholm Kongevej 11, 2970 Hørsholm, Denmark, E-mail: kiha@kvl.dk, [2]Katholieke Universiteit Leuven, Division Forest, Nature and Landscape, Celestijnenlaan 200E, 3001 Leuven, Belgium, [3]VITO, Boeretang 200, 2400 Mol, Belgium, [4]Swedish University of Agricultural Sciences, Box 7001, 750 07 Uppsala, Sweden, [5]Alterra Green World Research, P.O. Box 47, 6700 AA Wageningen, the Netherlands, [6]Karolinska Institute Microbiology and Tumorbiology Center (MTC), Box 280, 171 77 Stockholm, Sweden, [7]TNO, Environment and Geosciences,Laan van Westenenk 501, 7300 AH Apeldoorn, the Netherlands, [8]Faculty of Science, Institute of Environmental Biology, Sorbonnelaan 16, NL 3584 CA Utrecht, the Netherlands, [9]ECN, P.O. Box, 1755 ZG Petten, the Netherlands, [10]Carthago, Rotterdam, the Netherlands

Abstract. Afforestation objectives vary from one country to another and even within countries. Apart from the objectives, the specific conditions from a biophysical, environmental and socio-economic point of view should always be considered throughout the entire afforestation process, from policy decisions through location of the new forest, establishment and management, and the final utilisation of the forest. Decisions on how and where to afforest, and how much these decisions will affect the environmental impacts should ultimately be a compromise between the site quality in terms of climate, soil and preceding land-use, the initial goals set by planners and managers, and the stakeholders' preferences. The focus of AFFOREST has been on building knowledge and capacity to support decisions regarding afforestation of former arable land with respect to changes in C and N pools and fluxes and changes in

G.W. Heil et al. (eds.), Environmental Effects of Afforestation in North-Western Europe, 249–291.

water recharge. The guidelines in this chapter are based on literature reviews, the experimental data from chronosequences of afforested stands in Denmark, Sweden, and the Netherlands (described in Chapter 2, 3 and 4), and on the developed mechanistic metamodel (METAFORE) and the spatial Decision Support System (AFFOREST-sDSS) (Chapter 7, 8, 9 and 10). The structure of the guidelines is based on questions and corresponding answers under the main themes of water recharge, nitrate leaching, C sequestration, diversity of understory vegetation and complex questions involving more than one of the first three issues. Hopefully, the guidelines will be helpful and inspire landscape and forest planners in planning how and where afforestation should take place.

1. INTRODUCTION

In the earlier chapters, we presented some fundamental findings on the environmental effect of afforestation in north-western Europe and a new decision support tool for environmental impact assessment and planning based on these findings. In this chapter, we offer guidelines for afforestation of former arable land based on results from field measurements, modelling, and the use of the spatial Decision Support System (AFFOREST-sDSS). The guidelines are presented as questions and short answers, followed by more detailed clarifications. We start with single objective questions on water recharge, nitrate leaching, carbon sequestration and diversity of understorey vegetation, and we end with examples of multi-objective problem solving resulting from running the spatial DSS.

2. WATER RECHARGE

The water recharge (Q) of an area depends on the water balance, which can be expressed as:

$$Q = P - E - \Delta S \tag{1}$$

in which P is the precipitation, E the evapotranspiration (the loss of water to the atmosphere) and ΔS is the change in soil water storage. Water recharge consists of runoff, lateral drainage and leaching to the groundwater. Evapotranspiration occurs due to evaporation of rainfall from the canopy (interception), transpiration of the forest, and soil evapotranspiration.

2.1. Do new forests decrease water recharge relative to arable land?

It is widely accepted that trees generally use more water than any other vegetation. Specific research shows that water recharge in forests is less than under arable land since more water evaporates from higher canopies. When arable land is afforested the recharge therefore decreases. It is then a question of where and how to afforest in order to optimize the water recharge.

Mature trees use more water and their transpiration is larger than for newly planted young trees. Therefore, water recharge decreases during the transition from arable conditions over young trees to canopy closure where the decline levels off.

Literature data indicate that water recharge in afforested areas is lower than in clear-cut or agricultural areas. Observed differences in water recharge range from less than 100 mm yr^{-1} to more than 600 mm yr^{-1} and depend on the mean annual precipitation, the change in forest cover and the type of forest. The lowest reductions in water recharge are found for the conversion to open deciduous forests in dry areas. The highest values are found for dense coniferous forests in areas with a high annual rainfall (Bosch & Hewlett 1982; Sahin & Hall 1996).

The chronosequences studied within the AFFOREST project showed a decrease in annual water recharge of 20-230 mm yr^{-1} resulting from an increase in forest age over a period of 5 to 92 years. This decrease is less than indicated by the literature because AFFOREST field studies were mainly limited to areas afforested 5 to 30 years ago. When a clear-cut area or an agricultural area could be included as a reference the decrease in annual water recharge would increase by 50-200 mm. For example, at the Dutch oak site the decrease in water recharge was approximately 70 mm between the 4 and 18 year old forest and 130 mm between a grassland site and the 18 year old oak forest.

Studies on the change in water recharge during the growth of a forest stand are quite limited. Results from long-term monitoring studies indicate a quick decline in water recharge during the first 5 to 10 years after afforestation until the canopies have closed up, followed by a much slower decline when the trees grow older (Le Maitre & Versfeld 1997).

At the AFFOREST sites the water recharge follows this general pattern of a strong decline in water recharge in young stands followed by a slower decline when trees grow older (Figure 11.1).

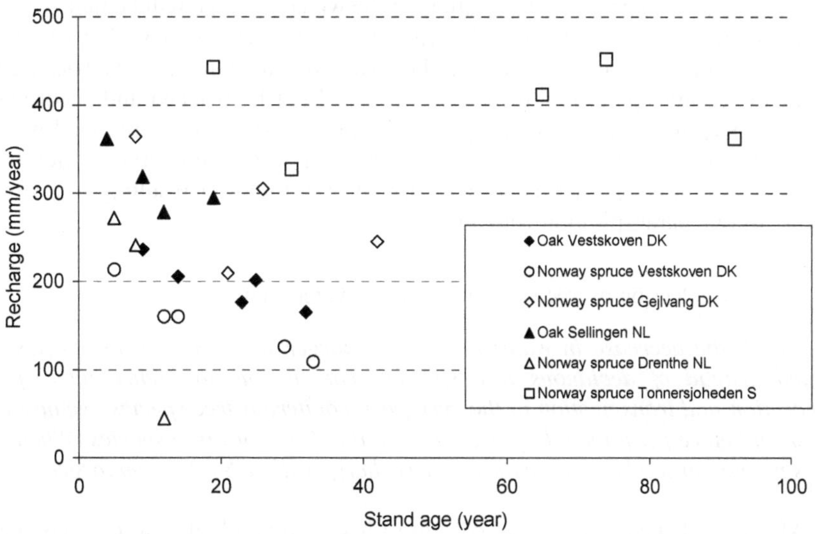

Figure 11.1. *Water recharge (mm yr^{-1}) at the five AFFOREST chronosequences of planted Norway spruce and oak as a function of age in Denmark (DK), Sweden (S) and the Netherlands (NL).*

The strongest decline in water recharge at the oak stands took place in the first 10 years. In the spruce stands the decline in water recharge is stronger and levels after 15-20 years. However, the decline in water recharge can be strongly influenced by planting density and management (See 2.4). For example, the decline in water recharge in the three Dutch spruce stands is extremely fast due to a high planting density and the absence of thinning.

2.2. Does the afforestation site selected influence the water recharge?

If it is the goal to maximize the water recharge and the choice is between afforestation of different soil types, a sandy site should be chosen to a clayey site. In a wet climate the reduction in water recharge following afforestation will be relatively higher than in a dry climate.

Water recharge varies from site to site due to differences in water holding capacity of the soil (sand vs. clay), the drainage conditions, the slope and the climatic conditions. Water recharge is normally higher from sandy soils than from clay soils due to the higher water holding capacity of clay soils (Finch et al. 1998). However, this effect may be overruled by differences in transpiration. For example, clay soils are often poorly drained, leading to transpiration reduction in wet periods due to anaerobiosis. In practice, the lowest leaching fluxes are often found on soils of intermediate texture where transpiration rates are neither reduced by drought nor by excessive wetness (van der Salm & de Vries 2000).

In general, the impact of afforestation on the water recharge will be higher in wet climates compared to dry climates (Bosch & Hewlett 1982; Sahin & Hall 1996). On dry sites the impact of afforestation will be lower compared to wet sites, because the potential evapotranspiration rates of the trees will not be reached and accordingly the differences in actual evapotranspiration between forest and arable land will be lower than under more favourable (e.g. somewhat wetter) conditions. Moreover, under dry or nutrient limited conditions growth will be lower resulting in a slower increase of evapotranspiration with age.

2.3. Does the selection of tree species influence water recharge?

In general, the decrease in water recharge is larger when coniferous species are planted instead of deciduous tree species. This is due to higher interception evaporation and transpiration in the evergreen coniferous tree species because of a higher and more permanent leaf area than in the deciduous tree species. When e.g. oak is planted instead of spruce the water recharge will be 80-250 mm larger.

The AFFOREST data showed that oak forest had a 80 mm higher water recharge in Denmark and a 250 mm higher recharge in the Netherlands compared to spruce forests (Chapter 3). However, differences between tree species are overruled by differences between the countries due to climatic effects. For example, water recharge in Sweden was higher compared to Denmark and the Netherlands, where precipitation was lower and evaporation was higher.

2.4. How will management of afforested sites influence water recharge?

Different management options clearly affect the water recharge. In order to optimize the water recharge the canopy of new forest should be kept as open as possible. Removal of old drainage pipes may increase around water recharge.

The loss of water due to evaporation of intercepted rainfall by the forest canopy is a major component in the water balance of forests. The amount of water lost by interception evaporation is strongly determined by the density of the forest. The losses will be low in open forests and high in dense forests. In AFFOREST, the losses by interception in the dense spruce forests in the Netherlands amounted to 400 mm yr^{-1}, whereas interception losses in the more open Swedish spruce forests were approximately 300 mm yr^{-1}. Furthermore, losses in oak forests are lower compared to spruce forests (see 2.3).

When thinning is performed the density is reduced and interception evaporation and transpiration decrease, which will increase the water recharge. Using close-to-nature forestry or continuous cover forestry with natural regeneration, water recharge will stay more constant over time.

Another factor that may lead to a decrease in water recharge is the density of the herb and shrub layer in the forest. Herbs and shrubs will add to the loss of water by evapotranspiration. In general, the highest water recharge will be found on bare soils and recharge will decrease with density and height of the vegetation.

In agricultural areas, pipe drains and artificial ditches to facilitate the use of machinery in wet seasons (spring and autumn) often artificially drain fields. Such drainage networks lead to an enhanced transport of soil water to streams and surface water and reduce the recharge to the groundwater. When such drainage systems gradually stop functioning after afforestation the recharge to the groundwater will increase. This may (partly) compensate the reduction in groundwater recharge due to the increased losses by evapotranspiration in forests compared to agricultural areas.

2.5. Concluding guidelines for water recharge

From existing knowledge about the water cycle in forests we have the following recommendations for afforestation where high groundwater recharge is wanted:

- The new forest should preferably contain deciduous tree species since these have a higher water recharge than coniferous species.
- Open forests (low basal area and species with a low crown density) yield a higher water recharge.
- Sites with sandy soils tend to have a higher water recharge than sites with finer textured soils. However, these differences may be overruled by regional hydrological conditions.

- Theoretically, removal or decline of old drainage pipes and ditches from agriculture may compensate for the negative effects on groundwater recharge of higher evapotranspiration in forests. However, the effect of removing drainage pipes has not been assessed.

3. NITRATE LEACHING

The N cycle in agricultural soils is an open cycle. Fertilisers (NPK) are supplied regularly in large amounts and approximately the same amount of N leaves the ecosystem by leaching or in harvested products. Management is intensive, characterised by frequent intervention, particularly soil work using heavy machinery and repeated application of pesticides. The amount of N bound in organic matter is high and soils have a high nitrifying capacity. Nitrogen leaching to seepage water and stream water is large since the soils often are 'saturated' with N and the vegetation cover is sparse during the wet season. On the contrary, old forests are characterised by a tighter N cycle where losses of N are low. Nitrification rates are low. Water from old forests is therefore generally of good quality with a low concentration of dissolved N compared to other land uses.

During afforestation major changes occur in the cycling and storage of N. Afforestation of former farmland is seen as a strategy to improve water quality, especially with regard to nitrate leaching. In this context, the challenge is to keep nitrate leaching from the new forests at a low level, despite the large N pool, which is a legacy of the former land use.

3.1. Does afforestation decrease nitrate leaching relative to arable land? How fast?

All available knowledge indicates that afforestation will cause lower nitrate leaching over a tree rotation compared to leaching from arable land, as a result of ceased fertilization and less soil disturbance in the new forest.

Afforestation will cause a rather fast reduction of nitrate concentrations within the first five years after planting since the demand for N is high in the early stage of stand development. However, high N status in the afforested soils and high N deposition may cause the stands to start leaching nitrate again after canopy closure when the demand for N decreases. The long-term nitrate leaching from old afforested stands is nevertheless expected to be lower than that from fertilised arable land.

A change in land use from agriculture to forestry implies that the high level of human intervention in the annual cycle of cultivating and harvesting crops is replaced by a much longer forest cycle with a lower interference. After afforestation, the soil remains more or less undisturbed apart from a possible soil preparation at the very beginning of the establishment and harvesting impacts at the end of the rotation period.

The large pool of N stored in the mineral soil in arable land will, to some extent, be redistributed to the living biomass or built into the organic matter pool in the forest floor. Afforestation of nutrient rich arable land is therefore likely to have a large influence on the organic N equilibrium resulting in a changing pattern of net gains and losses of N from the soil organic matter pool and the plant and microbial biomass fraction.

Such changes in soil conditions may enhance minerlaization of stored soil N and contribute further to the available N pool. Losses from terrestrial ecosystems mainly occur as dissolved N in seepage water. Nitrate seriously influences water quality and it is often the main form of mineral N leached. Since its adsorption to the soil is small, it is highly mobile and easily transported down the soil profile even at N-limited conditions (Gundersen & Rasmussen 1995; Stottlemeyer & Toczydlowski 1996).

Forests on former arable land have higher soil N status as a legacy of former fertilization. Because of this forests on former arable land may be more vulnerable to disturbance of the N cycle than old forests, resulting in enhanced leaching of nitrate. On the other hand, nitrate leaching may still be less than that from arable land. In both agricultural systems and forest ecosystems leaching of nitrate is induced by external input of N or by management practices or combinations of these:

- Increased input of N (N-deposition, fertilization, planting of N_2 fixing species)
- Decreased biological uptake (harvesting, clear-cutting, thinning, weed control, site preparation)
- Increased net minerlaization (liming, soil preparation)
- Decreased denitrification (decrease in groundwater level)

Concentrations of nitrate in soil water below the root zone is a pre-condition for nitrate leaching, and it indicates loss of nutrients from the ecosystem. In Denmark, higher concentrations of nitrate in soils below the root zone were observed in afforested land as compared to old forest ecosystems (Callesen et al. 1999) (Figure 11.2). Concentrations of nitrate in agricultural soils before afforestation were also higher than concentrations 10 years after afforestation on three sites in Denmark (Hansen & Vesterdal 1999).

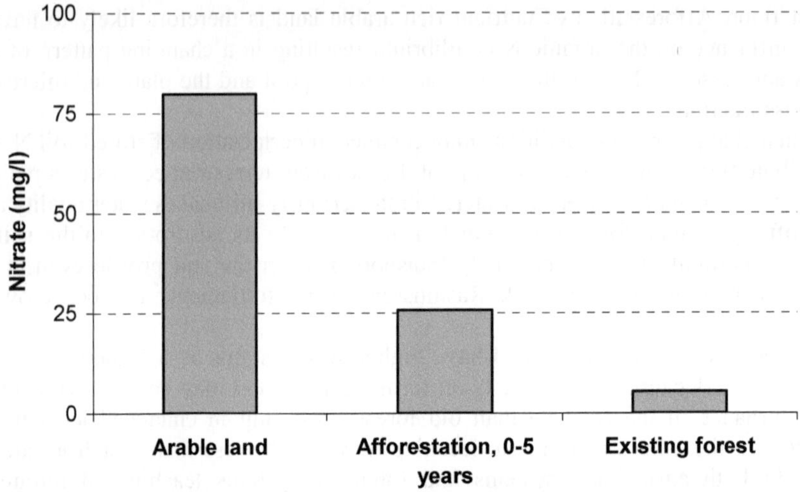

Figure 11.2. *Average nitrate concentrations (mg/l) in leaching water (at 75-100 cm depth) measured from 1986-1993 in arable land (>900 plots), afforestation areas (5 plots) and existing old forests (79 plots) in Denmark (Callesen et al. 1999).*

The effect on nitrate leaching of a change in land use from agriculture to forestry was evaluated using a simple modelling approach in the Netherlands (Rijtema & de Vries 1994). The change in land use led to a decrease in nitrate leaching regardless of the chosen tree species. Average nitrate leaching was estimated to be 74 kg nitrate-N ha^{-1} yr^{-1} on sandy soils under agricultural crops. This was reduced to approximately 11 kg nitrate-N ha^{-1} yr^{-1} under forest stands.

Recent studies of forests developed on former arable soils have shown that soils still have N cycling characteristics and rates of N minerlaization more comparable to cultivated soils than to soils under old forests - even 100 years after afforestation (Koerner et al. 1999; Magill et al. 1997; Jussy et al. 2000). These old arable soils have not been fertilised in the same way and with the same intensity, as we know it on the arable soils of today, but even so, the former land use is noticeable in the N cycling characteristics. The arable soils selected for afforestation today have been fertilised to a much larger extent. These soils often cannot accumulate more N, which means that the surplus of N deposited from the atmosphere is leached.

In the first years after afforestation, the large pool of available N can be converted to nitrate and leached. Leaching will most probably occur until the trees reach a certain size and ground vegetation has developed. Figure 11.3 shows average nitrate concentrations in soil water (75-90 cm depth) after afforestation at nine different sites in Denmark. High concentrations above the threshold value of 50 mg nitrate l^{-1} were only apparent in the first years after planting and the concentrations fell to a low level after approximately five years. Eight to 12 years after afforestation the stands were in good growth and the nitrate concentrations had fallen to much lower values within the range of nitrate concentrations in old forest soils.

Five years after afforestation the uptake of N in the trees increases remarkably since the trees have to build up their crown including branches, twigs and bark. The N demand of young forests is therefore believed to be greatest during canopy build-up and considerably decreases once canopy closure is reached (Miller & Miller 1988; Richter et al. 2000). Hereafter, the trees mostly grow in N-poor woody biomass. After canopy closure, deposition to the new forests will increase as a consequence of larger turbulence around higher trees. The input by atmospheric deposition will in many cases exceed the N demand of the trees and it is uncertain whether the new forests will be able to retain available N. If N cannot be retained in the soil there is a risk for renewed leaching of nitrate 20-30 years after afforestation.

Increased nitrate leaching with age has been observed for Sitka spruce in Wales 25 years after planting (Emmett et al. 1993) and with height in Denmark (Hansen 2003). In the AFFOREST chronosequences in Denmark and the Netherlands, leaching in Norway spruce only occurred after canopy closure at a stand age of 20 or more years (Figure 11.4). Leaching of nitrate from the oak stands increased with stand age in Denmark, however, leaching was high in the oak stands in the Netherlands right from the start after afforestation. No such patterns were apparent in the Swedish spruce chronosequence (12-92 years old) where no leaching took place because of a low throughfall deposition and less intensive former agriculture in the stands older than 60 years.

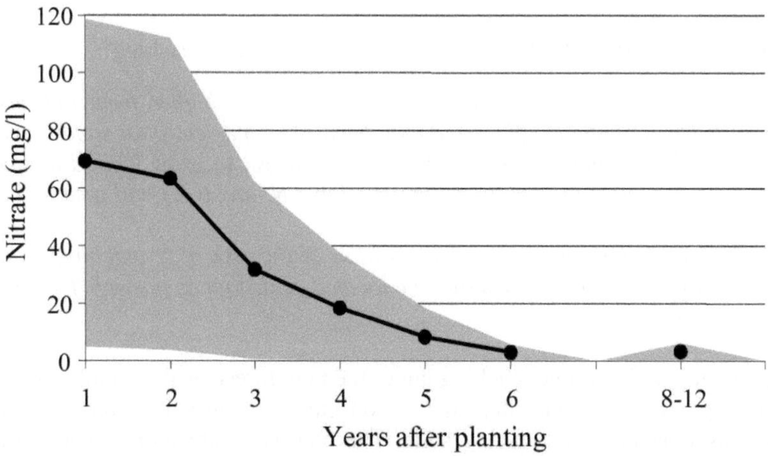

Figure 11.3. *Nitrate concentrations (mg/l) in 75-90 cm depth in new forests after afforestation on former arable land. Average of nine afforestation sites in Denmark. Shading indicates the standard deviation.*

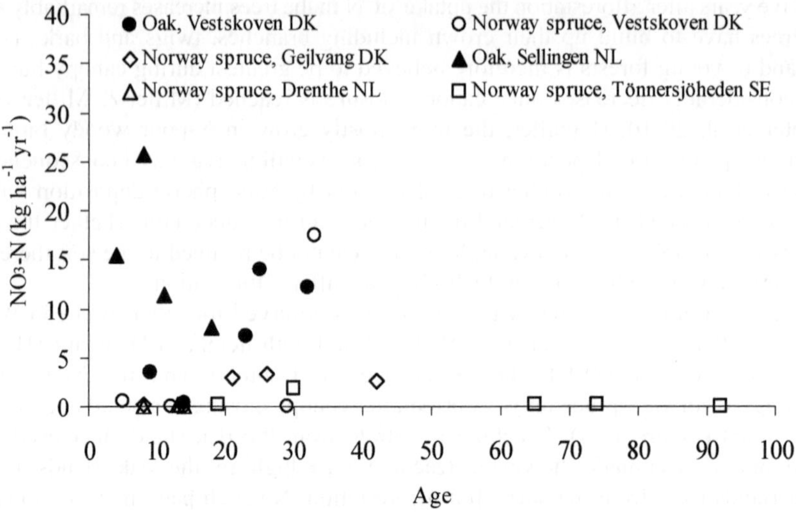

Figure 11.4. *Leaching of NO_3^--N (kg N ha^{-1} yr^{-1}) in stands of different ages in the AFFOREST chronosequences for oak and Norway spruce at sites in Denmark and the Netherlands.*

3.2. Does the selection of afforestation site influence nitrate leaching?

Nitrate leaching following afforestation will mostly be higher from nutrient-rich clayey soils when these are afforested than from more nutrient-poor sandy soils. The clayey soils already have enough N for the new forest to grow while the nutrient-poor soils will be able to immobilize the available N, and trees will take up most of the available N.

In areas with high atmospheric deposition of the risk of N leaching is higher. This is why afforestation areas should be located as distant as possible from local N emission sources.

A decisive factor for nitrate leaching under the new forests is the quality of the soil. Literature points to the fact that nitrate concentrations in soil water are lower in sandy soils than they are in clay soils. The clay soils are more N-rich and less additional N can be built into these soils. This means that when atmospheric deposition of N exceeds the amount used for building up the canopy a certain surplus of N is available for leaching. In contrast, the trees on nutrient-poor sandy soils with organic matter of high C/N ratio will be able to take up all available N in organic matter or by tree growth so that leaching of nitrate is lower or zero. The difference between N-rich and N-poor soils is also seen in their C/N-ratio where the N-poor sites have high C/N-ratios. The leaching of nitrate is in this way also connected with the C/N-ratio. At C/N-ratios less than 25 in the organic layer, literature almost always finds high nitrate concentrations in the deeper soil horizons (Gundersen et al. 1998).

Nitrate leaching is influenced by the atmospheric deposition of N. European data has shown that leaching of nitrate starts to increase at about 8-10 kg N ha^{-1} yr^{-1} (Kristensen et al. 2004). On average, the European forests leach approximately half of the deposited N (Gundersen et al. 1998) and the new forests will probably leach more since they are already very rich in N. The ammonium part of the N deposition is influenced by local N emission sources. Especially areas with intensive agriculture and livestock will experience a high input of ammonium N. To avoid high deposition of N to the new forests, afforestation areas should be located as distant as possible from local sources.

3.3. Does forest structure influence nitrate leaching?

Higher deposition has been observed in forest edges than in the interior forest, which is reflected in higher fluxes through the soil close to the edge (up to 50 meters). In order to lower the N deposition and leaching of nitrate in the new forests they should preferably consist of larger forest stands in connection with already existing forests.

The size and structure of the new forest will affect the deposition to the forest. For example, a much higher atmospheric deposition was observed at the forest edge in several studies. Indirectly, these variations in input to the forest might affect the soil properties as well. In southern Sweden, soil solution (30-40 cm) was sampled at different distances from the edge and into the forest (Påhlsson & Bergkvist 1995) in both a *Picea abies* (30 years and planted on former agricultural soils) and a *Fagus sylvatica* forest (approx. 100 years). The increased deposition to the soil at the spruce forest edge was reflected by higher element fluxes through the soil profile as well. The N flux through the soil decreased with distance from the edge over the first 25 m. At the edge, N fluxes were up to 6 times higher than in the interior of the stand. In the interior of the forest, most N was either taken up by trees or immobilised in the soil. In the beech forest, the edge effect was less pronounced in throughfall. This was also reflected by soil solution concentration, which was not correlated with the distance from the forest edge. Pore water chemistry profiles (greater than 2 m depth) were sampled by Kinniburgh & Trafford (1996) in the edge of a *Fagus sylvatica* stand in the southern part of England. Their analyses showed a dramatic increase in solutes close to the forest edge. This increase was seen 50 m into the forest.

Deposition and nitrate leaching in forest ecosystems are normally determined in the interior part of the forest. Thus, the large variations caused by forest edges are not accounted for. Today, a large part of the new forests are established as a mosaic of small, multifunctional plantations, rather than larger coherent forest areas, and leaching will be considerable from these "all-edge" forests. It is therefore recommended to plan and establish larger coherent forest areas in connection to already existing forests.

3.4. Does the selection of tree species influence nitrate leaching?

N deposition to coniferous tree species is higher than to deciduous tree species since conifers are evergreen and have surfaces that catch more deposition. Studies in old forests have shown that this extra input of N will result in larger leaching of nitrate beneath coniferous species. However, some studies contradict these findings, among these the AFFOREST chronosequences, where leaching was higher beneath oak than beneath Norway spruce.

When planting black and red alder and other N fixing trees it may result in additional leaching from afforested soils.

Individual tree species vary in their soil-forming properties. Different effects of tree species can be expected due to variability in litter quality, canopy architecture and canopy interception of atmospheric deposition, root structure, and rates of nutrient uptake and growth (Miles 1985).

N deposition in coniferous stands is approximately two-fold higher than in deciduous stands caused by the more efficient filtering effect of conifers due to their evergreen foliage and higher leaf area (Kristensen et al. 2004). An enrichment of throughfall for N compounds has been shown for paired comparisons of coniferous and deciduous stands (Brown & Iles 1991; van Ek & Draaijers 1994).

These observed differences in N deposition between the two tree types have been shown to create a higher output of nitrate from the soil under conifers when compared to deciduous tree species at the same site (de Vries & Jansen 1994; Rothe et al. 2002). However, observations from the European monitoring sites (level II sites) show that deciduous species have higher nitrate concentrations than conifers at similar input (Kristensen et al. 2004). One reason for the contrasting result from the European sites may be that the deciduous stands often grow on nutrient-rich soils (higher N status) that are more likely to leach.

In AFFOREST, the oak stands in Denmark and the Netherlands leached more nitrate (5-30 kg N ha^{-1} yr^{-1}) than the spruce forests (<5 kg N ha^{-1} yr^{-1} Figure 11.4). At Vestskoven in Denmark, the oak and spruce stands were planted on similar soils. Oak retained less N in biomass due to slower growth rate and less leaf biomass over the year, which left more N available for leaching.

N fixing trees like black and red alder have been used in silviculture as a tool to improve the soil fertility (Tarrant & Trappe 1971) and as an often-used nurse tree species in new plantations. The annual symbiotic N-fixation was estimated to be between 50 and 200 kg N ha^{-1} (Binkley et al. 1992; Bormann & DeBell 1981), which can result in leaching of 30-50 kg N ha^{-1} yr^{-1}. On agricultural soils where the soil N contents are high from the very beginning, afforestation using N-fixing species must be avoided since this further increases the risk of nitrate leaching.

3.5. How will management of afforested sites influence nitrate leaching?

When weeds are removed, the seedlings alone are unable to withhold available nitrate, which might result in increased nutrient losses. The most intensive preparation methods that disturb the soil the most, like deep ploughing, have been shown to trigger increased leaching. It is a balance between keeping the culture clean enough for the trees to survive and keeping N leaching to a minimum.

When stands are clear-cut plant uptake is suddenly disrupted and nitrate concentrations in soil water as well as leaching inceases. Within 2-5 years nitrate will return to preharvesting levels. Even thinning has been shown to increase nitrate leaching slightly.

When drainage pipes from former arable land use stop functioning after afforestation, soils revert to the original drainage regime. In case of more wet conditions, denitrification is significantly enhanced leading to a reduction in nitrate leaching.

Storage of N and leaching of nitrate from newly planted forest is dependent on the management practices applied prior to planting as well as practices during and at the end of a rotation.

3.5.1. Site preparation

Site preparation, frequently used when new forest stands are established, is carried out mainly to create a favourable environment that promotes fast and efficient establishment and good survival of seedlings. This is obtained by reducing the competing ground vegetation (e.g. grasses and herbs). Weed control in the beginning of a new culture greatly improves tree growth in a number of species (Chang & Peston 2000; Munson & Timmer 1995; Sutton 1995). Weed control includes a number of removal methods, which are given here in decreasing order of soil disturbance:

- Mechanical removal by patch scarification, trenching, mounding and ploughing (increases minerlaization)
- Chemical removal by the use of herbicides (decreases biological uptake)
- Mulching
- Competitive weed control

Deep ploughing is the preferred preparation for afforestation of former arable land since it suppresses weed establishment more efficiently than the other mechanical site preparation methods. A study by Matthesen & Kudahl (2001) compared the effect of different mechanical preparation methods on the vegetation cover of competing weeds and grasses in the first growing season on 3 afforestation sites in Denmark. Deep ploughing (down to 60 cm) reduced the cover percentage to approximately 50%, whereas trenching in between rows and agricultural ploughing (down to 20 cm) only reduced the cover 2-8%.

Use of herbicides before and after planting of seedlings is the most common way of controlling weeds and by far the most cost-effective method as well. Mulching involves covering the soil around trees with a cover material, which will prevent weeds from germination. Wood chips have been widely used as cover material, but also degradable plastic and cardboard have been used. Competitive weed control involves the use of other vegetation to take over and suppress the weeds, yet allow the seedlings to get enough light and water to grow. Especially, rye is used as competitive vegetation to weeds when planting on former agricultural soils. Also, nursery trees growing faster than the main tree species may help to earlier create a forest climate, hereby suppressing several weeds and preventing frost.

Mechanical weed control methods disturb the soil to some extent. The most intensive mechanical disturbances increase net N minerlaization, nitrification, and nitrate losses to seepage water (Vitousek & Matson 1985; Attiwill & Adams 1993). When weeds are removed, the seedlings alone are unable to withhold available nitrate (decreased uptake), which might result in increased nutrient losses (Ogner 1987a, b; Lundmark 1977; Lundmark 1988). On two Danish afforestation sites, Drastrup and Nørager, different weed control methods were examined (Pedersen et al. 2000; Gundersen et al. 2001). Agricultural ploughing (down to 20 cm) was compared to deep ploughing. At Drastrup, the nitrate concentration increased drastically after both agricultural and deep ploughing. The concentration of nitrate as well as leaching was highest in the deep ploughing treatment (200 mg l^{-1} corresponding to 100 kg N ha^{-1} yr^{-1}) as compared to the agricultural ploughing (100 mg l^{-1} corresponding to 40 kg N ha^{-1} yr^{-1}) (Figure 11.5).

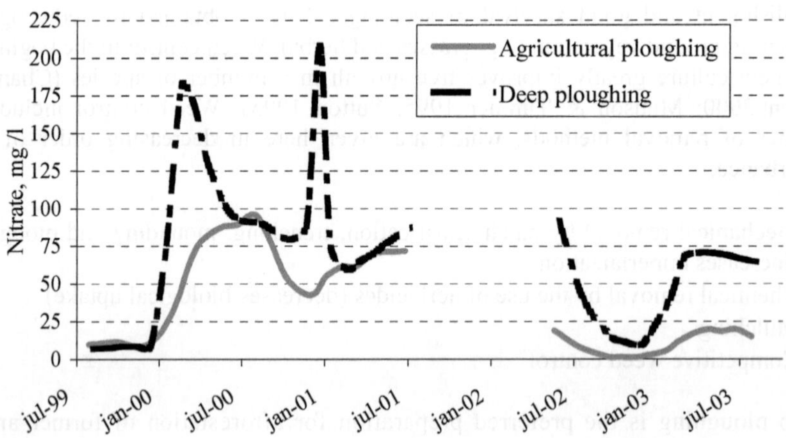

Figure 11.5. *Concentrations of nitrate (mg l^{-1}) in soil water (90 cm) using different site preparation methods on a sandy nutrient-poor soil at Drastrup in Denmark (Gundersen et al. 2001).*

At Nørager, results were comparable to those at Drastrup for the first leaching season (out of two) where 30-40 kg N ha^{-1} yr^{-1} were leached from the agriculturally ploughed plot and 90-130 kg N ha^{-1} yr^{-1} were leached from the deep-ploughed plot. During the second year at Nørager, the nitrate concentration in the deep-ploughed plot fell below the concentration in the agriculturally ploughed plot. In comparison, the average leaching from farmland in Denmark is estimated to be 106 and 55 kg NO_3-N ha^{-1} yr^{-1} from sandy and clayey soils, respectively (Grant et al. 2004).

At Drastrup, trees were also planted with large distance (2.5 x 2.5 m) and mulching with wood chips was performed. Here, the grass and weed cover was kept intact when planting, and it did not influence the concentration of nitrate (Gundersen et al. 2001), which was low already from the beginning because of a grass cover.

Mellert *et al.* (1998) found a close negative correlation between total vegetation cover and nitrate concentrations in soil solution (r^2=0.7). This is explained both by a higher total plant biomass (weeds and trees) on weedy areas and a higher N-accumulation in weeds compared to the tree (Smethurst & Nambiar 1989).

3.5.2. Thinning and harvesting

Canopy removal by thinning temporarily increases the amount of precipitation and sunlight reaching the forest floor, influencing the forest floor microclimate and thus the decomposition and minerlaization conditions. An increase in soil nitrate concentrations was visible in a thinning experiment in lodgepole pine where 60% of the trees were removed. Up to two years after thinning nitrate leaching was apparent but small and far from the 10-40 times increase in nitrate concentration found in the clear-cut stand (Knight et al. 1991). Bäumler & Zech (1999) also observed a small increase in nitrate concentrations after a 40% thinning. The concentrations were back to pre-cutting conditions after one year.

Like for thinning, removal of the whole canopy when harvesting increased the amount of precipitation reaching the forest floor and the light available for tree growth, leading to more favourable soil moisture and temperature conditions for decay microorganisms. (Piene & van Cleve 1978; Binkley 1984). However, for a clear-cut the disturbance is more severe since plant uptake is disrupted at the same time as minerlaization and nitrification increase. Furthermore, the outflow of run-off and seepage water is larger due to lower evapotranspiration (Knights et al. 1991; Qualls et al. 2000).

The effect of clear-cutting on nitrate leaching was followed in numerous old-forest studies. In general, the concentration of nitrate in soil and stream water increased after clear-cutting with peak nitrate concentrations the first three years after clear-felling. The nitrate concentrations will return to pre-cutting levels within relatively short time, normally 2-5 years. The effect of clear-cutting on leaching from originally afforested arable soils has not been examined yet. As these arable soils were N-saturated prior to afforestation it is hypothesised that nitrate leaching may be high after crown closure, at least in high deposition areas. A management strategy to prevent N leaching could be frequent thinnings, e.g. of whole trees for bioenergy purposes, to keep a high N sink at these sites. In this case, the removal of other nutrients with biomass must be balanced by the supply of other nutrients from soil and atmosphere or the loss of nutrients should be compensated by fertilization. Continuous cover forestry will also be a relevant option to avoid the disturbances connected with clear cutting that disrupts the N cycle.

3.5.3. Drainage

The drainage regime has large impact on soil N dynamics. Reduced drainage enhances the loss of nitrate through denitrification. Arable soils are often drained to ditches, and the natural drainage regime is usually more wet. Bad drainage can be due to slow infiltration of precipitation as in clay-rich soils or due to high groundwater tables. When drainage pipes from the former arable land use stop functioning after afforestation, soils revert to the original drainage regime. In case of more wet conditions, the availability of oxygen decreases and denitrification is significantly enhanced, which leads to a reduction in nitrate leaching.

3.6. Concluding guidelines concerning nitrate leaching

From existing knowledge about the N cycle in forests and afforested former arable land we have the following recommendations for afforestation where low nitrate leaching is a goal:

- Afforestation should preferably be performed as larger coherent forest areas in connection to already existing forest in order to decrease deposition caused by edge effects.
- The best place to afforest is far away from local emission sources.
- The new forests should preferably consist of deciduous tree species (e.g. beech and oak) since they have a lower deposition of N and a higher water recharge (see chapter 2) which mostly leads to lower nitrate concentrations in leaching water.
- N-fixing tree species like black and red alder should be avoided when afforesting former arable land.
- A fast establishment of both trees and weeds will cause the lowest nitrate leaching.
- As little site preparation and weed control as possible (accept some weeds) to avoid nitrate leaching or selection of weed control methods that will only disturb the system leniently.
- The crown cover should not be too closed. A relatively open forest established by medium thinning intensity is preferable since the deposition will be slightly decreased with decreased crown cover.
- Frequent thinnings for bioenergy can be evaluated in order to remove N from an area with high N status as for example former arable land. In this case, the removal of biomass must be balanced with the contribution of other nutrients from soil and atmosphere or the loss of other nutrients than N should be compensated by fertilization.
- Continuous cover forestry will be preferable at afforested sites as an alternative to clear cutting and the resulting disruption of the N cycle.

4. CARBON SEQUESTRATION

The land-use change from agriculture to forestry implies a shift from a landscape of low C density to one of higher C density. Carbon is stored in afforested systems as a result of the assimilation of CO_2 during photosynthesis and the subsequent accumulation of woody biomass. A certain amount of C is transferred each year to the soil with litterfall, sloughing of dead roots and C leaching from roots. The amount of C stored in soils after afforestation reflects the balance between input from the vegetation and the ongoing mineralization of organic C.

4.1. What is the potential of afforestation for carbon sequestration?

Afforestation of arable previously tilled land leads to constant or increasing soil C stocks. Studies show average rates of soil C sequestration between 0.3 and 0.8 t C ha^{-1} yr^{-1}. Carbon stocks were found to increase by in average 18% over a variable number of years.

The C storage in biomass depends on the growth rate of trees. Trees growing on former arable land have growth rates that are higher than expected from the parent material. Biomass C sequestration rates from 0-50 years are reported to range from 1.5-4.3 t C ha^{-1} yr^{-1}. Biomass C sequestration rates of the entire forest rotation (about 100 years) are more likely in the range of 1.5-2.0 t C ha^{-1} yr^{-1}.

Evidently, C storage mainly takes place in the above- and belowground biomass. Approximately 70% of the C stored in the new forest is sequestered here. The remaining is sequestered in soils.

There is substantial variation among studies regarding the relative contribution of soils and vegetation to total ecosystem C sequestration following afforestation. While there is always some sequestration in the biomass of growing forest trees, soils may gain C, experience no change in C or even loose C following afforestation (Guo & Gifford 2002; Vesterdal et al. 2002).

A recent review of many studies (Post & Kwon 2000) found that the average rate of soil C sequestration was 0.3 t C ha^{-1} yr^{-1} (range 0-3 t C ha^{-1} yr^{-1}) across different climatic zones. An analysis of 29 studies on afforestation of arable cropland showed that C stocks increase by in average 18% over a variable number of years (Guo & Gifford 2002). In the AFFOREST chronosequence studies on former cropland soils sequestered C at a rate ranging from 0 (no change) to 1.3 t C ha^{-1} yr^{-1} for the Dutch oak chronosequence (Table 11.1). The average rate of soil C sequestration was 0.8 t C ha^{-1} yr^{-1} across all sites.

Table 11.1. *Rates of soil C sequestration (t C ha^{-1} yr^{-1}) over 30-90 years in the AFFOREST chronosequences.*

Country	Soil type	Tree species	C sequestration
Denmark	Sandy clay	Oak, Spruce	~0
Denmark	Sand	Spruce	1.1
Sweden	Sand	Spruce	0.6
The Netherlands	Sand	Oak	1.3

Within the soil, C storage may occur because of the development of a forest floor – the organic layer consisting of dead leaves twigs etc. which blankets the mineral soil. However, additional C can be stored in the mineral soil, where it is more protected towards forest management and other environmental changes. In general, it takes longer (>30 years) for C sequestration to occur in the mineral soil (Paul et al., 2002). Such differences in allocation of sequestered soil C may be attributed to tree species, i.e. with oak accumulating less C than spruce in forest floors, and to the humus forms characteristic of different soil types.

The development in biomass C storage is basically the same as in any stand regenerating after disturbances like wind throw or clear-cutting. The storage depends on the growth rate of trees and possible understory species. Former arable soils differ from most forest soils in the fact that they are rich in nutrients, at least in the first decade following abandonment of agriculture. This is a legacy of frequent fertilization and liming. The nutritional demands of forest trees may therefore be more than met, and growth rates higher than expected from the parent material. In the literature, biomass C sequestration rates from 0-50 years are reported to range from 1.5-4.3 t C ha^{-1} yr^{-1} in boreal, cool temperate and subtropical climates (Vesterdal et al. 2002).

In the AFFOREST chronosequence studies rates of C sequestration of both oak and spruce across sites were quite similar for the first 40 years after afforestation with an average rate of 3.7 t C ha^{-1} yr^{-1}. The current growth rate of forest stands decreases with age, so biomass C sequestration rates of the entire forest rotation (about 100 years) is more likely in the range of 1.5-2.0 t C ha^{-1} yr^{-1}. Therefore, roughly 30% of the total stored C is sequestered in soils and the remaining 70% is sequestered in above- and belowground biomass (Figure 11.6).

In general, high short-term rates of biomass C sequestration is achieved in early succession tree species on nutrient-rich soils and in a climate with a long growth season in combination with ample amounts of moisture. In the longer term, however, more slow-starting late-succession tree species may provide higher rates of C sequestration. Also, other factors than growth rates matter for biomass C sequestration in a long-term perspective including several rotations of forest, e.g. effects of inherent soil properties, tree species and management practices must be expected to show up later.

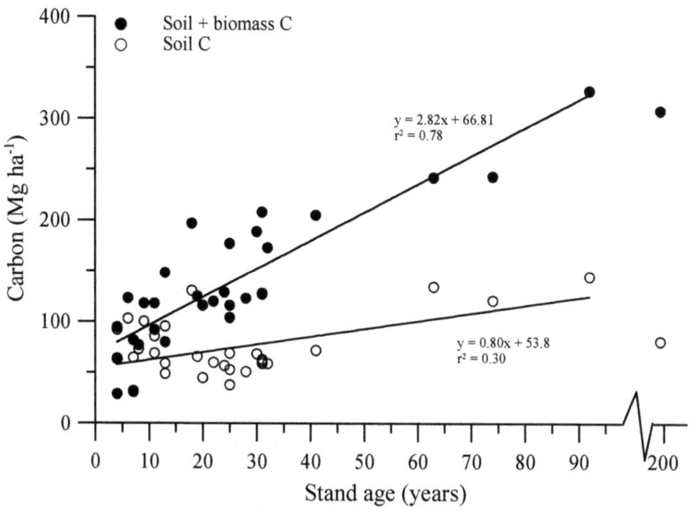

Figure 11.6. *Changes in total ecosystem C storage with time since afforestation in Denmark, southern Sweden and the Netherlands. The difference between the plotted lines indicates the relative contribution of biomass C sequestration.*

4.2. Does the selection of afforestation site influence the potential for carbon sequestration?

Storage of C in the forest floor is highest on nutrient-poor sandy soils. The effect of soil type on the storage of C in the mineral soil is not clear.
In the long-term, biomass C sequestration is expected to be highest at rich sites due to the higher increment rates. Afforestation of nutrient-rich soil types would therefore be the best option for biomass C sequestration.
Biomass growth appears to be the main sink for CO_2 following afforestation. This suggests that afforestation of rich soils is the optimal choice.

An important issue when planning afforestation projects is the selection of site, i.e. the effect on the mitigation potential of selecting marginal, nutrient-poor land versus more nutrient-rich arable land. In most cases land is available for afforestation because it is marginal for crop production. The soil type of the chosen land area influences the mitigation potential of an afforestation project, and the question is therefore which soil type will provide most C sequestration in soil and biomass?

There is good evidence that forest floors store more C when soils are poor, i.e. have sandy texture, low nutrient availability and low pH, and when soils are wet. This is because decomposition of litter proceeds very slowly in such soil types resulting in a thick humus layer, which blankets the mineral soil. It is less certain how C sequestration in the mineral soil is affected by soil type. In some cases, nutrient- and clay-rich soils have been suggested to store more C, because production of litter is high and because clay protects organic matter from decomposition. In other cases, poor soils seem to store more C than rich soils, which was explained by slow decomposition and complexation between C compounds and iron and aluminium in poor sandy soils. In the AFFOREST project there was an indication that sandy soils accumulated more soil C (including forest floor C) over 30-40 years than loamy soils.

Biomass production is known to vary considerably between sites for the same tree species due to soil nutrient status and climate. Accordingly, potential C sequestration in biomass is highly variable. However, biomass C sequestration at afforested sites will not differ that much over the first decades because abandoned arable soils are fertile for forest trees. Thus the influence of "native" soil type (parent material) may be masked by fertilization and liming during the previous land use and even sites with poor subsoil may have relatively high site index for trees. It is debated how long this fertilization effect will last, but in the long-term biomass C sequestration is expected to be highest at rich sites due to the higher increment rates. Afforestation of nutrient-rich soil types would therefore be the best option for biomass C sequestration in case this type of arable land is available.

The question is then, what soil type to afforest in order to maximise the C sequestration of the whole ecosystem? It is not possible to give a general answer to this question, as there seem to be different mechanisms controlling soil C sequestration. As mentioned before, biomass growth appears to be the main sink for CO_2 following afforestation. This suggests that afforestation of rich soils is the optimal choice. It is possible, however, that the soil contributes more when afforesting podsolized poor forest soils. It is thus recommended to select sites based on regional or local knowledge of the soil type and C stocks in such soil types under forest and crop production.

4.3. Does the selection of tree species influence the magnitude of carbon sequestration?

Tree species influence soil properties, but the influence of tree species on soil C sequestration is not well known. Some studies have indicated tree species differences, and others not.

Different tree species will affect C sequestration in biomass according to the relative growth potentials of the species. The longer-lasting and more continuous growth of oak and beech makes these species a good option on rich, loamy soil types.

In order to optimise C sequestration in afforestation areas, tree species with high rates of increment over a long life cycle should be selected.

Tree species can influence C sequestration because of different accumulation of living biomass and dead organic matter.

Tree species planted at the same soil type accumulate very different amounts of dead organic matter in the forest floor, i.e. some tree species also sequester more C than others in this part of the soil. For instance oak in Denmark sequesters less C (0.03 t C ha^{-1} yr^{-1}) in forest floors than Norway spruce (0.3 t C ha^{-1} yr^{-1}) over 30 years, and pine species have higher C sequestration rates (0.6 t C ha^{-1} yr^{-1}) than spruce. Other broadleaved tree species like ash and lime (*Tilia*) may sequester even less C in forest floors than oak.

There is little support for generalizations regarding tree species differences in mineral soil C sequestration. Some studies have indicated tree species differences, and others not. Based on experiences in the AFFOREST project we conclude that after about 30 years tree species like oak and Norway spruce do not influence the C stock in mineral soils differently. But in the long term this may not be the case.

The selection of tree species will affect C sequestration in biomass according to the relative growth potentials of the species. Thus, conifers may at sandy soils be superior to broadleaves in growth rate. However, the longer-lasting and more continuous growth of oak and beech makes these species a better option on rich, loamy soil types. Here spruce has high initial increment rates but a short-lasting life cycle, as it becomes unstable after less than 50 years.

The total ecosystem C changes in soil and biomass will be larger in conifers than broadleaved tree species in the short-term (<30 years since afforestation) because of higher rates of increment of the conifers within this age range and development of larger forest floors (Figure 11.7). It is difficult to compare tree species without taking the different management into account. For instance, rotation lengths are shorter for conifers, so biomass accumulation stops earlier and must start all over again. In the long-term, tree species selection should therefore be based on species that maintain high C stocks for a long period and continuously sequester C. In this case, stable, long-lived broadleaved species may provide the best alternative in spite of their lower rates of increment and forest floor accumulation.

In order to optimise C sequestration in afforestation areas, it is recommended to select tree species with high rates of increment over a long life cycle. Relevant tree species should also accumulate C in the soil. The specific selection of tree species must eventually be based on regional knowledge of tree species ecology at a given soil type.

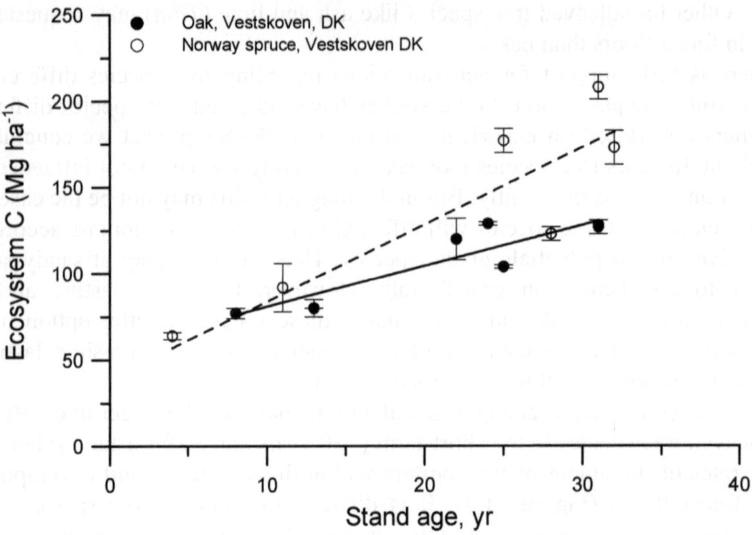

Figure 11.7. Example of ecosystem C change in Norway spruce and oak at Vestskoven, Denmark.

4.4. How will management of afforested sites influence carbon storage?

Continuous cover forestry using natural regeneration will have a higher average biomass C stock per ha and possibly also preserves more C in soils compared to traditional clear-cut/replanting systems.

The influence of thinning intensity on soil C storage is small. Reduced thinning intensity increases the biomass in the forest, thereby maintaining higher C stocks. High stocking levels (stems per ha) obtain maximum assimilation of CO_2 per area faster than when using lower stocking levels. Furthermore, harvest of excess biomass for bioenergy purposes in early thinnings will contribute to CO_2 mitigation.

When drainage pipes from the former arable land use stop functioning after afforestation, soils revert to the original drainage regime. In case of more wet conditions, C storage is significantly enhanced.

The answer to this question is basically the same for new and old forests. Management practices known to increase C stocks of the vegetation and/or the soil in existing forests would most probably also be relevant in afforestation management. Relevant management practices are, e.g. silvicultural system, thinning intensity, drainage regime and fertilization. Here we only address C stocks on afforested sites – C storage in wood products is not considered.

Traditionally, afforested stands are plantation forests where the silvicultural system is characterized by rotations initiated by planting and terminated by clear-cut harvesting. During recent years, continuous cover forestry (CCF) or nature-based forestry has come into focus. These systems aim at introducing the small-scale mosaic patterns found in natural forest in order to avoid costly large-scale clear-cuts and artificial reforestation efforts. The idea is that forest management should mimic the natural forest structure and dynamics where single trees die and create space for new trees. An important characteristic is the continuous cover or at least the smaller and less long-lasting openings in the forest canopy. This means that over time there is a higher average biomass C stock per ha in such forests compared to traditional clear-cut/replanting systems where 10-20 years per rotation is characterized by very low biomass C stocks. CCF possibly also preserves more C in soils as the input to soil is not interrupted by clear-cutting and replanting. As natural regeneration is the preferred method of regeneration, negative impact on soil C stocks by site preparation is also avoided. It is a challenge to develop first-generation afforested stands into more complex multi-aged stands. This issue was outside the scope of AFFOREST and is addressed elsewhere (Pommerening & Murphy 2004). However, for optimizing C stocks it is recommended to establish species-diverse forest stands that are able to perform natural regeneration or function as shelter for other tree species.

Reduced thinning intensity increases the biomass in the forest, thereby also maintaining higher C stocks. In the short term following afforestation, it would also be possible to use high stocking levels (stems per ha), i.e. obtain canopy closure and thus maximum assimilation of CO_2 per area faster than when using lower stocking levels. The influence of thinning intensity on soils was also explored in thinning trials in the AFFOREST project. There was no general evidence of different soil C storage. This is in line with other studies (de Wit & Kvindesland 1999). As strong thinning regimes are preferred in order to reduce N deposition and increase water recharge, reduced thinning intensity is not recommended for C sequestration in afforestation areas where water management is prioritized. Strong thinning regimes and use of wood for bioenergy purposes may consequently be a better option which contributes to CO_2 mitigation by substituting consumption of fossil fuel.

The drainage regime has large impact on soil C dynamics and tree growth. Reduced drainage is probably the main management parameter which may influence soil C stocks. Arable soils are often drained to ditches, and the natural drainage regime is usually more wet. Bad drainage can be due to slow infiltration of precipitation as in clay-rich soils or due to high groundwater tables. Wet conditions decrease the availability of oxygen, which hampers decomposition of soil organic matter. Wet soils therefore have significantly higher C stocks than well-drained soils. Up to five times more C may be found to a depth of 1 meter. Following afforestation, drains stop functioning over time, and soils revert to more wet conditions. It is therefore necessary to include these possible changes in the planning of afforestation projects in order to select suitable tree species for more wet soil conditions. For instance, oak is very tolerant with respect to drainage regime. In case suitable tree species are planted, the more wet conditions enhance C storage in the soil compartment significantly, thus increasing the CO_2 mitigation potential of afforestation. However, emissions of other greenhouse gases like N_2O and methane (CH_4) are enhanced in wet soils, and this possible negative effect on mitigation of greenhouse gases must be considered too.

4.5. Concluding guidelines concerning carbon sequestration

From existing knowledge of carbon stocks in biomass and soils in forests planted on former arable land we have the following recommendations for sequestration of C after afforestation:

- Forest trees sequester carbon faster than soils after afforestation. High short-term rates of biomass C sequestration are achieved in early succession tree species. In the longer term more slow-starting late-succession tree species may provide higher rates of C sequestration.
- Forest floors sequester C most rapidly after afforestation. However, in terms of permanency it is wiser to manage for sequestration of C in the mineral soil, where it is more protected toward forest management and other environmental changes. It takes longer (>30 years) for C sequestration to occur in the mineral soil.
- Tree species with high rates of increment over a long life cycle should be selected. Relevant tree species should also accumulate C in the soil, e.g. conifers. The specific selection of tree species must ideally be based on regional knowledge of tree species ecology at a given soil type.
- Afforestation of nutrient-rich soil types is preferable for biomass C sequestration. It is not possible to generalize regarding the potential of poor and rich soils for soil C sequestration. Local knowledge of C stocks in different soil types under forest and crop production can be consulted for decision support.
- Management of afforested stands should focus on use of biomass from thinnings for substitution of fossil fuels, increased rotation periods and the principles of nature-based management and continuous cover forestry. A change in drainage regime towards more natural, wet conditions when drains are impaired may significantly increase the C stock of the mineral soil.

5. DIVERSITY OF UNDERSTORY VEGETATION

In North-Western Europe, forest areas often serve a multitude of functions. Besides environmental and production purposes, nature development and recreation are objectives of afforestation. For both nature and recreation a diverse and abundant understory, consisting of valuable and typical forest species, is important.

As the tree layer is often planted and therefore quite artificial and the understory is left to spontaneous development, the question arises if it is feasible to expect the development of such a valuable understory to occur in a relatively short time span (if ever). Also, the differences in environmental conditions of the new forest areas due to anthropogenic influences as compared to ancient forests, will probably affect the development of a natural forest understory. However, some extra attention for the understory and active interference can stimulate and speed up the developments.

5.1. How diverse is the flora of current forests in North-Western Europe relative to arable land?

Arable land and recently planted forests are both not recognized for their high diversity in vascular plant species. However, diversity is not the only criterion in the determination of the value of an ecosystem. Due to the high amount of endangered species that occur on the forest floor of old forests, the rarity of well-developed understory systems in the north-western part of Europe and the vulnerability of the system, the new forest and its understory are potentially very valuable.

The exact value of the forest (understory) ecosystem compared to the current land use, but also compared to other possible land use types, depends on a combination of biological, recreational, landscape and economical factors and can be evaluated in a local, regional or larger context. Within the AFFOREST project, this consideration has not been addressed. The AFFOREST project focussed on the possibilities to steer the newly planted forests on arable land towards the highest system-specific diversity. Since the system-specific diversity is related to a certain combination of balanced environmental factors, most suitable for the species group of interest, the AFFOREST project focussed on the relation between plant performance of forest species and non-forest species and the environment. Recommendations for the afforestation of former agricultural land from a biodiversity point of view are based on that approach. So, it is not a high diversity for itself that is important, but the system-specific diversity that counts.

5.2. How is diversity of understory vegetation affected when arable land is afforested? How long does it take for the understory of afforested stands to be comparable to old forest sites?

In new forests planted on former agricultural soil, the vegetation is often characterized by a large proportion of nitrophilic, highly competitive species. The ancient forest species generally appear slowly in the understory vegetation. This slow development of a typical forest understory is ascribed to seed availability and environmental conditions.

The seeds of most forest species are short-lived. The chances of seeds reaching the new forest, depend on the distance between the forest and the seed sources in the neighbourhood, that is old-growth forest.

The most important environmental factors affecting the development of the forest understory are light and nutrients (especially N). Higher N availability in the soil, often found on former arable land, favour the early succession forest species while ancient forest species seem to respond more indifferently to the N supply.

A period of 50-100 years or even more before typical understory vegetation starts to build up after afforestation is not an unrealistic estimate. However, active intervention in the seed availability can speed up this process.

In Denmark, the understory vegetation of new forests planted on former arable land differing in age was characterized by use of a Forest Flora Index (Riis-Nielsen, unpublished data). This index describes the ecological habitat preference of species occurring in the vegetation records of the different forests, ranging from "not present in forests" to "confined to forest". The Forest Flora Index for each species was determined based on habitat descriptions (Hansen 1976). The mean Forest Flora Index determined in the AFFOREST chronosequences showed an increase with the age of the forest (Figure 11.8), however, a very slow increase. In literature, this slow development of a typical forest understory is ascribed to seed availability and environmental conditions (Verheyen & Hermy 2001; Honnay et al. 1999).

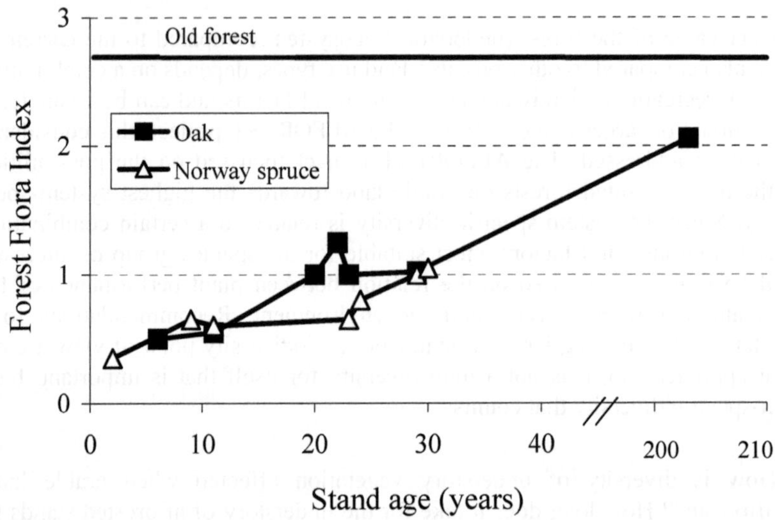

Figure 11.8. *Forest Flora Index (FFI) determined on the AFFOREST chronosequences in oak and Norway spruce at Vestskoven, Denmark. FFI was calculated based on habitat descriptions for each species (Hansen 1976). 0 = not present in forest, 1 = predominately outside forest, 2 = often present in forest, and 3 = confined to forest (Riis-Nielsen unpublished data).*

The seeds of most of the forest species are short-lived and the agricultural land-use often has a negative effect on the viability of the seed bank. Thus, the development of a valuable forest understory from seeds that have been stored in the ground since the disappearance of former forest can be ruled out. Therefore, the new forests are dependent on dispersal of seeds from seed sources outside the forest. The chances of seeds reaching the forest depend on the distance between the forest and the seed sources in the neighbourhood. Because the dispersal mechanism of many forest species is specialized in short-distance dispersal and because forest species tend to produce only few viable seeds, the development of the forest understory is often hampered by the availability of seeds.

The most important environmental factors affecting the development of the forest understory are light and nutrients (especially N). In a greenhouse experiment conducted within the AFFOREST project, the response of shade-tolerant forest species and species typical for younger or disturbed forests to different combinations of light and nutrients was tested. It is concluded that higher N availability in the soil, often found on former arable land, favoured the early successional forest species while ancient forest species seemed to respond more indifferently to the N supply. The effect of enlarged N availability was strongest at the higher light intensities. At very low light intensity, light was the most limiting growth factor and none of the species could take advantage of the high nutrient supply. Comparable results are found in other studies (Corré 1983). Ancient forest species are more adapted to low light and low N availability conditions, while early succession species are more adapted to high light and high N availability conditions.

It is difficult to give an indication of the time it takes for the understory of afforested stands to become comparable to old forest sites. If no special measures are taken to improve understory development, a period of 50-100 years or more (Brunet & von Oheimb 1998; Harmer et al. 2001) before a typical understory vegetation starts to build up is not an unrealistic estimate. However, by taking the appropriate measures, this process can be accelerated.

The development of the forest understory is dependent on the local situation and the design of the new forest. Indications of development time are often based on seed limitation and the time it takes for seeds to reach the forest. Active intervention in the seed availability can speed up this process. However, a slow development and a lengthy succession characterize the forest system. A natural forest goes through different developmental phases, each phase having its own characteristics and value and its own contribution to the construction of a balanced a-biotic and biotic system. Restoration defines a goal and a time frame for management measures, and at this point any threshold in the system becomes important (Heil 2004). The environmental factors of light and N availability determine the abiotic constraints, and seed availability determines the chance of assembling the system specific diversity.

5.3. Does the selection of afforestation site influence diversity of understory vegetation?

Selection of afforestation site does play an important role in the development of the forest understory. The site should preferably be planted directly bordering or in the vicinity of existing old-growth forest where the goal-vegetation is present. Sites with low N availability are preferable since N favours early successional species and increases their competitive strength.

The selection of afforestation site can affect both seed availability and environmental conditions (mainly N availability), which have been recognized as important determinants of the forest understory diversity.

To improve seed availability, the site should be in the vicinity of old-growth forest where the target vegetation is present and reproductive. In the most ideal situation, the new forest should be planted directly bordering the ancient forest. In other cases, suitable habitats between the forests, like hedges and shrubbery, can increase the colonization success (Butaye et al. 2001).

The stronger effect of N availability on non-forest species than on ancient forest species was already described in 5.2. The importance of competition was shown in the AFFOREST-field-experiment in a chronosequence of deciduous forests. In forests where no competition occurred, ancient forest plants performed well. In the younger forests where competition was severe due to higher light and nutrient levels, indications were found that the performance of the ancient forest species lagged behind. Also, in literature the suggestion is often put forward that N availability can favour early successional species in competitive situations. Most of the studies working with this hypothesis are aimed at unravelling the effect of N deposition on existing (ancient) forest systems. These studies clearly show that the performance of ancient forest species declines when the species are competing with fast-growing ruderals (Falkengren-Grerup 1993). The actual disappearance of ancient forest plants, and thus a decrease in biodiversity and value of the understory, is dependent on the duration and intensity of the competition and on the vitality of the ancient forest species.

In conclusion, preference should be given to sites that border existing old growth forests and sites with low N availability. If such sites cannot be found management measures will be necessary to increase seed availability and suppress competitive weed domination (see 5.5).

5.4. Does the selection of tree species influence diversity of understory vegetation?

The tree species sets the light environment of the understory vegetation and thus indirectly affects understory diversity. In addition, tree species influence diversity via their litter quality. Therefore, the selection of tree species is important for the development of the understory vegetation. Forest species are, in contrast to more competitive ruderal species, able to survive an environment with little light, which calls for selecting tree species with denser crowns.

In a natural situation, the occurrence of tree species is determined by the environmental conditions of the site (soil type, soil moisture etcetera); the understory composition, in turn, is associated with both the environment and the tree species. To develop a (semi-) natural forest, selection of the naturally occurring tree species is a prerequisite.

Different tree species are known to differ in the density of their crowns and thus in the relative amount of light that penetrates through their canopy (Leuschner & Rode 1999). The AFFOREST field experiment showed that forest species could survive rather low light levels. The results of the greenhouse experiment support this conclusion. The same experiments showed that non-forest species are more sensitive to light conditions. Growth rates, and thus competitive strength, decreased strongly with decreasing light.

Thus, to steer the development of the forest understory in the direction of the target vegetation, tree species with a relatively dense crown are preferred. However, light intensity on the forest floor can also be manipulated by the density of the trees, i.e. thinning intensity.

Another aspect that plays a role in the selection of tree species is important in the determination of the target vegetation type and the accompanying system specific diversity and species composition. Vegetation science of forest systems revealed a rigid combination of soil type, tree species and vegetation type (Stortelder et al. 1999). When selecting the tree species or, preferably, species combination, it is of importance to have a clear picture of the possible and desired target vegetation.

5.5. How will management of afforested sites influence diversity of understory vegetation?

Management can strongly affect the development of the species composition in the forest understory and can thus be used to steer the succession on the forest floor.

Seed availability could be improved by sowing the desired understory species. Also, flowering and seed production of existing plant populations in the new forest or in old forests in the surroundings could be stimulated.

The competition between ancient forest species and nitrophilic species can be controlled by active removal of competitive species, by grazing or by intervention in the light climate.

Management should aim at seed availability and at competition control. Two possibilities exist to increase seed availability. Firstly, sowing can provide an artificial supply of seed. Though strictly speaking not a management measure but more a measure of eco-engineering, but on condition that it is executed at the correct time, when environmental conditions are suitable for the species sown, a very effective measure. Secondly, flowering and seed production of existing plant populations in the new forest or in forests in the surroundings could be stimulated, for instance by thinning. Here, light is again an important factor. More light results in more flowering and higher seed production. Permanent deep shade leads to a stronger investment in survival of the existing population rather than expansion (Willems & Boessenkool 1999).

There are different options available to control competition and guard ancient forest species against exclusion by nitrophilic species. Firstly, active anthropogenic removal of competitive species is possible. Though effective, it is labour-intensive, environmentally unfit and produces a severe disturbance of a potentially stabile system and is therefore not very suitable. Secondly, several studies propose grazing as a measure for weed control, and interesting results have been reported (Papanastasis et al. 1995). Thirdly, an intervention in the light climate of the forest could be performed e.g. by applying a weak thinning regime, which allows fast canopy closure. The highly competitive non-forest species are, in contrast to forest species, not capable of surviving or competing in dense shade. However, permanent deep shade will also affect survival and, especially, expansion of ancient forest species (Willems & Boessenkool 1999). It can be concluded that seeding at a proper time, if the environmental conditions are favourable for ancient forest species, will be the most successful management measure to enhance the system specific diversity.

5.6. Concluding guidelines concerning understory vegetation

Based on existing knowledge we have the following recommendations in order to accelerate and steer the development of the understory vegetation in recently afforested arable land:

- Improve the seed availability of ancient forest species by:
  - selecting sites for the new forests adjacent to old-growth forests where the target species are present and reproductive
  - sowing the target species in the new forest
- Reduce the competition by non-forest species weeds by:
  - reduction of N availability to limit the competitive strength of weedy species
  - reduction of light availability to a level where the fast-growing species cease to grow but where ancient forest species can still survive
  - removal of fast-growing species by weeding or grazing.

6. COMPLEX QUESTIONS

Afforestation efforts are location specific and multi-objective. They are location specific since the benefits of the new forest in terms of e.g. biomass production, environmental services, and recreation strongly depend on the local site quality and its spatial organization in relation to its surroundings. They are multi-objective because they are based on decisions driven by multiple functions, with their respective claims to produce goods and services in a maximal way. Maximizing a single objective will often cause trade-offs for another objective. Addressing more than one forest function is therefore always a complex multi-objective optimization challenge, which implies insight in the complex interactions between the performance response of objectives to different site and management characteristics. In other words, the creation of a multipurpose forest should be a well thought compromise to meet the goals set by managers and stakeholders in an optimal way.

In the preceding sections, generalized and mono-objective guidelines were derived from field observations, literature review and expert knowledge. In this section, guidelines to questions with a higher level of complexity are given. Some questions are site specific but most questions are location specific and can be solved by linking the METAFORE model outputs to a spatial database. This is done in the AFFOREST-sDSS, the spatial Decision Support System of AFFOREST, which is designed as a user friendly tool that allows a flexible treatment of numerous afforestation questions. Also multi-objective questions setting weights or thresholds on the environmental impact categories C sequestration, nitrate leaching, and water recharge can be solved by the AFFOREST-sDSS.

Based on contacts with end users (national and local policy makers, land use planners, environmentalists, foresters) a range of policy relevant end-user questions were formulated (Table 11.2). Some of these questions are analyzed and solved in the following guidelines using the AFFOREST-sDSS. The selected questions illustrate the functionality of the sDSS and the possible guidance this tool offers. The selected questions and their solutions were saved as predefined cases in the AFFOREST-sDSS, which means that the solution can be followed step by step by running the predefined cases in the AFFOREST-sDSS software.

Table 11.2. *End user questions to the AFFOREST-sDSS. C= carbon sequestration; N= nitrate leaching; W= water recharge.*

Question	Country and Scale	Environmental impact category	Extra data need
Where are the best places in northwestern Europe to establish 100,000 hectares of new forest with maximum carbon sequestration as a goal?	NW EU 1 km^2	C	No
Where is the best place to afforest in order to reach the best possible environmental performance after 50 years given that the potential afforestation areas are situated adjacent to existing forest?	B (Flanders) 1ha	C+N+W	Forest map of Flanders
What is the difference in environmental performance (C, N, and H2O, respectively) after 100 years between high intensity and low intensity management for a planned afforestation area?	B (Flanders) 1ha	C+ N+W	Map of planned afforestation
The policy on afforestation in the Netherlands suggests planting 30,000 ha of oak forest. Where is the best place to afforest if carbon sequestration should be maximized, while at the same time nitrate leaching should not exceed the ground water standard of 50 mg/l?	NL 1ha	C+N+W	
Which tree species and which management strategies should Sweden use in each location of the territory to maximize C sequestration?	S 1ha	C	
For implementation of the EU Water Framework Directive, arable land in Denmark, assigned as groundwater protection area, is afforested with broadleaved forest	DK 1ha	C+N+W	Map of groundwater protection

mainly for recreational purposes. What is the environmental performance after 100 years?			areas
The government of Denmark decides to double the forested area by afforestation (intensive site preparation, medium stand tending) with oak over 100 years. After how many years is nitrate leaching below 5 kg N/ha/year on sandy soil? On clayey soil?	DK 1ha	N	Forest map of Denmark
A municipality has an objective to afforest an area with sandy soils in a way that produces as much clean drinking water as possible and as a second priority sequesters as much carbon as possible over 15 years. How should this be done in the best possible way?	DK 1ha	C+H+W	Map of area
A forest district has purchased marginal arable land (sandy soil) adjacent to existing forest. The aim is to obtain maximum wood production over 70 years using spruce and a low management intensity. What is the environmental performance?	DK 1ha	C+H+W	Map of purchased arable land

6.1. Where are the best places to afforest in north-western Europe in order to establish 100,000 hectares of new forest with maximum carbon sequestration as a goal?

This is a 'where' question with the afforestation strategy not being defined. It is wise to define a time horizon for evaluation, e.g. 50 years. It is solved in the AFFOREST-sDSS with the multicriteria option putting maximum weight on carbon. The solution was stored in the AFFOREST-sDSS as a predefined case named 'WHEREARmaxC'. The final result, showing the best places to afforest in order to reach the highest C sequestration, is seen in Figure 11.9.

This question needs to be solved in two steps. First the AFFOREST-sDSS selects the afforestation strategy resulting in maximum carbon sequestration for all available pixels (1 km^2) in the AFFOREST region (Sweden, Denmark, the Netherlands, and Flanders, Belgium). Available pixels are pixels with present agricultural land use. We look at the cumulative amount of carbon sequestered after 50 years. The result is a table with the best afforestation strategy per pixel. Since this output table is not sorted by C sequestration but by the pixel class value, a second step is needed in order to find the best 1000 km^2 (100,000 ha). In the second step, the output table is sorted by C and the first 1000 km^2 are selected. Based on this selection a new grid is created after converting to the 'grid' command of ArcView. The final result (Figure 11.9) is a map of the AFFOREST region highlighting the 100,000 ha where the cumulative C sequestration is the highest. A fair amount of these successful pixels are situated in the south of Flanders, Belgium (Color Plate 8). A mild climate with sufficient rainfall (800 mm), very fertile loamy soils but low initial soil carbon content under the current arable land use characterize this area. As a consequence, a high carbon sequestration potential in living biomass and in soil could be expected.

6.2. Where should we afforest in order to reach the best possible environmental performance after 50 years given that the potential afforestation areas are situated adjacent to existing forest?

This 'where' question was solved for Flanders, Belgium. An evaluation time of 50 years must be selected. Since the environmental performance is not specified, it is a multi-objective query with equal weight for the three environmental categories C sequestration, nitrate leaching, and water recharge. For the condition "adjacent to existing forest" an input map with the present forests in Flanders is needed. The solution is saved in the AFFOREST-sDSS as a predefined case named 'WHEREFlbestEP'. The final result, showing the best places to afforest in order to reach the best environmental performance, is seen in Color Plate 9 and Figure 11.9.

Similar to Question 18 we select a regional/national assessment but now for Flanders, Belgium (pixels 1ha). We also select a multi-criteria analysis but with equal weight for the three environmental categories C sequestration, nitrate leaching, and water recharge, because no specifications concerning environmental category was made. The environmental impact is calculated cumulatively for a period of 50 years, while no afforestation strategies are specified. For the condition "adjacent to existing forest" an additional GIS operation is needed before running the AFFOREST-sDSS, where the maximum distance to existing forest must be defined. We chose zones with a maximum width of 500m. These zones are entered as an ARCview shapefile in the AFFOREST-sDSS. The result is a map showing the best locations and the afforestation strategies there to obtain best possible environmental performance (Color Plate 9). In Figure 11.9 a detail of the result is shown with the existing forests (in grey), surrounded by a contour line at 500 m distance, indicating the potential afforestation zones. Within these zones, colored pixels are available for afforestation. The color indicates the afforestation strategy to follow in order to obtain the best environmental performance. In the area, we see pixels with afforestation strategy 21 and 12, being pine and beech afforestation respectively, both with high stand preparation intensity and low thinning intensity.

It is typical for optimizations where C sequestration plays a role that high intensity site preparation combined with low intensity thinning management give the best result, because it leads to higher biomass accumulation. However, in areas where spruce afforestation has the best environmental performance it goes in combination with a high thinning intensity. The reason is that for un-thinned spruce the trade-off in terms of decreased water recharge and increased nitrate leaching due to increased interception becomes too important.

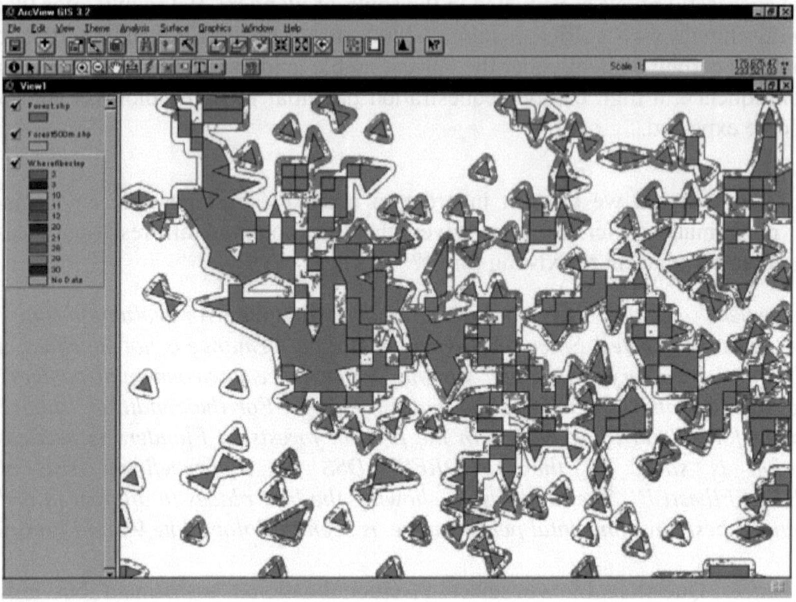

Figure 11.9. *Map of afforestation zones with best possible environmental performance adjacent to existing forests in Flanders (after zooming of Color Plate 8). Contour lines at 500 m distance, indicate the potential afforestation zones.*

6.3. The policy on afforestation in the Netherlands suggests planting 30,000 ha of oak forest. Where is the best place to afforest if carbon sequestration should be maximized, while at the same time nitrate leaching should not exceed the ground water standard of 50 mg/l?

This is a typical 'where' question solved in the AFFOREST-sDSS by using the 'Locate Afforestation Area' query with multicriteria option. In order to get a focused answer it is wise to specify a time horizon and the used afforestation strategy. The result was saved as a predefined case named WHERENLmaxCdrinkingnorm.

To solve this question we choose a time horizon of 50 years and an afforestation strategy of planting oak with intensive site preparation and low thinning intensity. There are two ways to approach this question.

The first one is similar to the case in question 18: a multicriteria approach with all weight on carbon. The result is an area with highest possible carbon sequestration but without information on nitrate leaching and water. This result must be saved as a shape file and brought in again in the sDSS for a second query. In this query the environmental performance of the area is addressed based on the given management strategy. After running the sDSS the end user must check in the output table which areas do not exceed the ground water standard. The output of nitrate leaching is not in concentrations but in absolute quantities (kg/ha). Concentrations can be obtained by dividing the N column by the W column and adjusting the units. Then the pixels that meet the requirement of 50 mg/l NO_3-N are selected. The table of this selection is then sorted by C, the first 30,000 ha are selected and a map is created.

The second approach is to select a multicriteria approach for the three Environmental Impact Categories (EIC). It must be ensured that carbon is maximized but that meanwhile the drinking water norm for nitrate is not exceeded. Putting large weight on carbon and little weight on nitrate and water recharge can do this. Instead of 30,000 ha we select 50,000 ha as a minimum in order to make sure that we obtain at least 30,000 ha in the end. The intermediate result is a map of the best pixels; approximately 70,000 ha were found (Figure 11.10). These pixels for sure have high carbon sequestration and low nitrate leaching. But do they meet the drinking water norm? To know this, the end user must check the output table in the same way as described in option A. The result is presented in Figure 11.11.

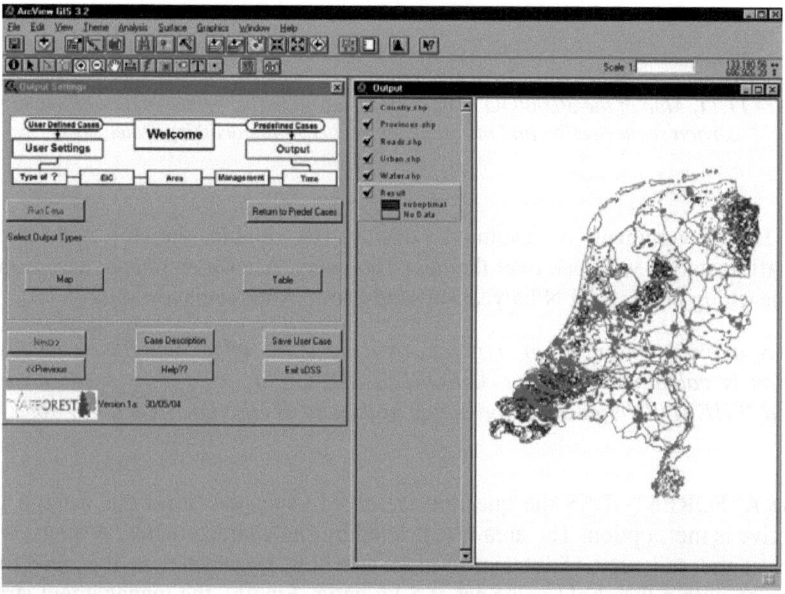

Figure 11.10. Selection of minimum 50,000 ha afforestation area with oak obtaining the maximum carbon sequestration, minimum nitrate leaching, and maximum water recharge.

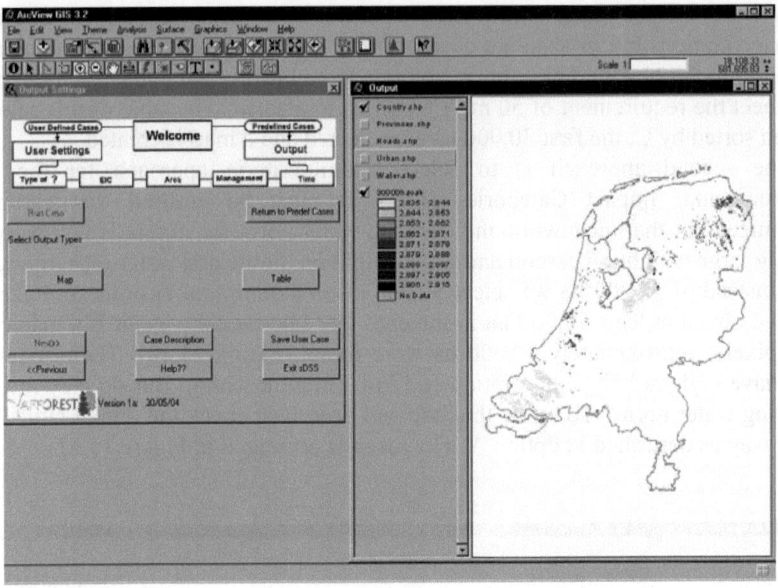

Figure 11.11. *Map of the 30,000 ha of oak afforestation in the Netherlands with the highest carbon sequestration and nitrate leaching below the drinking water norm.*

6.4. The government of Denmark decides to double the forested area by afforestation with oak over the next 100 years. After how many years is nitrate leaching below 5 kg N/ha/year on sandy soil? And on clayey soil?

This is a question on when a given environmental effect will be reached. The solution is calculated with the AFFOREST-sDSS and saved as a predefined case named WHENDKNleachsand for sandy soils and WHENDKNleachclay for clayey soils.

In the AFFOREST-sDSS the question is solved using the 'Find out when a given objective is met' option. The area is restricted by site characteristics. A query is built to select the soil type of interest. Nitrate leaching is selected as the objective to minimize with a threshold value set at 5 kg/ha/yr. Finally, the management strategy is selected. The results are presented in a time map (Figure 11.12 for sandy soils and Figure 11.13 for clay soils). The results for sandy soils show that the objective is never reached. The results for clay soils show that the objective is reached after a period between 1 and 32 years. Why it takes much longer at some places than at others is hard to interpret but it is mainly influenced by the local N deposition level.

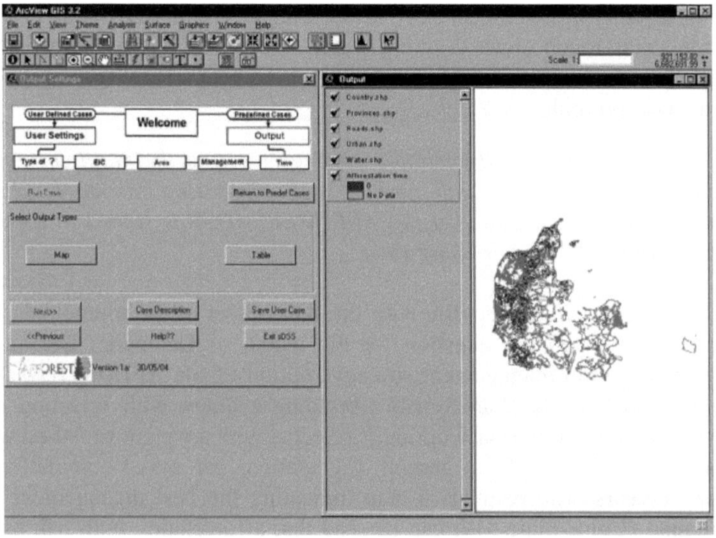

Figure 11.12. Map indicating when nitrate leaching is below 5kg/ha/yr for afforestation with oak on sandy soils. All agricultural land on sandy soils is in class0, which is without result. This means that the threshold value is never reached and nitrate leaching is always higher.

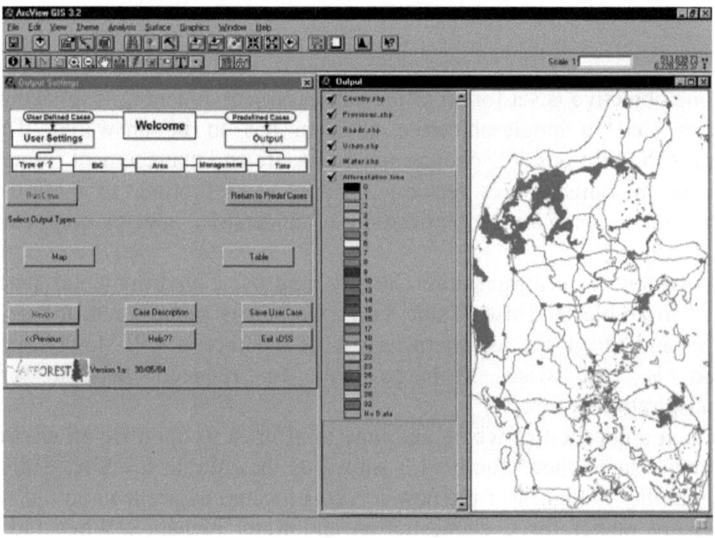

Figure 11.13. Map showing the time needed before nitrate leaching is below 5 kg/ha/year for afforestation with oak on clay soils.

6.5. A municipality has an objective to afforest an area with sandy soils in a way that produces as much clean drinking water as possible and as a second priority sequesters as much carbon as possible over 15 years. How should this be done in the best possible way?

This is a question on the afforestation strategy meeting the multiple objective of maximizing water recharge and carbon and minimizing nitrate leaching. The solution (Figure 16) calculated using AFFOREST-sDSS is stored in a predefined case called 'AFFSTRATDKdrinkingwater'.

The first step is to add a shapefile with the boundaries of the municipality. In this example, we answered the question for the whole of Denmark. The question is solved with the 'Find management strategy' option of the AFFOREST-sDSS. The area is restricted by site characteristics building a query with selection of sandy soils. Hereafter, the multicriteria option is selected with a weight of 0.4 on water and nitrate and a weight of 0.2 on carbon. Calculations are asked cumulatively for a period of 15 years. The result is a map indicating the best management strategy. From the map (Color Plate 9) it can be read that afforestation with oak using high intensity site preparation and high intensity thinning (afforestation strategy 1) is by far the most successful. By comparing this guideline with earlier ones, it becomes clear that by putting less emphasis on carbon sequestration and more weight on nitrate leaching and water recharge, the optimal thinning intensity moves from low to high.

7. CONCLUSIONS

If only one objective is set for an afforestation project, planning is less complicated. In such a case, a single-objective question has to be answered. Table 11.3 summarises the best possible choices regarding site and management for new forests in order to maximise water recharge, and C sequestration, to minimise nitrate leaching, and to encourage colonization of understory species characteristic for ancient woodlands.

It is evident that afforestation largely improves environmental performance compared to arable land use (Table 11.3). Carbon is sequestered, nitrate leaching decreases, and forest species characteristic of old forests will slowly appear. The exception is that water recharge will be reduced because of higher evapotranspiration in forests.

It is then a question of where and how to afforest to optimise all environmental effects at the same time. Table 11.3 shows us that nitrate leaching decreases and seed availability is highest if the new forest is located adjacent to an old forest, but this does not matter for C sequestration and water recharge. When old drainage pipes from former arable use are impaired or removed, it is positive for water recharge, nitrate leaching, and C sequestration. Selection of deciduous species is also favourable for the environmental impact categories being considered. In combination with intensive thinning it will significantly reduce interception evaporation. However, there may be a trade-off in the short term for carbon storage due to the slower initial growth rates of deciduous trees. But in the long-term, the greater stability towards disturbances such as windstorms and the possibility to perform continuous cover forestry can compensate this.

Afforestation is often based on decisions driven by multiple objectives. When multi-purpose demands are involved, it is difficult and involves many pitfalls to plan the new forests in a proper way while meeting the array of natural, political, and socio-economic conditions. Maximising a single objective will often cause trade-offs for another objective. It is therefore a complex optimisation challenge to address more than one objective. It is necessary to know the interactions between the performance of individual services provided by afforestation and different site and management characteristics.

A range of possible complex questions concerning environmental effects of afforestation have been posed (Table 11.2) and the solution to part of them, solved using the AFFOREST-sDSS, have been shown here. The questions are relevant examples formulated based on contacts with end users (national and local policy makers, land use planners, environmentalists, and foresters). Two general conclusions can be made based on the use of the AFFOREST-sDSS. Initially in the planning process, the 'where' question is more important than the 'how' question, which means that the environmental performance in terms of carbon sequestration, nitrate leaching, and groundwater recharge is more influenced by the site of afforestation than by afforestation management. In other words, climate, soil and N deposition level play a more important role than the tree species and the management strategy chosen when afforestation is performed.

Once an afforestation area is selected, the 'how' question becomes of major importance. When the management strategy is optimised there seems to be a trade-off between carbon sequestration on the one hand and groundwater recharge on the other. Measures leading to high carbon sequestration (i.e. fast growing tree species, intensive stand preparation, and low thinning intensity) at the same time lead to low groundwater recharge. This relation is rather obvious, and different authors have demonstrated the relationship between net primary production and evapotranspiration. The relationship between these two environmental impact categories and nitrate leaching is less obvious, but in general, high nitrate concentrations in groundwater often accompanies low recharge and high carbon sequestration. Such situations will be typical for fertilised and unthinned spruce stands. Reversely, older heavily thinned broadleaved stands will have lower nitrate concentrations in the groundwater due to lower N deposition and lower interception evaporation, while recharge is higher and carbon sequestration lower in such old stands.

Table 11.3. *Overview of afforestation strategies in order to optimise the environmental impacts.*

	Water recharge	Nitrate leaching	Carbon sequestration	Characteristic forest understory species
What happens after afforestation compared to arable land?	• Decreases during first 10-20 years • Thereafter constant	• Fast reduction during first five years • Increases again after canopy closure (~20 years)	• Average rates of C sequestration range from 0.3-0.8 t C ha^{-1} yr^{-1} in soils and from 1.5-4.3 t C ha^{-1} yr^{-1} in biomass	• It takes 50-100 years or more before typical characteristic forest understory species start to develop • Seed availability, light and nutrients determine the time scale
Which afforestation sites should be selected?	• Sandy soils	• Nutrient-poor sandy soils • Low N deposition areas away from local sources • Larger forest fragments connected to former forest	• Nutrient-rich soils • Moist soil conditions	• Bordering or in vicinity of old-growth forest = seed availability • Low N availability
Which tree species should be selected?	• Deciduous species preferable to conifers	• Deciduous species or mixtures of species • No use of N fixing species	• Tree species with high increment rates over a long life cycle • Species with slowly decomposing litter	• Deciduous trees with high density of overstory = low light levels
Which management practices will optimise?	• Low density of overstory • No artificial drainage • Low density of understory = bare soils	• Little site preparation • No mechanical weed control • Fast establishment • Weak thinning regime and no clear-cutting • No artificial drainage	• Continuous cover forestry • Weak thinning regime • No artificial drainage • Stronger thinning regimes in productive sites where wood products can substitute fossil fuels	• Seed availability improved by sowing • Anthropogenic removal of competitive species • Grazing

8. REFERENCES

Attiwell, P. M., & Adams, M. A. (1993). Tansley Review No. 50. Nutrient cycling in forests. *New Phytologist, 124,* 561-582.

Bäumler, R., & Zech, W. (1999). Effects of forest thinning on the streamwater chemistry of two forest watersheds in the Bavarian Alps. *Forest Ecology and Management, 116,* 119-128.

Binkley, D. (1984). Does forest removal increase rates of decomposition and nitrogen release? *Forest Ecology and Management, 8,* 229-233.

Binkley, D., Sollins, P., Bell, R., Sachs, D., & Myrold, D. (1992). Biochemistry of adjacent conifer and alder conifer stands. *Ecology, 73*(6), 2022-2033.

Bormann, B. T., & DeBell, D. S. (1981). Nitrogen content and other soil properties related to age of red alder stands. *American Journal of Soil Science Society, 45,* 428-432.

Bosch, J. M., & Hewlett, J.D. (1982). A review of catchment experiments to determine the effect of vegetation changes on water yield and evapotranspiration. *Journal of Hydrology, 55,* 3-23.

Brown, A. H. F., & Iles, M. A. (1991). Water chemistry profiles under four tree species at Gisburn, NW England. *Forestry, 64*(2), 169-187.

Brunet, J., & von Oheimb, G. (1998). Colonization of secondary woodlands by Anemone nemorosa. *Nordic Journal of Botany, 18,* 369-377.

Butaye, J., Jacquemyn, H., & Hermy, M. (2001). Differential colonization causing non-random forest plant community structure in a fragmented agricultural landscape. *Ecography, 24,* 369-380.

Callesen, I., Rauland-Rasmussen, K., Gunderssen, P., & Stryhn, H. (1999). Nitrate concentrations in soil solutions below Danish forests. *Forest Ecology and Management, 114,* 71-82.

Chang, S. X., & Preston, C. M. (2000). Understorey competition affects tree growth and fate of fertilizer-applied ^{15}N in a coastal Bristish Columbia plantation forest: 6-year results. *Canadian Journal of Forest Research, 30,* 1379-1388.

Corré, W. J. (1983). Growth and morphogenesis of sun and shade plants. III. The combined effects of light intensity and nutrient supply. *Acta Botanica Neerlandica, 32,* 277-294.

de Vries, W., & Jansen, P. C. (1994). Effects of acid deposition on 150 forest stands in the Netherlands. 3. Input output budgets for sulphur, nitrogen, base cations and aluminium. Wageningen, the Netherlands, DLO-Winand Staring Centre for Integrated Land, Soil and Water Research, Report 69.3, 60 pp.

de Wit, H. A., & Kvindesland, S. (1999). Carbon stocks in Norwegian forest soils and effects of forest management on carbon storage. 14, 1-52. Ås, Norsk Institutt for Skogforskning & Institutt for Skogfag, NLH. Rapport fra Skogforskningen - Supplement.

Emmett, B. A., Reynolds, B., Stevens, P. A., Norris, D. A., Hughes, S. Görres, J., & Lubrecht, I. (1993). Nitrate leaching from afforested Welsh catchments – Interactions between stand age and nitrogen deposition. *Ambio, 22,* 386-394.

Falkengren-Grerup, U. (1993). Effects on beech forest species of experimental enhanced nitrogen deposition. *Flora, 188,* 85-91.

Finch, J. W. (1998). Estimating groundwater recharge using a simple water balance model – sensitivity to land surface parameters. *Journal of Hydrology, 211,* 112-125.

Grant, R., Blicher-Mathiesen, G., Jensen, P. G., Pedersen, M. L., Clausen, B., & Rasmussen, P. (2004). Landovervågningsoplande 2003. NOVA 2003. Danmarks Miljøundersøgelser. 118 s. http://faglige-rapporter.dmu.dk

Gundersen, P., & Rasmussen, L. (1995). Nitrogen mobility in a nitrogen limited forest at Klosterhede, Denmark, examined by NH_4NO_3 addition. *Forest Ecology and Management, 71,* 75-88.

Gundersen, P., Emmet, B. A., Kjønaas, O. J., Koopmans, C., & Tietema, A. (1998). Impact of nitrogen deposition on nitrogen cycling: a synthesis of NITREX- data. *Forest Ecology and management, 101,* 37-55.

Gundersen, P., Friis, E., & Hansen, K. (2001). Nitratudvaskning fra skovrejsning og vedvarende græsarealer 1998 – 2001. Drastrup projektet. Aalborg Kommune og Forskningscentret for Skov & Landskab, Hørsholm, 30 sider. (In Danish).

Guo, L. B., & Gifford, R. M. (2002). Soil carbon stocks and land use change: a meta analysis. *Global Change Biology, 8,* 345-360.

Hansen, K. (1976). Ecological Studies in Danish Heath Vegetation. *Dansk Botanisk Arkiv, 31*(2), 1-118.

Hansen, K. (Ed.) (2003). Næringsstofkredsløb i skove - Ionbalanceprojektet. Forest & Landscape Research nr. 33. Skov & Landskab, Hørsholm. 300 pp.
Hansen, K., & Vesterdal, L. (1999). Skovrejsning - nitratudvaskning, jordens pH og kulstofbinding. Videnblade Skovbrug nr. 4.6-9, Forskningscentret for Skov & Landskab, Hørsholm. 2 pp.
Harmer, R., Peterken, G., Kerr, G., & Poulton, P. (2001). Vegetation changes during 100 years of development of two secondary woodlands on abandoned arable land. *Biological Conservation, 101,* 291-304.
Heil, G. W. (2004). Modelling of Plant Community Assembly in Relation to Deterministic and Stochastic Processes. In V. M. Temperton, R. J. Hobbs, T. Nuttle, & S. Halle (Eds.), Assembly Rules and Restoration Ecology, pp. 230-245. Washington: Island Press, London: Covelo.
Hermy, M. (1994). Effects of former land use on plant species diversity and pattern in European deciduous woodlands. In T. J. B. Boyle & C. E. B. Boyle (Eds.), Biodiversity, Temperate Ecosystems and Global Change, pp. 123-144. Berlin Heidelberg: Springer-Verlag.
Honnay, O., Hermy, M., & Coppin, P. (1999). Impact of habitat quality on forest plant species colonization. *Forest Ecology and Management, 115,* 157-170.
Jussy, J.-H., Colin-Belgrand, M., & Ranger, J. (2000). Production and root uptake of mineral nitrogen in a chronosequence of Douglas-fir (Pseudotsuga menziesii) in the Beujolais Mounts. *Forest Ecology and Management, 128,* 197-209.
Kinniburgh, D. G., & Trafford, J. M. (1996). Unsaturated zone pore water chemistry and the edge effect in a beech forest in sourthern England. *Water, Air, and Soil Pollution, 92,* 421-450.
Knight, D. H., Yavitt, J. B. & Joyce, G. D. (1991). Water and nitrogen outflow from lodgepole pine after two levels of tree mortality. *Forest Ecology and Management, 46,* 215-225.
Koerner, W., Dambrine, E., Dupouey, J. L., & Benoit, M. (1999). $\delta^{15}N$ of forest soil and understory vegetation reflect the former agricultural land use. *Oecologia, 121,* 421-425.
Kristensen, H. L., Gundersen, P., Callesen, I., & Reinds, G. J. (2004). Atmospheric N deposition influences soil nitrate concentration differently in European coniferous and deciduous forests. *Ecosystems, 7,* 180-192.
LeMaitre, D. C., & Versfeld, D. B. (1997). Forest evaporation models: relationships between stand growth and evaporation. *Journal of Hydrology, 193,* 240-257.
Leuschner, C., & Rode. M. W. (1999). The role of plant resources in forest succession: changes in radiation, water and nutrient fluxes, and plant productivity over a 300-yr-long chronosequence in NW-Germany. *Perspectives in Plant Ecology, Evolution and Systematics, 2,* 103-147.
Lundmark, J. E. (1977). Marken som del av det skogliga ekosystemet. (The soil as part of the forest ecosystem). *Sveriges Skogsvårdsförb. Tidskrift, 2,* 108-122. (In Swedish).
Lundmark, J. E. (1988). Skogsmarkens ekologi, del II, tillämpning (The forest soil's ecology, part II, application). Skogsstyrelsen Jönköping. (In Swedish).
Magill, A., Aber, J., Hendricks, J., Bowden, R., Melillo, J., & Steudler, P. (1997). Biogeochemical response of forest ecosystems to simulated chronic nitrogen deposition. *Applied Ecology, 7,* 402-415.
Matthesen, P., & Kudahl, T. (2001). Skovrejsning på agerjord – ukrudtsudviklingen. Videnblade Skovbrug nr. 4.2-4, Forskningscentret for Skov & Landskab, Hørsholm. 2 pp.
Mellert, K.-H., Kölling, C., & Rehfuess, K. E. (1998). Vegetationsentwicklung und Nitrataustrag auf 13 Sturmkahlflächen in Bayern. *Forstarchiv, 69,* 3-11.
Miles, J. (1985). The pedogenic effects of different species and vegetation types and the implications of succession. *Journal of Soil Science, 36,* 571-584.
Miller, H. G., & Miller, J. D. (1988). Response to heavy nitrogen applications in fertilizer experiments in British forests. *Environmental Pollution, 54,* 219-231.
Munson, A. D., & Timmer, V. R. (1995). Soil nitrogen dynamics and nutrition of pine following silvacultural treatments in boreal and Great Lakes-St. Lawrence plantations. *Forest Ecology and Management, 76,* 169-179.
Ogner, G. (1987a). Glyphosate application in forest-Ecological Aspects. II. The quality of water leached from forest soil lysimeters. *Scandinavian Journal of Forest Research, 2,* 469-480.
Ogner, G. (1987b). Glyphosate application in forest-Ecological Aspects. V. The water quality of forest brooks after manual clearing or extreme glyphosate application. *Scandinavian Journal of Forest Research, 2,* 509-516.
Påhlsson, A.-M., & Bergkvist, B. (1995). Acid deposition and soil acidification at a southwest facing edge of Norway spruce and European bech in south Sweden. *Ecological Bulletins, 44,* 43-53.
Papanastasis, V., Koukoura, Z., Alifragis, D., & Makedos, I. (1995). Effects of thinning, fertilisation and sheep grazing on the understory vegetation of Pinus pinaster plantations. *Forest Ecology and Management, 77,* 181-189.

Paul, K. I., Polglase, P. J., Nyakuengama, J. G., & Khanna, P. K. (2002). Change in soil carbon following afforestation. *Forest and Ecology Management, 168,* 241-257.

Pedersen, L. B, Riis-Nielsen, T., Ravn, H. P., Dreyer, T., Krag, M., Nielsen, A. O., Matkovski, A., & Sunde, P. B. (2000). Alternativer til pesticidsprøjtning i skovkulturer. *Skoven, 8,* 355-359. (In Danish).

Piene, H., & Van Cleve, K. (1978). Weight loss of litter and cellulose bags in a thinned white spruce forest in interior Alaska. *Canadian Journal of Forest Research, 8,* 42-46.

Pommerening, A., & Murphy, S. T. (2004). A review of the history, definitions and methods of continuous cover forestry with special attention to afforestation and restocking. *Forestry, 77,* 27-44.

Post, W. M. & Kwon, K. C. (2000). Soil carbon sequestration and land-use change: processes and potential. *Global Change Biology, 6,* 317-327.

Qualls, R. G., Haines, B. L., Swank, W. T., & Tyler, S. W. (2000). Soluble organic and inorganic nutrient fluxes in clearcut and mature deciduous forests. *Soil Science Society of America Journal, 64,* 1068-1077.

Richter, D. D., Markewitz, D., Heine, P. R., Jin, V., Raikes, J., Tian, K., & Wells, C. G. (2000). Legacies of agriculture and forest regrowth in the nitrogen of old-field soils. *Forest Ecology and Management, 138,* 233-248.

Rijtema, P. E., & de Vries, W. (1994). Differences in precipitation excess and nitrogen leaching from agricultural lands and forest plantations. *Biomass and Bioenergy, 6* (1/2), 103-113.

Rothe, A., Huber, C., Kreutzer, K., & Weis, W. (2002). Deposition and soil leaching in stands of Norway spruce and European Beech: Results from the Höglwald research in comparison with other European case studies. *Plant and Soil, 240,* 33-45.

Sahin, V., & Hall, M. J. (1996). The effects of afforestation and deforestation on water yields. *Journal of Hydrology, 178,* 293-309.

Smethurst, P. J., & Nambiar, E. K. S. (1989). Role of weeds in the management of nitrogen in a young Pinus radiata plantation. *New Forests, 3,* 203-224.

Stortelder, A. H. F., Schamimee, J. H. J., & Hommel, P. W. F. M. (1999). De vegetatie van Nederland. Deel 5. Plantengemeenschappen van ruigten, struwelen en bossen. Leiden: Opulus Press.

Stottlemyer, R., & Toczydlowski, D. (1996). Modification of snowmelt chemistry by forest floor and mineral soil, Northern Michigan. *Journal of Environmental Quality, 25,* 828-836.

Sutton, R. F. (1995). White spruce establishment: initial fertilization, weed control, and irrigation evaluated after three decades. *New Forests, 9,* 123-133.

Tarrant, R. F., & Trappe, J. (1971). The role of Alnus in improving the forest environment. *Plant and Soil, 35* (special volume), 335-348.

Van der Salm, C., & De Vries, W. (2000). Soil acidification in loess and clay soils in the Netherlands. *Water, Air, and Soil Pollution, 120,* 139-167.

van Ek, R., & Draaijers, G. P. J. (1994). Estimates of atmospheric deposition and canopy exchange for three common tree species in the Netherlands. *Water Air, and Soil Pollution, 73*(1-4), 61-82.

Verheyen, K., & Hermy, M. (2001). The relative importance of dispersal limitation of vascular plants in secondary forest succession in Muizen Forest, Belgium. *Journal of Ecology, 89,* 829-840.

Vesterdal, L., Rosenqvist, L., & Johansson, M.-B. (2002). Effect of afforestation and subsequent forest management on carbon sequestration in soil and biomass. In Hansen, K. (Ed.), Literature review for AFFOREST: Planning afforestation on previously managed arable land - influence on deposition, nitrate leaching, precipitation surplus, and carbon sequestration pp. 63-88. http://www.fsl.dk/afforest/html/filearchive.asp?id=1

Vitousek, P. M., & Matson, P. A. (1985). Disturbance, nitrogen availability, and nitrogen losses in an intensesively managed loblolly pine plantation. *Ecology, 66,* 1360-1376.

Willems, J. H., & Boessenkool, K. P. (1999). Coppiced woodlands and their significance for herbaceous plant species conservation, pp. 188-196. In Dong Ming & M. J. A. Werger (Eds.), A spectrum of ecological studies. In honour of Professor Zhang Zhangcheng at his 70th birthday. Southwest China Normal University Press.

CHAPTER 12

THE LESSONS LEARNED FROM AFFOREST – A SYNTHESIS

K. HANSEN[1], B. MUYS[2], G.W. HEIL[3] and L. VESTERDAL[1]

[1]Forest and Landscape Denmark, KVL, Hørsholm Kongevej 11,2970 Hørsholm, Denmark, E-mail: kiha@kvl.dk, [2]Katholieke Universiteit Leuven, Division Forest Nature and Landscape, Celestijnenlaan 200E, 3001 Leuven, Belgium, [3]Faculty of Science, Institute of Environmental Biology, Sorbonnelaan 16, 3584 CA Utrecht, the Netherlands

Abstract. AFFOREST was research project in the EU 5th Framework Programme for Research & Technological Development (Energy, Environment and Sustainable Development theme). The project ran for four years during the period 2000-2004 and included partners from four countries: Belgium, the Netherlands, Sweden and Denmark. Knowledge has been obtained from literature and existing data sets, completed with field observations in afforestation chronosequences on former agricultural soils. Two oak (*Quercus robur*) and four Norway spruce (*Picea abies*) chronosequences were investigated. The main products are a geographical database, a mechanistic carbon/nitrogen/water process-based metamodel, a GIS-model based on the metamodel and the spatial information, and finally a spatial decision support system (AFFOREST-sDSS) for scenario analyses and decision support. The AFFOREST-sDSS is able to test and compare different scenarios of afforestation and to propose optimal solutions for afforestation planning. On the basis of these scenario analyses and the performed field work, guidelines for optimal afforestation management were developed.

G.W. Heil et al. (eds.), Environmental Effects of Afforestation in North-Western Europe, 293–306.

1. BACKGROUND FOR AFFOREST

Forests originally covered most of the European continent. Over the last several thousand years, forests were cleared for growing and cultivating crops for food production. The land use patterns known today were formed. In many European countries, agricultural management has been intensified over the last century leading to increased production but also to negative environmental effects. Today, a surplus of agricultural production has caused decisions and changes in the EU Common Agricultural Policy (CAP) and important areas of agricultural land have been taken out of production. Therefore, several EU countries have decided to work towards conversion of agricultural land to forestry in some areas. Priorities in planning the geographic location and management of these areas are among others groundwater protection (quality and quantity) and article 3.3 of the Kyoto Protocol stipulating that carbon (C) sequestration after afforestation may be accounted as a sink in national greenhouse gas budgets. Other incentives for afforestation are the increased demand for recreational areas and commitments to maintain, enhance and protect biological diversity. Afforestation of today therefore aims at serving multiple purposes, whereas afforestation of the past primarily aimed at increasing the wood production.

2. THE AFFOREST PROJECT IN SHORT

AFFOREST was a project in the EU 5th Framework Programme for Research & Technological Development (Energy, Environment and Sustainable Development theme). The project ran for four years from May 2000 and included 6 partners from Belgium, the Netherlands, Sweden and Denmark. The focus of AFFOREST was on building knowledge and capacity to provide support for decisions regarding afforestation on former arable land with respect to changes in C and N pools and changes in water fluxes (Chapter 1) and to assist managers to optimise the location of new forests according to these environmental criteria.

New field data were collected in chronosequences of forests established on former arable land within the last 0-90 years in order to better quantify and understand changes in the water, C and nitrogen (N) cycles (Chapter 2, 3 and 4). Six afforestation chronosequences were investigated, of which two with oak (*Quercus robur*) and four with Norway spruce (*Picea abies*). A total of 31 forest stands were investigated.

These and additional literature data from published experiments were used to calibrate, validate and simplify mechanistic models describing the water, C and N cycles. Here, the vegetation model CenW (C, energy, nutrients and Water) (Chapter 7) as well as the detailed Nutrient Cycling and Soil Acidification Model (NUCSAM) (Chapter 8) were applied. Spatial interactions were included in the calculation of N deposition with the Eutrend model (Chapter 5). In addition, a new understory model was developed to test the possibilities for the development of the forest field layer after afforestation of former agricultural land, and to show the interactive effect of light and N on the potential understory plant performance (Chapter 6).

To support the decisions on afforestation on former arable land in north-western Europe, the project developed a mechanistic metamodel (METAFORE) (Chapter 9) based on CenW, NUCSAM and Eutrend, as well as a fully operational spatial Decision Support System (AFFOREST-sDSS) (Chapter 10), useful for scenario analysis and environmental impact assessment. In addition, a set of guidelines were derived (Chapter 11). The output generated by METAFORE is captured in a database, which can be queried by the AFFOREST-sDSS routines in order to obtain useful and reliable answers on complex questions of the end-user on where, how, how long and how much to afforest to reach a certain environmental performance. These decision support tools were then used to calculate a number of real-life scenarios at the regional policy maker level and at the local management level. Some of these scenarios were implemented as pre-defined cases in the AFFOREST-sDSS and were further explained in the tutorial for users and the published guidelines.

3. SCIENTIFIC ACHIEVEMENTS AND MAIN DELIVERABLES

The AFFOREST project has made various scientific achievements and provided a number of main deliverables in order to improve knowledge on the environmental impacts of afforestation on former arable land. All principal AFFOREST products are available for downloading at the AFFOREST project website www.sl.kvl.dk/afforest. Also, a self-instructive CD-ROM including the final AFFOREST-sDSS along with a user tutorial and other end products of AFFOREST has been produced.

3.1. Literature review

A literature review on environmental effects of afforestation was written in the initial phase of the project. Literature was screened in order to find suitable data for use in AFFOREST. The report describes the impact of tree species, stand structure, closeness to the forest edge, and different management options on dry deposition, nitrate leaching, groundwater resources, C sequestration and field layer biodiversity in connection to afforestation. Also, the literature report contains a review on Decision Support Systems (DSS), spatial Decision Support Systems (SDSS), and spatial analysis and modeling techniques.

3.2. Field experiments in chronosequences of afforested stands

It is difficult to directly measure the long-term effect of afforestation on former arable land having only little time. One would need 50-100 years in order to study a whole rotation period. Instead, time can be substituted with space using a chronosequence approach. The chronosequence approach will give a first indication of the trends in time after afforestation of former arable land. Most of the field work in AFFOREST was performed in chronosequences. The term chronosequence, as we use it, is a series of forest stands planted in different subsequent years on abandoned arable land. All stands are planted on similar soils in the same area. All stands in a chronosequence are exposed to the same climatic conditions and the same air quality regime. However, older plantations have obviously experienced a different previous agricultural regime and a different air quality in their early growth stages than recently planted forests. In view of the long timescales of the processes being considered (C sequestration, nitrate leaching and groundwater recharge), the field measurements were intended to contribute to the existing literature rather than to be the sole inputs underpinning the modelling runs.

Chronosequence study sites were found, equipped and monitored in Sweden, Denmark and the Netherlands. Three chronosequences were studied in Denmark: two spruce chronosequences up to 30 and 40 years on very contrasting soil types (sandy and clayey) and one oak chronosequence up to 30 years including an old oak stand of about 200 years. Two chronosequences were studied in the Netherlands: oak and spruce on similar sandy soils up to 30 years. One chronosequence of Norway spruce up to 90 years was studied in Sweden.

3.2.1. Carbon sequestration in biomass and soils

The overall aim was to provide better estimates of C sequestration in living biomass and in the soil as a result of afforestation activities. Biomass and soil C assessments were carried out in all chronosequences.

There was a clear positive effect of afforestation on the total C sequestration. For the afforested ecosystem as a whole, i.e. biomass and soil C combined, C sequestration in soil comprised about one third and C sequestration in biomass comprised about two thirds of the total C sequestration in the afforested ecosystem. In the short term (30-40 years), total C sequestration was higher in Norway spruce than in oak whereas soil type did not clearly influence the rate of C sequestration.

Biomass C sequestration ranged between 2.7 and 4.6 Mg C ha^{-1} yr^{-1} for stands younger than 45 years with no clear influence of different site characteristics. This suggests that the effect of parent material (texture, nutrient status, drainage class) on growth in the first 45 years appears to be masked by the soil enrichment, which is a legacy of former agriculture. Due to nutritional homogeneity of former arable soils general equations may therefore possibly be used to estimate biomass C sequestration during the first decades following afforestation. Later in the rotation of afforested stands (older than 45 years), biomass C sequestration differs more, which could be attributed to different management, tree species-specific growth patterns and less influence of former fertilization.

For soils, there is good evidence that afforestation of former arable cropland leads to either constant or increasing total soil C storage. In the short term studied in most detail (30 years), tree species selection has little influence on total soil C sequestration. However, forest floors on top of the mineral soil were clearly influenced by tree species. In stands younger than 40 years, C sequestration in the forest floor is higher under Norway spruce than under oak. Soil C stocks in the two Danish spruce chronosequences suggested that afforestation of nutrient-poor sandy soils results in larger forest floor C sequestration and also larger total soil C sequestration than afforestation of nutrient-rich clayey soils.

3.2.2. Water recharge

In each chronosequence, hydrological conditions were measured monthly during a period of one to two years. Changes in interception evaporation were investigated by measuring rainfall and throughfall fluxes. Soil hydrological fluxes were modelled using the dynamic simulation model SWAP. The model has been validated on measured throughfall data and soil water contents.

Afforestation had a clear influence on the water recharge of the considered sites. Water recharge was generally lower under Norway spruce compared to oak. In the spruce stands, 5–30% of the incoming precipitation leads to water recharge to surface and groundwater. In the oak stands, 20–35% of the precipitation left the system as water recharge. In general, water recharge declined with an increase of the stand age. In the oak stands, leaching decreased from 35 to 20% of the precipitation in the first 30 years. In the Norway spruce stands, the decline in water recharge was approximately 10-20% (100-150 mm).

The decline in water recharge was mainly caused by an increase in interception evaporation by the trees. In the oak stands, interception losses increased by more than 100 mm (from 10% of the precipitation in the youngest stands to 20% at an age of 30 years). In the spruce stands, interception evaporation increased with 100 to 200 mm and interception was responsible for 20 to 40% of the water losses.

3.2.3. Nitrogen deposition and nitrate leaching from the root zone

Bulk precipitation and throughfall were measured for approximately two years in all six chronosequences. Soil solutions of the mineral soil were sampled by suction cup lysimeters at monthly intervals at all forest stands in all chronosequences. The simulated hydrological fluxes were used to calculate leaching fluxes at the bottom of the root zone.

Throughfall deposition of N increased with tree height. Tree height and age were strongly related but for longer time scales tree height was a better predictor of deposition. The "filtering" capacity of Norway spruce was larger than for oak trees mostly related to a larger roughness of needles compared with leaves as well as the fact that spruce trees keep their foliage during winter. The potential for canopy exchange and N uptake of Norway spruce was higher than for oak. Part of the total N deposition was captured by the trees and taken up in the canopy, thereby reducing the throughfall deposition of N. The highest calculated total N deposition was 34 kg ha^{-1} yr^{-1} in the oldest oak stand at Sellingen in the Netherlands. Throughfall deposition of base cations was significantly enhanced compared to bulk deposition, due to dry deposition as well as exchange in the canopy, especially for K. The patterns for inert elements (e.g. Na and Cl) were consistent but for species that can be actively taken up by the canopy, such as NH_4^+, the pattern was less robust.

The measured monthly soil solution nitrate concentrations were generally below the drinking water standard of 50 mg dm^{-3} NO_3^-. Afforested sites varied substantially in their ability to retain N in the ecosystem. Differences in N retention and nitrate leaching among sites was the result of a complex combination of factors such as N deposition, soil type, soil N status, tree species, successional status of the vegetation and hydrology. In general, a shift in land use from agriculture to forestry lead to improved soil water quality (decreased nitrate concentrations) and to decreased export of nitrate from the root zone to deeper soil levels.

Over approximately 35 years following afforestation, nitrate leaching from the root zone was generally higher below oak (5-25 kg ha^{-1} yr^{-1}) than below Norway spruce (0-18 kg ha^{-1} yr^{-1}). These findings are in contrast with several studies, showing lower nitrate leaching below deciduous species as compared to conifers. The long-term effect (rotation period) of tree species on nitrate leaching in plantation forests can not be predicted by our experiments. However, the results indicate that in spruce stands on nutrient-rich soils (high N-status), nitrate leaching and concentrations of nitrate may increase substantially as the stands mature.

Compared to nutrient-poor sandy soils, nutrient-rich clayey soils appeared more vulnerable to disturbance of the N cycle and to increased N deposition, possibly leading to N saturation and enhanced nitrate leaching.

3.3. Field layer vegetation experiments and modelling

To test the influence of light and nutrients on the performance of herbaceous species that are expected to occur in different phases of forest development after afforestation, a garden experiment was performed. Three different levels of shade i.e.: 2-3% daylight, 8-10% daylight, and approximately 60% daylight were applied. Within the shade treatments, a fertilizer treatment was applied. Species behaviour was analyzed in terms of total biomass production, Specific Leaf Area and root-shoot ratio. In the field, an inventory was made on species distribution in different forest types.

In the garden experiment, a large effect of light availability was found on total biomass production, Specific Leaf Area and root:shoot ratio. All species showed the same pattern in their response to light but the absolute values differed between the species from different phases of forest development. Species from young forests (i.e early successional species) in general showed a higher plasticity and higher absolute values of Specific Leaf Area and root:shoot ratio than species more common to later stages of forest succession (i.e. ancient forest species). An effect of fertilizer treatment was only apparent in the higher light treatments.

It can be concluded that species from early phases in forest succession are better capable of adjusting their biomass allocation pattern to different light and nutrient conditions than are species commonly occurring in so-called ancient forests. This can lead to a competitive advantage at higher light availability. However, at low light conditions, none of the species was capable of profiting from a higher fertilizer treatment and the ancient forest species, specialized in growing at low light availabilities, performed best.

This experiment provided data for a field layer model simulating and comparing the performance of field layer species from different phases of forest development. The model included a photosynthesis and N acquisition module, and on basis of these, a growth module and a plastic biomass allocation module. Different plant species could be defined by calibrating the values of the Specific Leaf Area and root:shoot ratio, which are the two plant traits that showed to be imperative in acclimation to light and nutrient availability and that differ between the species of interest. The output, total plant mass at the end of the growing season, has been used to evaluate the performance of the different species. At low light and low N availability the ancient forest species showed a higher biomass production than the pioneer species. Increased N availability had a positive effect on plant production but not at very low light levels. The effect of N availability was stronger for the pioneer species. These results were found for both a coniferous forest and a deciduous forest. In the deciduous forest however, increased N availability had a negative effect on the total biomass production of the ancient forest species. The results of the model simulations emphasize the importance of a design and management of new forests on former agricultural land aimed at a low light availability and removal of excessive N.

3.4. Process-based modelling

Field data from the chronosequences and additional data from published studies were used to calibrate, validate and simplify mechanistic models describing the water, C and N cycles. These models were further developed for the AFFOREST project based on existing models. The models were applied to simulate the effects of afforestation in other circumstances than those for which calibration was done.

3.4.1. Deposition

The initial deposition of N on the AFFOREST region was calculated using the Eutrend model and country specific emission data for both NH_3 and NO_x as well as land use and roughness length maps. The emissions used originated mostly from national emission databases. Specific roughness length and land use maps were constructed using the original Corine database with land use classes on a 250x250 m resolution (Flanders, the Netherlands and Denmark) and Swedish land use information on a resolution of 1x1 km. After reclassification of the initial land use data, the 250x250 m resolution data were aggregated to 1x1 km based on dominant land use. Based on the 250x250 m land use map a roughness map was constructed. For each of the individual land use classes a specific roughness length was used.

When certain locations (grid cells on a map) are afforested, the existing N deposition has to be corrected for land use changes. The Eutrend model calculated the enhanced deposition in afforested pixels and the diminished deposition in the remaining pixels. For the correction of the initial deposition due to growing forest four different cases were distinguished, depending on the initial land use (pasture or arable) and the tree species used for afforestation (coniferous or deciduous). The calculated N deposition was compared with measured deposition in the AFFOREST chronosequence sites. The comparison showed fairly good results (except for Sweden) and the results were decided to be satisfactory to include in the metamodel METAFORE. The resulting N deposition after these corrections is used within the AFFOREST-sDSS to describe the new deposition situation after a certain amount of years after afforestation.

3.4.2. *Vegetation*

An overview was made of forest models that could be used for the AFFOREST project. Based on the criteria for the final product of the project, the CenW model, a forest growth model with linked C, energy, nutrient and water cycles was chosen.

The original CenW-model was changed to run on fortnightly time steps. Photosynthetic C gain was calculated based on light absorption, temperature, soil water status and foliar N concentration. Some C is lost in respiration and the remainder utilized for growth, with allocation to different plant organs determined by plant nutrient status, tree height and species-specific allocation factors. Nitrogen can be taken up from the mineral N pool. Nitrogen can be supplied by external sources or from the decomposition of organic matter. The nutrient cycle is closed through litter production by the death of trees, or by shedding of plant parts, such as roots, stems and, most importantly, foliage. This way C and N are transferred to the soil to form organic matter. Organic matter is eventually decomposed, thereby releasing CO_2 to the atmosphere. Any N in excess of microbial requirements can enter the pool of mineral N from where it can be taken up by plants. The prototype model based on CenW was developed for coniferous trees. However, the model was extended for deciduous trees by including growth buds in which C can be stored during winter time, and from which re-growth is initiated in spring.

The model was validated against data from the AFFOREST chronosequences. To test the model, patterns of water use, N and C dynamics simulated by the model were compared against the observed data from literature. To run the simulations, the model was initialized and thereafter was completely run by external driving variables. The model used rainfall, net radiation and minimum and maximum temperatures recorded. The simulation of both C sequestration in the vegetation and of tree height was evaluated to be satisfactory.

3.4.3. *Soil*

The detailed Nutrient Cycling and Soil Acidification Model (NUCSAM) was chosen for modelling soil. The model has been made available in a PC-version for application within the project. The model has been extended with a feature for graphical presentation of the model results together with the observations. Available results from the chronosequence measurements have been processed, evaluated and transformed into model input for all sites. The resulting water balances (calculated using the hydrological model SWAP – see above) were used for the calculation of input-output budgets for N and as input for the model NUCSAM. Parameterisation and calibration of the model NUCSAM was performed for all chronosequences. At individual plots the agreement between observed and simulated changes in soil solution chemistry was reasonably good.

3.5. The metamodel METAFORE

It appeared that existing complex mechanistic models as described above required too many input data that are generally unavailable for application on a regional scale. Therefore, a new simplified metamodel was derived by the name of METAFORE. The AFFOREST metamodel METAFORE is an integrated and simplified mechanistic model. METAFORE joins together four detailed process-based models into one metamodel because the original models were too large and complex to run for many pixels with restricted data availability. The components are a deposition model based on the Eutrend model, a soil model based on the models SMART and NUCSAM, a water model called WATBAL and finally a vegetation model derived from CenW. In order to speed up the processing time of the AFFOREST-sDSS, METAFORE was pre-run for a large number of cases to produce large lookup tables which could be accessed by the AFFOREST-sDSS. The metamodel can be run in stand-alone mode as well, and allows the effects of a number of parameters on the environmental issues of concern to be explored rapidly. A user-friendly graphical interface (GUI) has been developed for the metamodel.

In this design, interaction between the model components is the responsibility of the metamodel framework. Model components never interact directly with each other, but ask their input data from the metamodel framework and present their results to this framework. Individual model components are thus not dependent on specific implementations of the other model components: they solely rely on interface definitions. The model components work with different time steps. The deposition and soil components are designed to work with a yearly time step, while the water and vegetation components work with a monthly time step. The metamodel framework bridges this gap in time steps, by implementing two loops. A set of additional variables is used to calculate yearly values from the monthly results.

METAFORE is used in two modes within the AFFOREST-sDSS system. The first mode is the use of METAFORE to create the sDSS tables as a basis for finding optimal solutions to user questions. METAFORE calculates for each location in the GIS database the environmental performance for each afforestation strategy. This run is needed only once to create all sDSS tables. Once these tables are created the sDSS uses the tables for evaluating user questions. The task of the AFFOPREST-sDSS is now simply searching through these tables. In this design, only changes in the input data (meaning changes in site specific characteristics) require a new model run.

The second mode of operation of METAFORE is a direct simulation of a specific afforestation strategy for a specified location. This means that the metamodel reads site-specific data for that location and simulates the soil-water-plant processes for the location. Detailed reports and graphs of this simulation are produced, and the user can change any of the (location specific) site characteristics or afforestation strategy and perform sensitivity analysis in this manner. Whenever a map is visible in the system this mode of operation is available, so any pixel may be selected and analyzed in detail.

METAFORE was calibrated and validated for the AFFOREST region in three steps. Firstly, the metamodel was validated on the chronosequence data from all three countries and on literature data and it was able to describe the vegetation growth and the soil chemistry adequately. Secondly, METAFORE was validated against simulations with the detailed process models. Some predefined cases were constructed, in which soil characteristics, climate variables and afforestation strategy were defined and the simulation results of the detailed process models were similar to the results of the metamodel. Lastly, the metamodel was calibrated against existing knowledge of experts (rules of thumbs). A large number of simple questions about ecosystems and the answers to the questions were used to score the agreement amongst experts on the validation of the statement. METAFORE was evaluated against this list, and results were satisfactory.

3.6. Geo-referenced database

A spatial geo-referenced database was built. The AFFOREST geographical database provides the basic information used to model the development of the environmental conditions as a consequence of different afforestation alternatives. It covers data on current land use, the soil type, climate conditions, N deposition and latitude. The original data were combined into classes, feasible for the AFFOREST-sDSS. The final database is available at two different resolution scales. At the European scale, the spatial resolution is 1 km^2 and at the national scale the spatial resolution is 1 hectare. The purpose of this is to provide a good base for answering relevant questions at different spatial scales. At the European scale, one single map file covers all four countries, whereas at the national scale data are kept in separate map files. The database is provided in Arc/Info grid format. Metadata have been collected for all input data sources for documentation of the database and these are included in the help system of the AFFOREST-sDSS.

The European geo-referenced database was constructed based on the four national AFFOREST databases. This involved i) to project the national databases into a common geographical reference system and ii) to resample the data to a 1-km cell size. The reference system used in the AFFOREST European database is based on Lambert's azimuthal projection, which represents the area extent correctly and is an ETRS89 compliant reference system.

3.7. The AFFOREST-sDSS

The envisaged users of the AFFOREST-sDSS are afforestation policy planners in national and regional administrations and local managers of afforestation projects. To accommodate for these two distinct categories of users, two corresponding levels of spatial resolution have been identified in the AFFOREST-sDSS.
The Graphical User Interface (GUI) is formed as a set of dialog windows and help texts that guide the user through the system and help to structure his/her afforestation questions into one of the fourteen type questions the system can handle. There are five ways to navigate through the AFFOREST-sDSS:

- Predefined cases can be chosen. These are examples where specific afforestation problems are solved. These scenarios serve as tutorial cases for users to better understand the system and solve their own specific questions.

- 'Where' questions: All variables can be set except for the definition of the initial system.
- 'How much' questions: All variables can be set except for the definition of the environmental performance (water recharge, C sequestration, nitrate leaching).
- 'How' questions: All variables can be set except for the definition of the afforestation strategy (tree species choice, site preparation, stand tending).
- 'Time' questions: All variables can be set except for the definition of the evaluation time (0-100 years).

According to the type of question the results can either be viewed as maps, tables or both. A tutorial has been developed, which guides the novice user through the possibilities in the AFFOREST-sDSS.

AFFOREST-sDSS is able to test and compare different scenarios of afforestation and to propose optimal solutions for afforestation planning. The system will support the prioritizing of alternative locations for afforestation measures, propose optimal solutions for afforestation planning, and support the development of guidelines. The AFFOREST-sDSS shows a promising road to the integrated valorisation of knowledge and technical capabilities to better integrate environmental concerns in the planning and management of human interventions in the rural environment.

3.8. Guidelines on afforestation

Guidelines have been produced on recommended practices for local afforestation in relation to management, and recommended strategies for regional afforestation policies in relation to pursued environmental objectives (water recharge, nitrate leaching, C sequestration and field layer vegetation). The guidelines are based on both experimental data from the chronosequence studies and output from the AFFOREST-sDSS. These guidelines should help landscape and forestry planners to establish and manage new forests planted on former agricultural land with less environmental effects on the groundwater quality and quantity.

4. APPLICABILITY OF THE AFFOREST PRODUCTS

The work in AFFOREST has improved the knowledge of C sequestration in afforested agricultural. As a contribution to mitigation of rising atmospheric CO_2 concentration, countries can choose whether or not to include afforestation and reforestation activities in the national accounting of greenhouse gas budgets under the Climate Convention and the Kyoto Protocol. The work on land use change in AFFOREST primarily focused on improving the basic knowledge of contributions of afforested cropland to C sequestration. These results will help to bridge the gap between currently existing knowledge and policy demands and support the reporting of C sequestration due to afforestation according to article 3.3 and due to forest management according to article 3.4 of the Kyoto protocol. Due to the fragmentation of afforestation areas at the landscape level, national forest inventories may not be able to monitor afforestation adequately by random sampling. It may therefore be necessary to draw on other data sources. AFFOREST has provided key figures for changes in soil and biomass C pools and has delivered tools to estimate the impact of afforestation. These products may be valuable as a supplement to plot-based national forest inventories.

Nitrogen pollution of water bodies in the European Union has increased substantially in recent decades. This is largely a result of an intensification and expansion of intensive agricultural activities and industries. Currently, critical loads for N are considered exceeded over most of Europe. The Nitrates Directive (91/676/EEC), adopted by the Council of the European Communities, concerns the protection of waters against pollution caused by nitrates from agricultural sources. Likewise, the Directive (98/83/EEC) seeks to establish requirements for the quality of drinking water. Part of the motivation for planning large afforestation programs in north-western Europe are to decrease the current leaching of nitrate to the groundwater and hereby ensure cleaner drinking water, lower the water processing costs and ultimately guarantee the health of the inhabitants. Our knowledge on nitrate leaching after afforestation of former arable land has been greatly improved through the work of AFFOREST. Through sampling in the chronosequences it has been shown that afforestation in north-western Europe, the most intensely N-polluted area of Europe, decreases nitrate leaching significantly compared to the arable land use replaced. The strategic impact and social relevance is therefore a possibility to improve the quality of the groundwater, and hereby improve the European environment as well as provide EU citizens with cleaner drinking water.

Since time was substituted with space in our chronosequence measurements, the trends in time after afforestation of former arable land is only an indication of what the future environmental impact will be. These trends are very distinct. However, repeated sampling in the chronosequences about ten to twenty years after the first sampling would allow for testing the experimental results obtained in AFFOREST. This will provide real evidence of the directional change for each stand in the chronosequences and will also test the predictive value of the chronosequence approach.

The AFFOREST-sDSS allows scientific results to be translated into policy-relevant guidelines and supports decision-making about afforestation strategies in particular places. It provides an approach for determining where and how to afforest former arable land in an optimised way according to the four impact categories discussed in this book: nitrate leaching, groundwater recharge, C sequestration and field layer biodiversity. AFFOREST has produced a set of tools which could potentially be used by end users (forest managers, regional and national planners etc.) in their effort to plan the new forests planted on former agricultural soils in the best possible way for the environment. End users were involved in defining the questions subject to research and how the results were to be presented.

Restrictions on the use of underlying data may limit access to the fine spatial scales in some countries, but overall the tool is available for use in the countries of origin. A certain level of technical expertise will be necessary to use the tools. It needs to be stressed that the role of these tools is to support decision making, not to make the decisions, as decision makers will need to use other criteria as well when planning afforestation. Therefore, the output of the AFFOREST-sDSS is complementary to and should be used together with other sources of information like empirical evidence, expertise and scientific literature.

The AFFOREST-sDSS as well as the metamodel METAFORE are useful tools for policy development within the four countries concerned (Belgium, Denmark, the Netherlands and Sweden). For other countries, or for the EU as a whole, the AFFOREST-sDSS could not be used directly as environmental conditions differ. However, the formal structure of the sDSS would be directly transferable, and the approach used in AFFOREST could be used to develop similar decision support systems in the EU in general or in other countries with similar questions on environmental impact assessment.

It is evident that afforestation largely improves environmental performance compared to the arable land use. Carbon is sequestered, nitrate leaching decreases, and forest species characteristic of old forests will slowly appear. The exception is that water recharge will be somewhat reduced because of higher evapotranspiration in forests. Therefore, current plans on continued afforestation in Europe are likely to overall improve the environment significantly.

Afforestation is often based on decisions driven by multiple objectives. When multi-purpose demands are involved, it is difficult and involves many pitfalls to plan the new forests in a proper way while meeting the array of natural, political, and socio-economic conditions. Maximising a single objective will often cause trade-offs for another objective. It is therefore a complex optimisation challenge to address more than one objective. It is necessary to know the interactions between the performance of individual services provided by afforestation and different site and management characteristics. A further development of the AFFOREST-sDSS to also include decisions on biodiversity, recreation and afforestation adapted to climate change would be a challenge and a preferable tool when planning the forests of tomorrow.

5. THE AFFOREST PROCESS

The AFFOREST project involved co-operative work between seven institutions in four countries. It required expertise in a wide range of disciplines from plant ecology to information technology. All sub-projects were interconnected, and failure in one might have jeopardised the whole outcome. Therefore, all partners have been depending on each other in an integrated chain of work. This need has created a strong interdisciplinarity and fully integrated products and it has been unique and refreshing to work with experts from different fields and to experience that we really worked as ONE group having creative, open-minded discussions.

The European added value of the consortium is significant, as the project has required the expertise of the contributors, their data, and their process-based models to establish a critical mass to address the problem at the European scale. It is the combination of complementary expertise and resources available from the European-wide group of different organisations that has permitted the development of the AFFOREST-sDSS. Thus, AFFOREST has contributed to science and technology co-operation between academic research communities and end-users such as policy makers, agencies responsible for pollution control, and forest and land-use planners since it addresses issues of concern to all.

COLOR PLATES

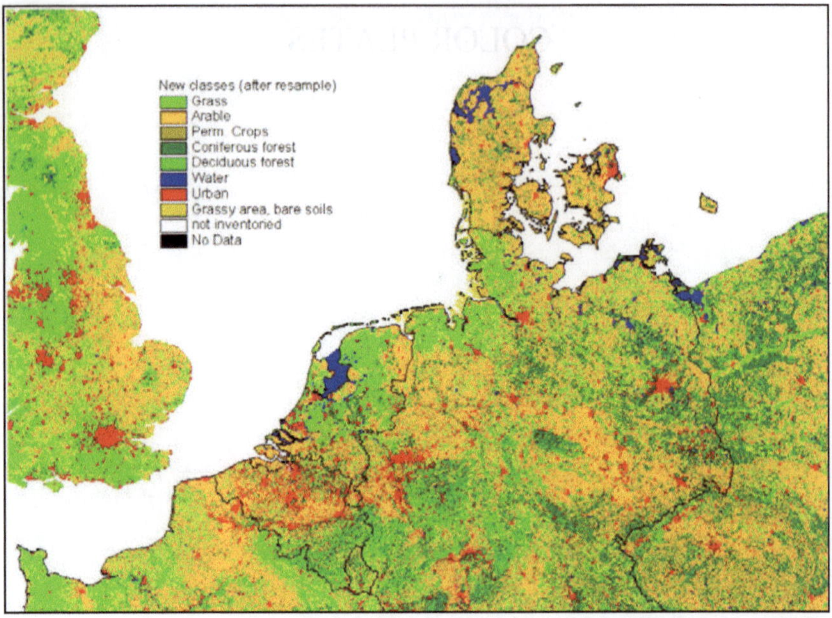

Color plate 1. *Overview of the land use classes after reclassification of Corine data.*

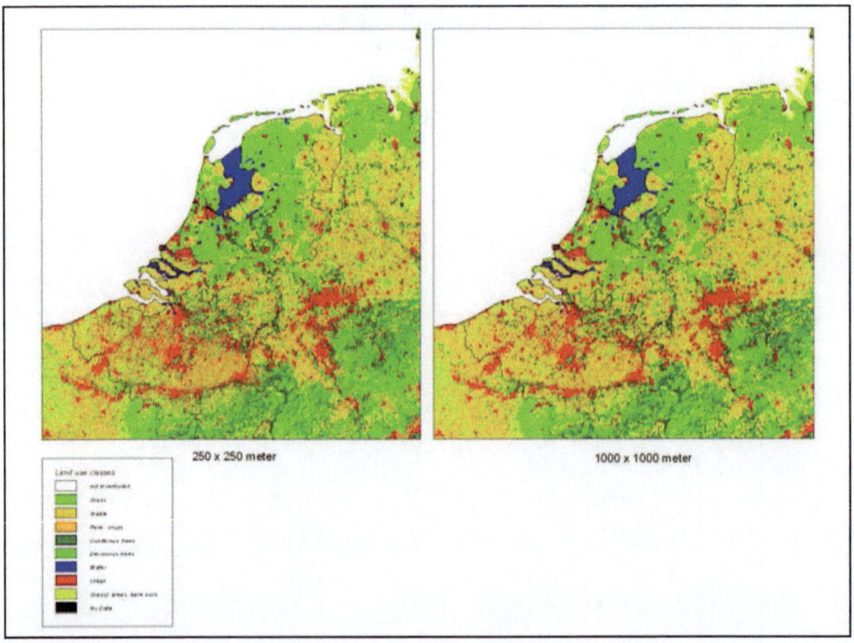

Color plate 2. *Overview of the land use map for the Netherlands and Flanders after aggregation from 250x250 m to 1x1 km.*

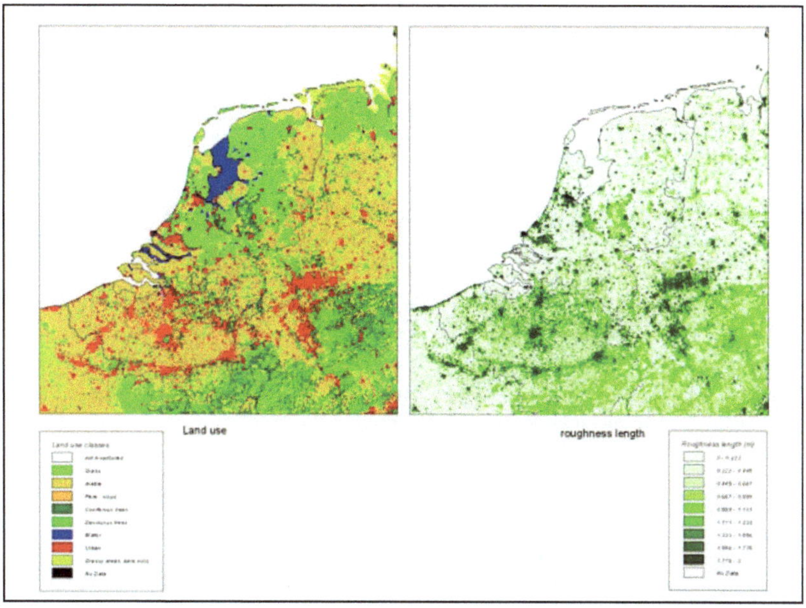

Color plate 3. *Land use and corresponding roughness length map for Flanders and the Netherlands.*

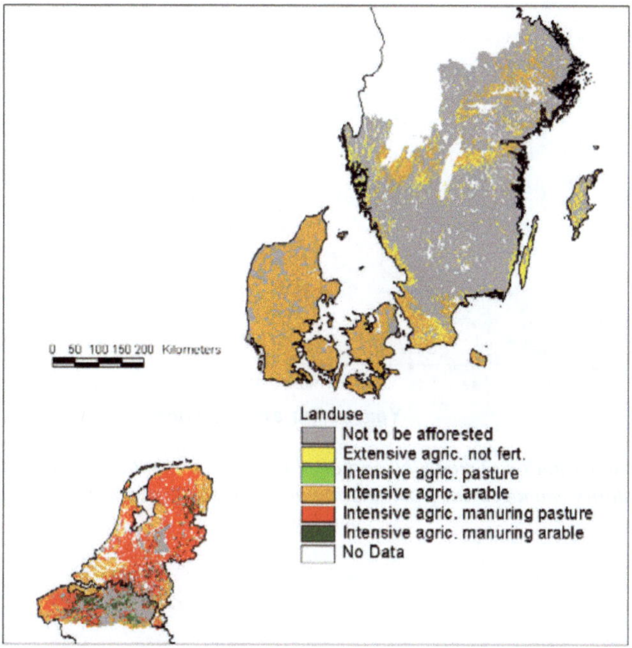

Color plate 4*. Land use map from the low resolution European AFFOREST database.*

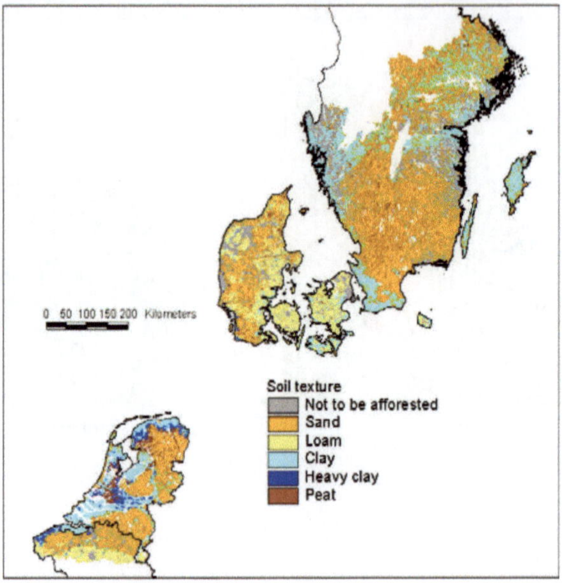

Color plate 5. *Soil texture map from the low resolution European AFFOREST database.*

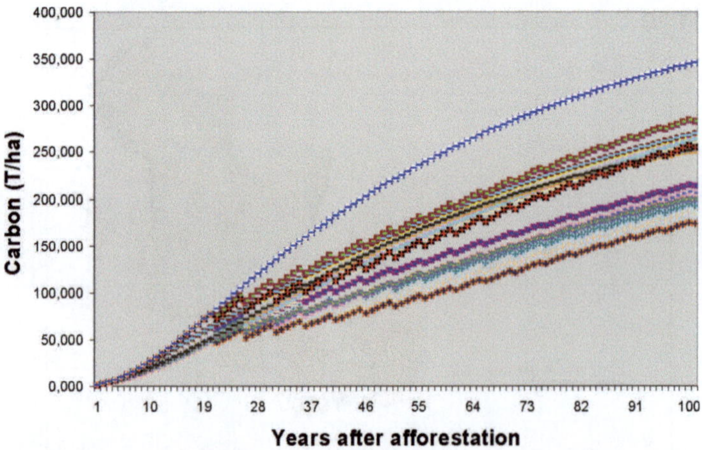

Color plate 6. *Evolution of modelled carbon storage for one initial system under 36 types of afforestation management (combination of tree species, site preparation and stand tending)*

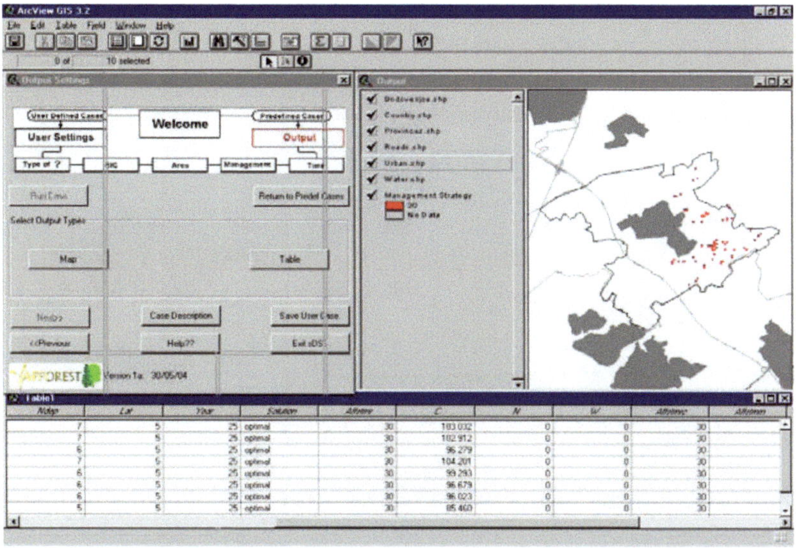

Color plate 7. *Example of output from AFFOREST-sDSS.*

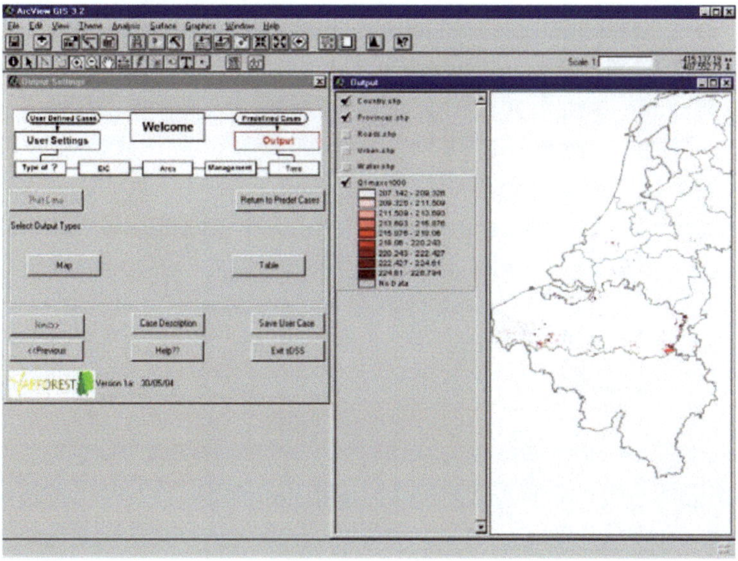

Color plate 8. *Map with the best 100,000 ha of the AFFOREST region for cumulative carbon sequestration after 50 years following the best possible afforestation strategy (only part of the region is visible).*

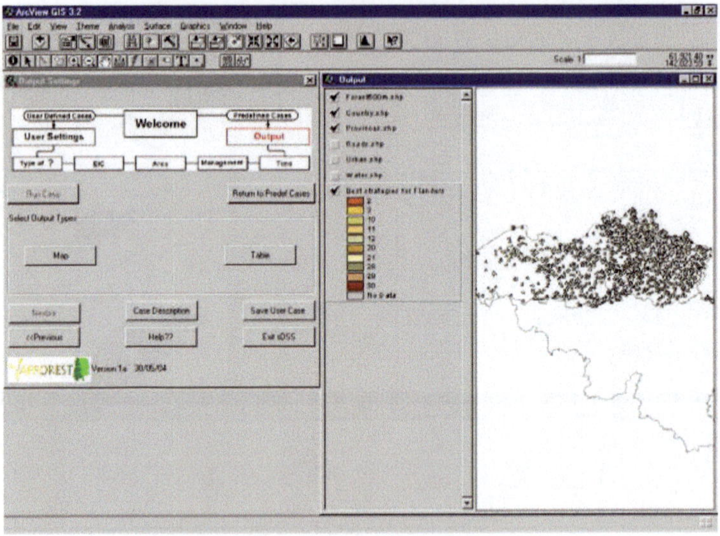

Color plate 9. *Map of locations in Flanders, Belgium, adjacent to existing forests with the best possible environmental performance 50 years after afforestation.*

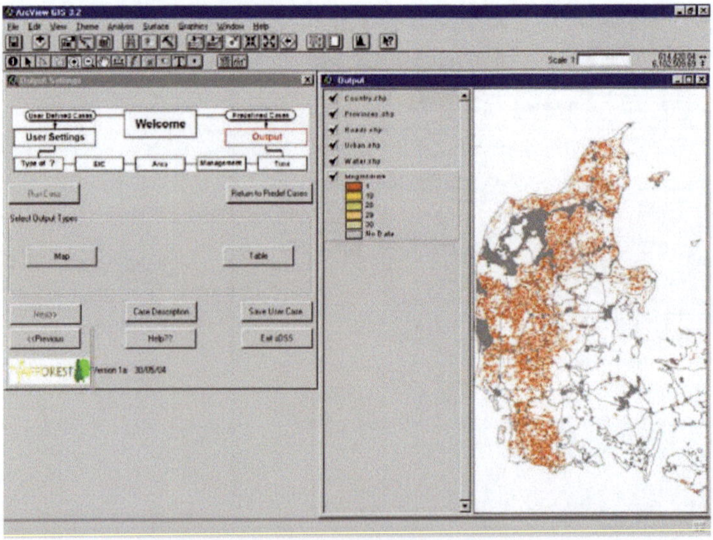

Color plate 10. *Best management strategy for afforestation on sandy soil in Denmark with maximum clean drinking water as a first priority and maximum carbon sequestration as a second priority after 15 years.*

INDEX

MIX
Papier aus verantwortungsvollen Quellen
Paper from responsible sources
FSC® C105338

If you have any concerns about our products,
you can contact us on
ProductSafety@springernature.com

In case Publisher is established outside the EU,
the EU authorized representative is:
Springer Nature Customer Service Center GmbH
Europaplatz 3, 69115 Heidelberg, Germany

Printed by Libri Plureos GmbH
in Hamburg, Germany